# ADVANCES IN PROTEIN CHEMISTRY

Volume 74

## Mechanisms and Pathways of Heterotrimeric G Protein Signaling

# ADVANCES IN PROTEIN CHEMISTRY

EDITED BY

FREDERIC M. RICHARDS
Department of Molecular Biophysics
and Biochemistry
Yale, University
New Haven, Connecticut

DAVID S. EISENBERG
Department of Chemistry and Biochemistry
Center for Genomics and Proteomics
University of California, Los Angeles
Los Angeles, California

JOHN KURIYAN
Department of Molecular and Cellular Biology
University of California, Berkeley
Berkeley, California

VOLUME 74

## Mechanisms and Pathways of Heterotrimeric G Protein Signaling

EDITED BY

STEPHEN R. SPRANG
Department of Biochemistry
University of Texas Southwestern Medical Center
Dallas, Texas

ELSEVIER

AMSTERDAM • BOSTON • HEIDELBERG • LONDON
NEW YORK • OXFORD • PARIS • SAN DIEGO
SAN FRANCISCO • SINGAPORE • SYDNEY • TOKYO
Academic Press is an imprint of Elsevier

Academic Press is an imprint of Elsevier
84 Theobald's Road, London WC1X 8RR, UK
Radarweg 29, PO Box 211, 1000 AE Amsterdam, The Netherlands
Linacre House, Jordan Hill, Oxford OX2 8DP, UK
30 Corporate Drive, Suite 400, Burlington, MA 01803, USA
525 B Street, Suite 1900, San Diego, CA 92101-4495, USA

First edition 2007



ISBN: 978-0-12-034288-4
ISSN: 0065-3233

For information on all Academic Press publications
visit our website at books.elsevier.com

Pinted and bound in USA
07 08 09 10 11 10 9 8 7 6 5 4 3 2 1

# CONTENTS



## Structural Basis of Effector Regulation and Signal Termination in Heterotrimeric G$\alpha$ Proteins

STEPHEN R. SPRANG, ZHE CHEN, AND XINLIN DU



## How Do Receptors Activate G Proteins?

WILLIAM M. OLDHAM AND HEIDI E. HAMM



## Some Mechanistic Insights into GPCR Activation from Detergent-Solubilized Ternary Complexes on Beads

TIONE BURANDA, ANNA WALLER, YANG WU, PETER C. SIMONS, SEAN BIGGS, ERIC R. PROSSNITZ, AND LARRY A. SKLAR

## Activation of G Protein–Coupled Receptors

Xavier Deupi and Brian Kobilka



## Kinetic Analysis of G Protein–Coupled Receptor Signaling Using Fluorescence Resonance Energy Transfer in Living Cells

Martin J. Lohse, Carsten Hoffmann, Viacheslav O. Nikolaev, Jean-Pierre Vilardaga, and Moritz Bünemann



## Regulation of Rho Guanine Nucleotide Exchange Factors by G Proteins

Paul C. Sternweis, Angela M. Carter, Zhe Chen, Shahab M. Danesh, Ying-Fan Hsiung, and William D. Singer

# PREFACE

The molecular mechanism that underlies eukaryotic signal transduction through heterotrimeric G proteins is of ancient origin and persists in all kingdoms of life. In their most primitive forms, G proteins require nothing but their intrinsic nucleotidase activity to assume the property of a two-state control device that is activated by GTP binding and inactivated by GTP hydrolysis. The most fundamental cellular processes, notably protein synthesis and membrane trafficking, involve G protein domains as checkpoint control modules. The ability of G proteins to specifically recognize—in a GTP-dependent fashion—specific conformational states of effectors is thermodynamically equivalent to stabilizing them. Acting on various effectors, G proteins thereby control catalytic processes. The family of mammalian G proteins encompasses hundreds of G proteins, diverse in sequence and function, which recognize a host of effectors. All potential points of control—the rate and cellular location at which GDP is exchanged for GTP, and similarly, that of its hydrolysis—are exploited in the evolution of a host of regulatory proteins: Guanine nucleotide Exchange Factors (GEFs), Guanine nucleotide Dissociation Inhibitors (GDIs), GTPase Activating Proteins (GAPs), among others. With the aid of such accessory proteins, both GTP- and GDP-bound states of G proteins can be harnessed to control directional processes, such as nuclear import or export. Preserved in all members of the G protein superfamily is a switching mechanism, built within and around the GTP-binding site, that couples GTP binding and hydrolysis to conformational change, and which is recognizable even in the remotely homologous motor ATPases.

The ''heterotrimeric'' G$\alpha$ proteins that mediate signal transduction in eukaryotes use this mechanism as well, but have traversed a distinct pathway of molecular evolution. Activated by agonist-dependent, seven-transmembrane plasma membrane receptors (G protein-Coupled Receptors, or GPCRs, heterotrimeric G proteins are transducers of extracellular chemical stimuli. Intact G protein heterotrimers, each composed of a GDP-bound G$\alpha$ subunit, and a heterodimer of intimately associated G$\beta$ and G$\gamma$ subunits, does not itself act on effectors. Rather it is an inert species in which the two signal transducers, G$\alpha$ and G$\beta\gamma$ are mutually sequestered by virtue of their strong GDP-dependent interaction. By catalyzing GTP-for-GDP exchange on G$\alpha$, leading to the release or rearrangement of G$\beta\gamma$, GPCRs liberate both components to pursue distinct, but frequently intertwined signaling agenda. The mechanisms by which the two species cooperate in subsequent signaling events remain a significant, largely unexplored problem.

This volume focuses on recent developments in the efforts of many laboratories to understand the kinetic and structural basis of G protein signaling. In Chapter 1, my colleagues and I review structural and mechanistic aspects of signal transduction by $G\alpha$ proteins. Structural studies conducted over the last ten years have shown that $G\alpha$ subunits have evolved topographically distinct sites for effectors, GTPase activating proteins and, as described by Oldham and Hamm in Chapter 2, for receptors. Signaling through $G\alpha$ proteins may involve little more than simple effector binding, but may also involve control of effector activity through complex allosteric effects. Some effectors, through their intrinsic GAP activity, control the lifetime of the signaling state of the $G\alpha$ by which they are regulated. Others function as GTP-dependent organizing centers for G protein components. Crystallography, guided by a variety of biochemical and genetic techniques, has provided mechanistic insight into some of these systems. The chemical and conformational events that accompany signal termination, whereby a tight-binding $G\alpha$ substrate (GTP) is converted to an equally tight-binding product (GTP), are still rather poorly understood, although headway is being made to delineate the chemical mechanisms of GTP hydrolysis. How hydrolysis is energetically and kinetically coupled to the conformational transition that reduces affinity for effectors is more complex than one might imagine.

Relatively few $G\alpha$-controlled effector systems have been well studied, let alone identified. The number of G protein-regulated pathways seems to be sparse relative to the myriad GPCRs that shape the cellular and molecular context in which those pathways are elaborated. In Chapter 6, Sternweis and colleagues describe the family of GEFs for Rho GTPases. Individual members of this family follow distinct, $G\alpha$-specific modes of regulation that are influenced by components of other signaling pathways. Like the phospholipases $\beta$, RhoGEFs are GAPs for the $G\alpha$ proteins by which they are regulated, suggesting the potential for steady-state interactions with receptors at the plasma membrane.

GPCRs continue to perplex and amaze those investigators who bravely seek to understand their mode of action. Unexpected complexity has been discovered through the application of new methods and techniques to study the kinetic and conformational behavior of GPCRs, both *in vivo* and *in vitro*. The three-dimensional structure of a complex between an agonist-bound GPCR with a G protein heterotrimer remains an elusive prize. Yet, it will almost certainly be the case that any crystal structure will reveal only one of a manifold of states. In Chapter 2, Oldham and Hamm review our current understanding of the G protein–GPCR interface, and the molecular basis of G protein activation by receptors. Insights come from a variety of experimental approaches, including mutagenesis, antibody mapping, chemical cross-linking, the use of fluorescence probes, electron microscopy, site-specific spin labeling, and importantly, X-ray and electron diffraction.

Conformational intermediates in the pathway to receptor activation are clearly evident in the spectroscopic studies of the $\beta_2$-adrenergic receptor described by Deupi and Kobilka in Chapter 4. Receptors are exquisitely sensitive to differences in the chemical structures of ligands that occupy their agonist-binding sites. In light of molecular genetic, biochemical, and crystallographic information, ligand-induced spectroscopic changes can be interpreted in terms of receptor structure. Several laboratories have begun to show, for a variety of receptors, that such receptor states are functionally distinct, and may be differentially stabilized in a ligand-specific manner at the agonist-binding site and perhaps at other allosteric sites in the receptor.

The complex kinetic behavior that accompanies the binding of G proteins to receptors is revealed by sensitive new approaches. Buranda *et al.* describe, in Chapter 3, how detergent-solubilized G proteins and receptors can be assembled on beads in the presence of agonists and partial agonists, and how flow cytometry can be used to report on the thermodynamics of their interaction. Flow cytometry also yields the kinetics of G protein and ligand dissociation from receptor:agonist: G protein ternary complexes in solution. In Chapter 5, Lohse and colleagues show how fluorescence resonance energy transfer can be used to follow the kinetics of receptor stimulation by agonists, and subsequent G protein activation, in real time and in living cells. Together these studies provide kinetic parameters for a series of events: agonist-induced conformational changes within the receptor (milliseconds), nucleotide-induced dissociation of the receptor from the nucleotide-free heterotrimer (sub-second), and the dissociation of G$\alpha$ from G$\beta\gamma$. For at least some G proteins, the latter may not occur within the time frame of rapid signaling, or may involve re-arrangement of the subunits with half times in the range of seconds, rather than their complete dissociation.

The contributors to this volume have shown how far our understanding of the molecular mechanisms of G protein signaling has come, but also how much remains to be discovered. What is the chemo-mechanical mechanism of G protein activation by receptors? What are the relevant states of receptors; how are they differentially stabilized by ligands; what is the functional role of receptor dimerization in the modulation and control of G protein activation, or the coordination of other signaling functions. Do G protein ''signalosomes'' exist? If so, what (in addition to the obvious!) are their components; how tightly are they integrated—kinetically and structurally—and how and where are they assembled at the plasma membrane? Beyond these "classical" concerns, a new research field has recently emerged with the discovery of receptor-independent regulatory pathways for Gα in cytokinesis and perhaps other cellular processes. This is an exciting field. For me, in no small part due to the remarkable colleagues with whom I have shared the adventure: Al Gilman, who made it possible, Elliott Ross, Paul Sternweis, Alan Smrcka, Steve Lanier, and Roland Seifert among others, together with talented students and fellows who have, as they say, "done the work", not to mention much of the thinking. Most importantly, I express my gratitude to the contributors to this volume, who have so thoughtfully illuminated this fascinating area of research.

Stephen Sprang
University of Texas Southwestern Medical Center
University of Montana

# STRUCTURAL BASIS OF EFFECTOR REGULATION AND SIGNAL TERMINATION IN HETEROTRIMERIC G$\alpha$ PROTEINS

By STEPHEN R. SPRANG, ZHE CHEN, AND XINLIN DU

Department of Biochemistry, University of Texas Southwestern Medical Center, Dallas, Texas 75390



## Abbreviations

| | |
|---|---|
| ApCpp | adenosine 5′-($\alpha\beta$-methylene)-triphosphate |
| ATP$\alpha$S | adenosine 5′-($\alpha$-thio)-triphosphate ($R_p$) |
| cGMP | cyclic guanosine 3′,5′-monophosphate |
| CHD | cyclase homology domain |
| DEP | disheveled/Egl-10/plextrin homology domain |
| DH | Dbl homology |
| FTIR | Fourier transform infrared Raman |
| G$\alpha$ | G protein $\alpha$ subunit |
| GAP | GTPase-activating protein |
| GDI | guanine nucleotide dissociation inhibitor |
| GEF | guanine nucleotide exchange factor |
| GPCR | G protein–coupled receptor |
| GppNp | guanosine 5′-($\beta\gamma$-imino)-triphosphate |
| GRK | G protein receptor kinase |
| GTP$\gamma$S | guanosine 5′-($\gamma$-thio)-triphosphate |
| KIE | kinetic isotope effect |

*ADVANCES IN PROTEIN CHEMISTRY, Vol. 74*
DOI: 10.1016/S0065-3233(07)74001-9





0065-3233/07 $35.00

| | |
|---|---|
| LARG | leukemia-associated RhoGEF |
| LBHB | low barrier hydrogen bond |
| mAC | membrane adenylyl cyclase |
| MM | molecular mechanics |
| NF1-333 | neurofibromin fragment (residues 1–333) |
| PLC$\beta$ | phospholipase C$\beta$ |
| PDE | cGMP phosphodiesterase |
| PH | plextrin homology |
| QM | quantum mechanics |
| RGS | regulator of G protein signaling |
| RH-RhoGEF | RGS homology Rho guanine nucleotide exchange factor |
| rmsd | root mean square deviation |
| SOS | Son of sevenless |
| $VC_1$ | $C_1$ domain of type V mAC |
| $IIC_2$ | $C_2$ domain of type II mAC |

## Abstract

This chapter addresses, from a molecular structural perspective gained from examination of x-ray crystallographic and biochemical data, the mechanisms by which GTP-bound G$\alpha$ subunits of heterotrimeric G proteins recognize and regulate effectors. The mechanism of GTP hydrolysis by G$\alpha$ and rate acceleration by GAPs are also considered. The effector recognition site in all G$\alpha$ homologues is formed almost entirely of the residues extending from the C-terminal half of $\alpha 2$ (Switch II) together with the $\alpha 3$ helix and its junction with the $\beta 5$ strand. Effector binding does not induce substantial changes in the structure of G$\alpha$•GTP. Effectors are structurally diverse. Different effectors may recognize distinct subsets of effector-binding residues of the same G$\alpha$ protein. Specificity may also be conferred by differences in the main chain conformation of effector-binding regions of G$\alpha$ subunits. Several G$\alpha$ regulatory mechanisms are operative. In the regulation of GMP phospodiesterase, G$\alpha$t sequesters an inhibitory subunit. G$\alpha$s is an allosteric activator and inhibitor of adenylyl cyclase, and G$\alpha$i is an allosteric inhibitor. G$\alpha$q does not appear to regulate GRK, but is rather sequestered by it. GTP hydrolysis terminates the signaling state of G$\alpha$. The binding energy of GTP that is used to stabilize the G$\alpha$:effector complex is dissipated in this reaction. Chemical steps of GTP hydrolysis, specifically, formation of a dissociative transition state, is rate limiting in Ras, a model G protein GTPase, even in the presence of a GAP; however, the energy of enzyme reorganization to produce a catalytically active conformation appears to be substantial. It is possible that the collapse of the switch regions, associated with G$\alpha$ deactivation, also encounters a kinetic barrier, and is coupled to product (Pi) release or an event preceding formation of the GDP•Pi complex. Evidence for a catalytic intermediate, possibly metaphosphate, is discussed. Gα GAPs, whether exogenous proteins or effector-linked domains, bind to a discrete locus of Gα that is composed of Switch I and the N-terminus of Switch II. This site is immediately adjacent to, but does not substantially overlap, the Gα effector binding site. Interactions of effectors and exogenous GAPs with Gα proteins can be synergistic or antagonistic, mediated by allosteric interactions among the three molecules. Unlike GAPs for small GTPases, Gα GAPs supply no catalytic residues, but rather appear to reduce the activation energy for catalytic activation of the Gα catalytic site.

## I. Introduction and Scope

Heterotrimeric G proteins, the molecular transducers of GPCRs, are composed of a guanine nucleotide-bound Gα, together with a heterodimer formed by two protein components, Gβ and Gγ. Nearly 20 human Gα proteins, the products of 17 distinct genes, have been identified. Gα proteins are members of the Ras GTPase superfamily, and are composed of an ~220-residue Ras-like α/β core domain into which is inserted an ~120-residue α helical module that is unique to the family of Gα proteins. Gβ subunits (encoded by 6 genes) are ~340-residue proteins, composed of a characteristic WD-40 amino acid sequence repeat manifested in a seven-blade, β-propeller fold. The ~75-residue Gγ subunits (12 isoforms) are extended, largely helical molecules that interact tightly with Gβ over an extended surface; the high affinity of the interaction, and the extensive hydrophobic surface areas of the uncomplexed subunits, renders the Gβγ heterodimer effectively nondissociable under physiological conditions. The structural characteristics of these molecules have been reviewed in some detail (Hamm, 1998; Sprang, 1997).

Both Gα and Gβγ components of heterotrimeric G proteins function as signal transducers, and the targets of their action are collectively referred to as G protein effectors. Effectors include mAC and PLCβ that catalyze the synthesis of diffusible small molecule second messengers, respectively, cyclic AMP and inositol-1,4,5-trisphosphate with diacylgycerol (Gilman, 1987; Rhee, 2001). Inward rectifying potassium channels (Sadja *et al.*, 2003) and voltage-gated calcium channels (De Waard *et al.*, 2005) are, respectively, activated and inhibited by G protein βγ components. Heterotrimeric G proteins also activate Rho GTPases (Sternweis *et al.*, this volume) and participate in nonreceptor-mediated regulation of cytokinesis (Willard *et al.*, 2004). Some of these systems have been accessible to study through *in vitro* biochemistry and structural biology, and consequently, it has been possible to formulate some of the physical and structural principles that govern the mechanisms of G protein regulation. In this chapter, we consider discoveries of the last 10 years that have led to a structural understanding of effector regulation by the G$\alpha$ components of heterotrimeric G proteins. This narrative only begins to describe the structural complexity of G protein–signaling mechanisms. It has become increasingly evident that most, if not all, G protein–mediated regulation occurs in the context of multiprotein signaling complexes, with emergent kinetic and regulatory behavior not expected from the properties of the isolated components (Abramow-Newerly *et al.*, 2006; Biddlecome *et al.*, 1996; Peleg *et al.*, 2002; Ross and Wilkie, 2000). The techniques available to study the structures of such dynamic complexes are limited, and much remains to be revealed that lies beyond the scope of this discussion.

Modulation of effector activity by G proteins (G protein signaling) is a cyclic process (Gilman, 1987). The basal, inactive state of the transducer in this system is the G protein heterotrimer itself, composed of a high affinity complex between a G$\beta\gamma$ heterodimer and a GDP-bound G$\alpha$ subunit. While bound together in this complex, neither component is capable of interacting with effectors. Specific engagement of agonist-stimulated GPCRs with the heterotrimer results in the exchange of GDP for GTP at the catalytic site of G$\alpha$, and full or partial disengagement of G$\alpha$ from G$\beta\gamma$ (Bunemann *et al.*, 2003; Frank *et al.*, 2005; Gilman, 1987). The intrinsic rate at which GDP dissociates from G protein heterotrimers is extremely low (Higashijima *et al.*, 1987c), and GPCRs increase this rate by several orders of magnitude. The mechanism by which GPCRs catalyze this, the activation step in the G protein regulatory cycle, is not fully understood; current models are discussed by Oldham and Hamm (this volume). $Mg^{2+}$ is invariably coordinated to G$\alpha$ and GTP, and therefore, "G$\alpha$•GTP" implicitly refers to the complex with $Mg^{2+}$, unless stated otherwise. G$\alpha$•GTP and G$\beta\gamma$ may act on different effectors or act either independently or synergistically on the same effector. A signaling cycle is terminated by G$\alpha$-catalyzed hydrolysis of GTP, which results in dissociation of G•GDP from effector and reassociation with G$\beta\gamma$. The intrinsic rate of GTP turnover by G$\alpha$ proteins is low—on the order of 2 $min^{-1}$, but can be accelerated 1–3 orders of magnitude by GTPase-activating domains within the effector, or by exogenous GTPase activating proteins (GAPs) that contain RGS domains (Ross and Wilkie, 2000). It is now apparent that G protein signaling involves greater spatial and temporal complexity than is suggested by this simple scheme. There is evidence that G$\alpha$:effector complexes are preassembled on the plasma membrane in signaling complexes composed of a G protein, agonist-stimulated GPCR, effector, and possibly a GAP (Dowal *et al.*, 2006). G protein activation (nucleotide exchange on Gα) and signal termination (GTP hydrolysis) can become synergistically coupled, leading to rapid cycling with high, steady-state effector activation (Biddlecome *et al.*, 1996; Dowal *et al.*, 2001; Ross and Wilkie, 2000; Zhong *et al.*, 2003).

## II. A Selective Survey of Gα Protein Structure and Function

The vertebrate Gα proteins form a closely related family of structurally conserved proteins. The most similar are more than 90% identical in amino acid sequence, while the most divergent differ by more than 60%. The taxonomy and evolution of the family have been reviewed (Gilman, 1987; Wilkie and Yokoyama, 1994). Four classes of Gα proteins have been described. The members of each class are related in sequence and, to differing extents, act on the same sets of effectors (Morris and Malbon, 1999). The Gs subfamily is represented by the prototypical "stimulatory" Gαs—expressed as short and long splice variants—the classic activator of all membrane-intrinsic adenylyl cyclase isoforms (Gilman, 1987; Hanoune and Defer, 2001; Sunahara *et al.*, 1996). Early biochemical studies of this system established the paradigm for G protein signaling, even before the distinct molecular components involved in the pathway were defined (Cassel and Selinger, 1978; Cassel *et al.*, 1977). The first structural insights into the mechanism of effector activation by Gα were also gained from crystallographic studies of Gαs complexes with the catalytic domains of adenylyl cyclase (Tesmer *et al.*, 1997b). Gαolf, stimulated by the family of olfactory receptors, is closely related to Gαs, and also stimulates adenylyl cyclase (Schild and Restrepo, 1998).

The Gi family includes Gαi, Gαo, Gαt, and Gαz. Gαi isoforms, of which three have been characterized, were first described as inhibitors of type V and type VI adenylyl cyclase (Chen and Iyengar, 1993; Taussig *et al.*, 1993). The mechanism of this inhibition is biochemically well understood, but its structural basis is suggested only by mutagenic and modeling studies. Together, Gαs and Gαi exert reciprocal, antagonistic but noncompetitive control over the catalytic activity of adenylyl cyclase. Gαi proteins participate in a rich repertoire of signaling functions. Among the more intriguing of these are nonreceptor-mediated cytokinesis pathways that involve cytoplasmic GDIs and GEFs (Afshar *et al.*, 2004; Hess *et al.*, 2004; Lanier, 2004; Tall *et al.*, 2003; Willard *et al.*, 2004). The two isoforms of retinal transducin (Gαt) activate cGMP PDE on stimulation of visual rhodopsins by binding to the inhibitory subunit of the PDEγ. Structural studies in Paul Sigler's laboratory have illuminated the mechanism of transducin-mediated sequestration of the inhibitory PDE$\gamma$ subunit in the context of RGS9, a GAP for G$\alpha$t (Slep *et al.*, 2001).

Gq family members (G$\alpha$q, G$\alpha$11, G$\alpha$14, and G$\alpha$15/16) are all activators of PLC$\beta$ isoforms (Rhee, 2001; Singer *et al.*, 1997), although functional diversity within this family is appreciable (Hubbard and Hepler, 2006). PLC$\beta$ is the first G$\alpha$ effector demonstrated to function also as a GAP (Berstein *et al.*, 1992), and is subject to dual and possibly synergistic regulation by G$\alpha$q and G$\beta\gamma$ (Smrcka and Sternweis, 1993). G$\alpha$q also binds to G protein receptor kinases, which mediate initial steps in GPCR desensitization and downregulation by GRKs (Krupnick and Benovic, 1998; Pitcher *et al.*, 1998). G$\alpha$q does not appear to modulate GRK activity, but may help to localize GRK to the membrane. G$\alpha$q is itself sequestered, together with G$\beta\gamma$, at the plasma membrane by GRK. Structural analysis of GRK2 complexes with G$\beta\gamma$ and G$\alpha$q provide the first view of a G protein–signaling complex as it might assemble at the plasma membrane (Tesmer *et al.*, 2005).

The G12 family proteins G$\alpha$12 and G$\alpha$13 are regulators of a family of RhoGEFs (Hart *et al.*, 1998). Effectors within this family exhibit GAP activity toward G$\alpha$12/13 (Kozasa *et al.*, 1998); recent biochemical and structural studies, described here and in the chapter by Sternweis *et al.* in this volume, have begun to suggest how GAP and effector-stimulatory activity are coupled in this family of effectors.

Three-dimensional structures have been now determined for at least one representative of each of the four classes of G$\alpha$ subunits: G$\alpha$t (transducin) (Noel *et al.*, 1993) and G$\alpha$i1(Coleman *et al.*, 1994), G$\alpha$s (Sunahara *et al.*, 1997b), G$\alpha$12 and G$\alpha$13 (Chen *et al.*, 2005; Kreutz *et al.*, 2006), and G$\alpha$q (Tesmer *et al.*, 2005). The G$\alpha$ fold is highly conserved, such that the structures of these molecules, in equivalent nucleotide-bound states, can be superimposed with an rmsd of less than ~1.5 Å in the positions of corresponding C$\alpha$ atoms. Comparison of GDP-bound and GTP analogue-bound G$\alpha$t and G$\alpha$i revealed the conformational transitions that occur when GTP is hydrolyzed (Lambright *et al.*, 1994; Mixon *et al.*, 1995). As in Ras, local conformational transitions occur within two short linear segments of G$\alpha$ that flank the GTP-binding site, and, following the nomenclature established for Ras, are designated as Switch I and Switch II (Figs. 1 and 2 for residue and sequence information). Both segments contain sequence motifs that are conserved in the G protein superfamily (Sprang, 1997). Residues within Switch I and Switch II are involved in $Mg^{2+}$ coordination, GTP binding, and the catalytic mechanism.

The physical basis of the kinetic linkage between GTP hydrolysis and conformational change arises from the involvement of Switch I and Switch II in both processes. The P-loop, or Walker A motif (Walker *et al.*, 1982)

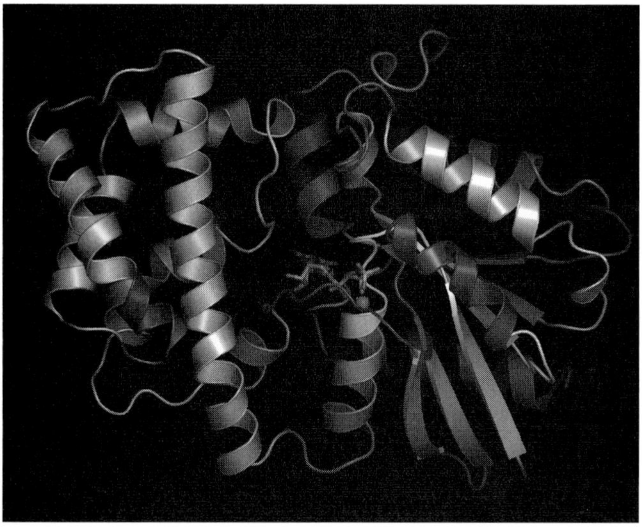

FIG. 1. Tertiary structure of Gαi1 bound to the nonhydrolyzable GTP analogue GppNp; α helices are depicted as coiled ribbons and β strands as arrows in this and subsequent figures. Switch I and Switch II are labeled and colored deep blue; Switch III and Switch IV are colored cyan. The P-loop is yellow. The GTP analogue GppNp is shown as a stick figure, with atoms color-coded: green, carbon; nitrogen, blue; oxygen, red; and phosphorus, yellow. $Mg^{2+}$ is depicted as a magenta sphere. Figures were produced with the molecular graphics program PyMOL™ (http://www.pymol.org). (See Color Insert.)

which enfolds the α and β phosphates of the guanine nucleotide, adopts the same conformation in all Gα proteins, and its structure is not affected by GTP hydrolysis or the binding of effectors or regulatory proteins. Switch I contains conserved residues that participate in catalysis and, together with the P-loop, in $Mg^{2+}$ binding. Residues in Switch II are also involved in these functions and also form the binding pocket for the γ phosphate of GTP. GTP hydrolysis results in an order-to-disorder transition in Switch II of Gαi and Gα12, while in Gαt, this element adopts an alternative but apparently stable conformation. Switch III, which is not a locus of conformational change in Ras, corresponds to a loop-helix segment that contacts Switch II through ionic and polar interactions. Switches I–III are within or in proximity to Gα effector-binding sites. Hence, the conformational changes that occur within these segments on GTP hydrolysis are responsible for the reduction in affinity of Gα for effectors. The helical domain also contains a segment, Swich IV, that was noted to undergo a nucleotide-dependent conformational change in Gαi (Mixon *et al.*, 1995). Because it is remote from the nucleotide, Gβγ, or effector-binding sites, the

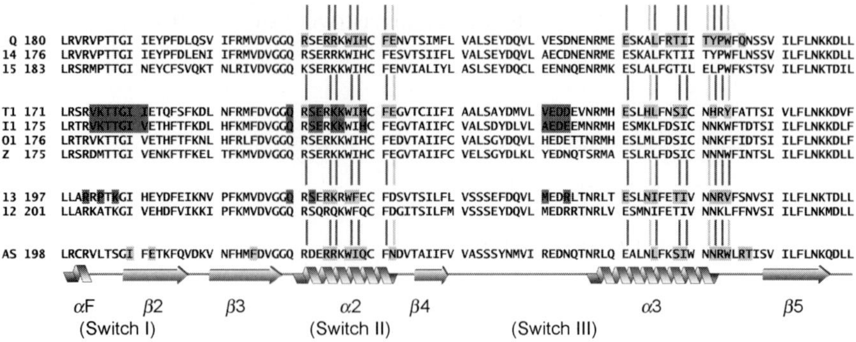


FIG. 2. Primary structure of the effector/GAP-binding region of selected human Gα proteins: Gαq (GenBank accession number AAM12610), Gα14 (AAM12617), Gα15 (AAM12618), Gαt1 (AAM12622), Gαi1 (AAM12619), Gαo1 (AAM12609), Gαz (AAM12613), Gα13 (AAM12616), Gα12 (AAM12614), and Gαs (AAM12612). Vertical bars denote highly conserved (dark green bars) or nonconserved (light green bars) residues that are involved in effector interactions in at least two Gα:effector complexes for which structural data are available. Residues shaded green are in contact with effector domains; residues shaded red contact GAP domains; magenta shading indicates contact with both domains. Note that some of the contacts in the α3-β5-binding region involve residues derived from Gαi in the crystal complexes of Gαt-i:PDEγ:RGS9 and Gα13-i: rgRGS (see text and legend to Fig. 3). (See Color Insert.)

mechanistic significance of changes in Switch IV was unclear. However, Switch IV was found to form part of the binding site for GoLoco/GPR peptides that inhibit the release of GDP from Gαi/o proteins (Kimple *et al.*, 2002).

Gα proteins are either or both thio-palmitoylated or N-terminally myristoylated, and Gγ are isoprenylated at their C-termini (Casey, 1994; Wedegaertner *et al.*, 1995). These modifications can serve as membrane anchors, but can also affect the affinity of G protein interactions with effectors and regulators. N-terminal myristoylation of Gαi/o, for example, promotes high-affinity binding to Gβγ (Linder *et al.*, 1991) and to adenylyl cyclase (Taussig *et al.*, 1993), while palmitoylation of Gαz and Gαi reduces affinity for the GAP proteins GzGAP and RGS4, respectively (Tu *et al.*, 1997). Heterologously expressed, N-terminally myristoylated Gαi1 is soluble and monomeric (Linder *et al.*, 1991). Therefore, it is possible that the myristoyl group is shielded by intramolecular contacts with protein side chains. If so, myristoyl groups, rather than simply tether Gα to the membrane, may dock within the protein and modulate its structure. This in fact appears to be the case for recoverin, where the myristoyl group is buried in the $Ca^{2+}$-free state, but is extruded on $Ca^{2+}$ binding (Ames *et al.*, 1995;

Tanaka *et al.*, 1995). The N-terminal myristoyl group of the cAMP-regulated protein A kinase catalytic unit is buried in a deep hydrophobic pocket and anchors the N-terminal residues that lie outside of the catalytic core (Zheng *et al.*, 1993). Our understanding of the structural basis of Gα: effector/regulator interactions is gained largely from crystallographic studies conducted within unmodified proteins. Consequently, these interactions are described with the caveat that lipid modifications are likely to have important, if unknown, consequences.

## III. Mechanisms of Effector Recognition and Regulation by Gα•GTP

Like all members of the G protein superfamily, Gα proteins in the GTP-bound state have high affinity for effectors; in the GDP-bound state, affinity for effectors is diminished. The reverse is true for Gβγ, which binds Gα•GDP tightly. This selectivity exploits the high affinity of Gα for both GTP and GDP, but extracts a severe penalty in the form of a substantial activation energy for GTP hydrolysis and GDP release; hence, the low steady-state rate of GTP hydrolysis by Gα. Effectors are structurally and functionally diverse. They may engage Gα•GTP as regulatory binding partners or as allosteric regulators. As described in Section V, effectors can also affect the lifetime of the GTP-bound state of Gα.

Our current understanding of the structural basis of Gα-effector interactions is largely based on crystal structures of four Gα:effector complexes: Gαs bound to the catalytic domains of adenylyl cyclase (Tesmer *et al.*, 1997b); the ternary complex of Gαq, Gβγ, and GRK2 (Tesmer *et al.*, 2005); Gαt bound to PDEγ in the presence of the RGS domain of RGS9 (Slep *et al.*, 2001); and Gα13 bound to the N-terminal rgRGS domain of p115RhoGEF (Chen *et al.*, 2005) (Fig. 3). Thus, the structure of an effector complex has been obtained for each of the four classes of Gα. In the latter two, the Gα protein incorporates residues from Gαi to allow recombinant expression in bacteria. The structural data demonstrate, first, that all effectors bind to a common recognition surface of Gα, although individual effectors also may target sites outside of this surface. Second, effector binding does not appear to induce substantial (<1 Å rmsd at Cα-positions) changes in the structure of Gα•GTP; hence, the effector-binding site is preformed in the GTP-bound state. Third, the interactions between effectors and Gα are fairly class-specific; Gαi•GTP and Gαt•GTP are both able to bind PDEγ (the former with 100-fold less affinity) (Skiba *et al.*, 1996), but do not act as adenylyl cyclase activators. Exceptions to this rule can be

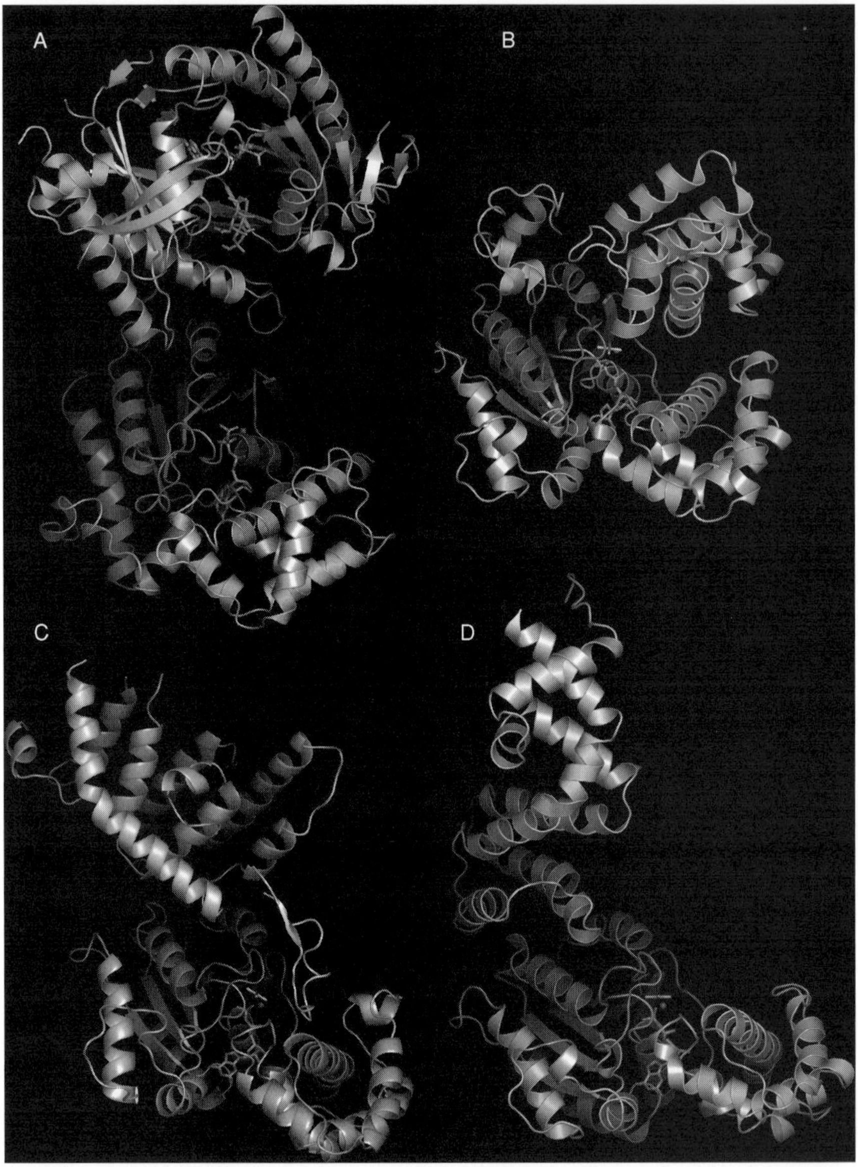


FIG. 3. Gallery of Gα:effector:GAP complexes. The Gα subunits, colored according to the scheme in Fig. 1, are similarly oriented in the four panels. Effector domains are colored pale green, and GAP domains light tan; Panel A, complex among Gαs•GTPγS and the $VC_1$ and $IIC_2$ domains of mammalian mAC. The latter is shown bound to forskolin and a nonhydrolyzable analogue of ATP (protein Databank ID 1CJT)

found: one member of the p115RhoGEF family is stimulated by both Gα13 and Gαq, but others are activated only by Gα12/13 (Booden *et al.*, 2002).

Two of the Gα:effector complexes that we describe in this chapter involve interactions between the Gα protein and RGS-like domains of the effector. As discussed in greater detail in Section V, many RGS domains exhibit GAP activity toward Gα. This is true for the major family of RGS domains whose members share substantial sequence similarity. The RGS-like domains of p115RhoGEF and GRK2 belong to separate and distinct subgroups that bear only remote sequence similarity to the larger family of RGS GAPs, and bind to Gα in the mode characteristic of effectors.

### A. *A Common Gα:Effector Interface*

The effector recognition site of Gα is formed almost entirely of the residues extending from the C-terminal half of α2 (Switch II) together with the α3 helix and its junction with the β5 strand (Figs. 2 and 4). In its interaction with adenylyl cyclase, Gαs also binds residues in the β2–β3 hairpin that follows Switch I. Several of the Switch II effector-binding residues also form contacts with Gβ in the G protein heterotrimer. Therefore, Gβγ and effector binding to Gα are mutually exclusive (Lambright *et al.*, 1996; Wall *et al.*, 1995, 1998). In some effector:Gα complexes, elements well outside of the α2, α3 locus of Gα participate in effector binding, as is suggested for adenylyl cyclase and p115RhoGEF, discussed below. Structural information regarding these latter complexes is limited by the use of truncated proteins, as in the case of p115RhoGEF or minimal catalytic domains, as for adenylyl cyclase, in order to obtain protein complexes amenable to crystallization. Nevertheless, involvement of the α2, α3 locus is the common and perhaps minimal requirement for formation of a Gα:effector complex. In contrast to Gα proteins, effectors are structurally diverse, although the Gα-binding regions of those that have been characterized are composed of α helical domains or α helical segments (Fig. 3). Adjacent parallel helices of the adenylyl cyclase $C_2$ domain embrace Switch II/α2 of Gαs in a scissors grip. A pair of antiparallel helices in the RGS-like domain of GRK2 packs against α2 and α3 of Gαq to

(Tesmer *et al.*, 1999); Panel B, the complex among Gαt-i•GDP•$Mg^{2+}$•$AlF_4^-$, PDEγ, and the RGS (GAP) domain of RGS9 (1FQJ) (Slep *et al.*, 2001); Panel C, the complex between Gαt-i•GDP•$Mg^{2+}$•$AlF_4^-$ and the rgRGS domain of p115RhoGEF (1SHZ) (Chen *et al.*, 2005); Panel D, the interaction between Gαq•GDP•$Mg^{2+}$•$AlF_4^-$ and the RGS-like domain of GRK2 in the ternary complex among Gαq, GRK2, and Gβγ (2BCJ) (Tesmer *et al.*, 2005). (See Color Insert.)

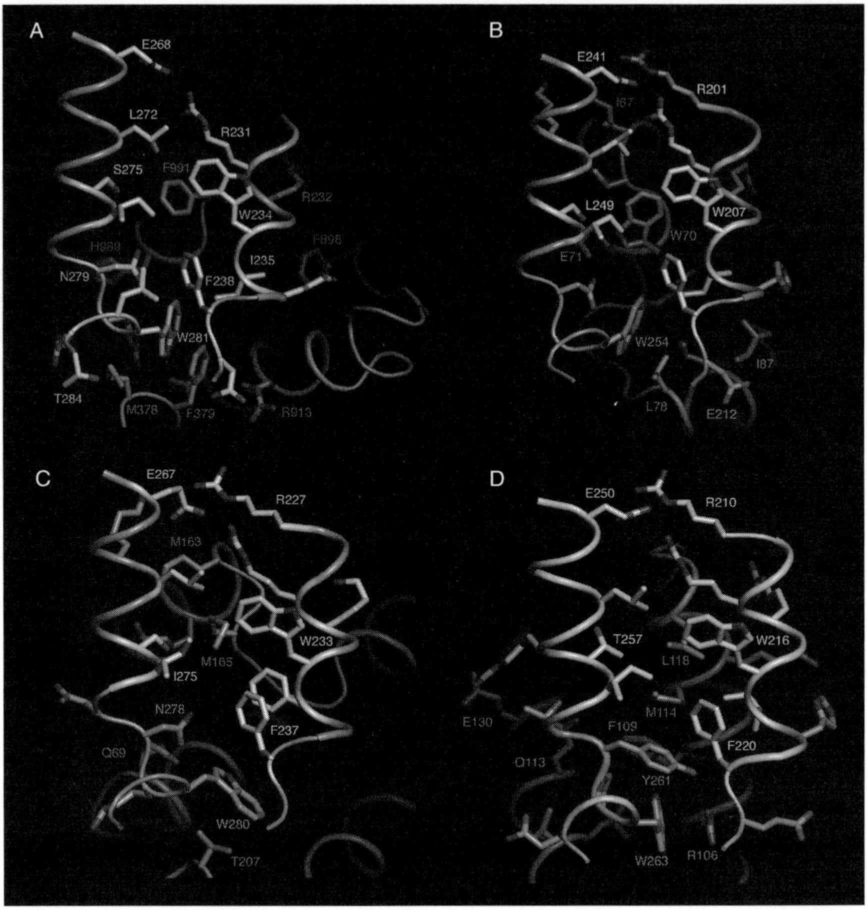


FIG. 4. Gα:effector interactions at the effector-binding surface common to Gα proteins. Cα trace of Gα and effector is rendered as a smooth coil, with side chain atoms of selected residues depicted. Shown in the same orientation in each panel are the α2 (Switch II) and α3 helices (including the loop connected to β5) of Gα, looking toward the effector, and colored orange. The side chains of effector-contacting residues of Gα are shown and conserved residues (Fig. 2) are colored light tan. The effector region of the Gα-binding partner is colored blue. Selected side chains within 4 Å of the Gα-binding surface are shown. Atoms are color-coded as in Fig. 1, except that carbon atoms assume the color assigned to the backbone. The single letter code is used to identify residue types. The complexes shown in panels A–D correspond to those in the corresponding panels of Fig. 3. (See Color Insert.)

generate a loose four helix bundle. The three short antiparallel helical segments of PDEγ thread a serpentine path on the surface of Gα, delving into the trough between α2 and α3. The RGS-like domain of p115RhoGEF presents a complex surface, composed of three α–α hairpins that form an array of contacts at both ends of the Gα13 effector-binding motif. Although the RGS-like domains of GRK2 and p115RhoGEF are structurally similar, they bind differently to Gα.

The common effector-binding surface of Gα proteins is defined by about 15 well-conserved residues located on α2 and α3 (Figs. 2 and 4). The side chains of the two helices are knit together by ion pair and van der Waals interactions in the GTP-bound conformation to create a rigid matrix of amino acid side chains into which the effector docks. Mutagenesis and functional studies demonstrated that residues at this locus are key to Gαs activation of adenylyl cyclase (Berlot and Bourne, 1992), Gαi1 inhibition of adenylyl cyclase (Grishina and Berlot, 1997), binding of Gαt to PDEγ (Natochin *et al.*, 1998a; Skiba *et al.*, 1996), Gαq activation of PLCβ (Arkinstall *et al.*, 1995; Venkatakrishnan and Exton, 1996) and binding to GRK2 (Day *et al.*, 2004), and Gα13 interaction with p115RhoGEF (Grabocka and Wedegaertner, 2005). Mutagenesis experiments have also identified residues that do not appear at the Gα:effector interfaces revealed in crystal structures. Inconsistencies between structural and functional studies can be attributed either to nonlocal effects of mutations on the structure and dynamics of the protein complex or to the incomplete nature of crystal complexes in which not all functionally relevant effector domains are present.

The common effector-binding surface of Gα contains a cavity, and presents a knob, that complement corresponding features of opposite curvature on the surface of the effector. The Gα cavity is formed by a constellation of residues on the inward-facing surfaces of α2 and α3, many of which are conserved among Gα sequences (Figs. 2 and 4). Into this cavity, effectors insert one or more bulky hydrophobic or aromatic residues, which typically project from the exposed junction between two helices. PDEγ presents a tryptophan; the $C_2$ domain of type II adenylyl cyclase, a phenylalanine; GRK2, a leucine–methionine pair; and p115RhoGEF, a methionine into the α2–α3 pocket of Gα. Mutation of these residues reduces the affinity of adenylyl cyclase for Gαs (Yan *et al.*, 1997), and abolishes binding of PDEγ and GRK2 to Gαt and Gαq, respectively (Otto-Bruc *et al.*, 1993; Slepak *et al.*, 1995; Sterne-Marr *et al.*, 2003).

The Gα effector-recognition knob is formed by residues near the α3–β5 junction. In all Gα isoforms except for Gα13 and Gα12, a tryptophan residue (Trp 234 in Gαs) constitutes the "knob" that fits into a complementary cavity in the effector. Mutation of the α3 tryptophan and surrounding residues at Switch II abrogates PDE$\gamma$ binding to G$\alpha$t (Faurobert *et al.*, 1993; Natochin *et al.*, 1998a; Skiba *et al.*, 1996) and binding of GRK2 to G$\alpha$q (Lodowski *et al.*, 2005). Because effectors differ in tertiary structure, the shape and composition of the cavity that accepts the G$\alpha$ knob is variable; likewise, structural and sequence variation among G$\alpha$ isoforms is also greatest in this region of the effector-binding locus (Figs. 2 and 4), and may be a key to the selectivity of G proteins toward their effectors. A sense of the structural origin of selectivity may be gained by noting the steric conflicts that would occur in models of nonphysiological G$\alpha$:effector complexes (viz., G$\alpha$s•GTP with PDE$\gamma$)—a thought experiment that is easily done after superimposing coordinates of all experimentally determined G$\alpha$:effector complexes, using only the $\alpha 2$,$\alpha 3$-$\beta 5$ atoms of the G$\alpha$ proteins. Most steric conflicts between G$\alpha$ and mismatched effectors would be confined to the C-terminus of $\alpha 3$, near its junction with $\beta 5$. In some instances, residues near the C-terminus of $\alpha 2$ would also generate steric conflicts with inappropriate effectors. Hence, the $\alpha 3$ locus of G$\alpha$ is a significant determinant of effector selectivity, as corroborated by mutagenesis analysis of the G$\alpha$t:PDE$\gamma$ interaction (Natochin *et al.*, 1998a). Similarly, the specificity of G$\alpha$q toward PLC$\beta 1$ requires residues 243–276 at the junction of $\alpha 3$ and $\beta 5$ (Venkatakrishnan and Exton, 1996).

Different effectors may recognize distinct subsets of effector-binding residues of the same G$\alpha$ protein. For example, mutation of residues 259–261 in $\alpha 3$ of G$\alpha$q does not block stimulation of PLC$\beta$. However, Tyr 261 of G$\alpha$q is located at the interface with GRK2, and mutation of this residue impairs binding (Fig. 4C) (Tesmer *et al.*, 2005). Specificity may also be conferred by differences in the main chain conformation of effector-binding regions of G$\alpha$ subunits. For example, G$\alpha$i1 and G$\alpha$s bind at different sites on their common effector adenylyl cyclase. However, G$\alpha$i differs from G$\alpha$s at only two positions within the $\alpha 2$, $\alpha 3$ effector-binding surface. Model building suggests that the exclusion of G$\alpha$i from the G$\alpha$s-binding surface of AC is not due to amino acid sequence differences between the two G$\alpha$ proteins. Rather, it appears that differences in the spatial relationship between the $\alpha 2$ and $\alpha 3$ motifs of G$\alpha$i1 and G$\alpha$s account for the high selectivity of the two G proteins for their respective binding sites on AC (Sunahara *et al.*, 1997b). There is evidence from mutagenesis that G$\alpha$i1 may use residues in the $\alpha 4$-$\beta 6$ loop, outside of the consensus effector-binding site in its recognition of type V and VI adenylyl cyclases, but not in the $\alpha 3$-$\beta 5$ loop (Grishina and Berlot, 1997). This finding has yet to be confirmed by structural data.

Although G$\alpha$ proteins exhibit considerable selectivity, effector-binding surfaces can nevertheless tolerate substantial variation. That is why it has been possible to use chimeric G$\alpha$ proteins in structural studies of G protein: effector complexes. Some G$\alpha$ isoforms are difficult to express in heterologous systems. This is true of Gαt, Gαq, and Gα12/13. However, in the course of experiments undertaken to identify the site of effector binding, Skiba *et al.* (1996) found that a Gαt-i chimera could be expressed in high yield as a soluble protein in *E. coli.* This protein, in which residues comprising and flanking the entire α2, α3-β5 effector-binding region were derived from Gαi1, proved to be of great use in subsequent structural studies of Gαt, and notably, its complex with RGS9 and PDEγ (Slep *et al.*, 2001). We also created a well-expressed chimeric protein in which the helical domain and switch regions of Gα13 were substituted for the corresponding residues of Gαi1 (Chen *et al.*, 2005). The Gα13-i chimera served as a substrate for GAP activity of p115RhoGEF *in vitro,* and formed a stable complex with the rgRGS domain of the latter that could be crystallized. Because of the chimeric nature of the G proteins used to form complexes with PDEγ and rgRGS, portions of the effector-binding surfaces contributed by α3 were derived from Gαi rather than Gαt or Gα13, respectively, the appropriate partners for the respective receptors. There are five such substitutions in the Gαt-i:PDEγ complex and seven in Gα13-i:rgRGS complex. Most are conservative, but demand different packing or hydrogen-bonding interactions. Nevertheless, the "inappropriate" Gα residues are accommodated at the effector interface. The chimeric Gα proteins have 2–5% of the affinity of the corresponding wild-type Gα proteins for their effector partners as measured by direct binding (Gαt-i with PDEγ) or GAP activity (rgRGS domain of p115RhoGEF toward Gα13-i) (Chen *et al.*, 2005; Skiba *et al.*, 1996). Accommodation of chimeric Gα proteins by effectors requires distribution of binding energy over the Gα:effector recognition surface. From the data available, no simple spatial or consensus sequence rules have emerged that account for the specificity of Gα:effector interactions.

### B. *Functional Consequences of Gα:Effector Binding*

Effector binding, per se, is the most direct output of Gα activation, and this event may be sufficient to account for the mechanism of PDE activation, in which Gαt binds and sequesters the inhibitory subunit of the enzyme complex. Thus, the C-terminal residues of PDEγ responsible for inhibition of the catalytic PDEαβ heterodimer are sequestered in the interaction with Gαt (Brown, 1992; Lipkin *et al.*, 1988). In other systems, Gα are allosteric activators or inhibitors, or act in more subtle and less well-understood roles as organizing centers of G protein–signaling complexes.

#### 1. *Adenylyl Cyclase*

Vertebrate mACs are regulated by a variety of mechanisms, and most of these are isoform-specific (Cooper, 2003; Hanoune and Defer, 2001; Sunahara and Taussig, 2002; Sunahara *et al.*, 1996). However, all nine isoforms of mAC are activated by Gαs•GTPγS (Mou *et al.*, 2005; Tesmer *et al.*, 1997b, 1999, 2000; Zhang *et al.*, 1997). The findings, in the context of biochemical data, have been reviewed (Hurley, 1999; Simonds, 1999; Tesmer and Sprang, 1998). A detailed mechanistic understanding of mAC activation by Gα proteins remains elusive, and much that is believed is based on informed speculation from limited structural and biochemical data. Even less well understood is the nature of synergistic or antagonistic interactions with other regulators such as Gi-family Gα proteins and Gβγ.

The catalytic core of mAC and other class III cyclases is formed by a pair of homologous ~220-residue CHDs, as reviewed by Sinha and Sprang (2006). In the intact mAC enzymes, the CHDs are separated by putative multiple-pass membrane-spanning domains. The N- and C-terminal CHDs, $C_1$ and $C_2$, respectively, are arranged with respect to each other by a twofold axis of symmetry. The catalytic site is shared between the two domains, and therefore the possibility exists for two, dyad-related catalytic sites. The class III cyclases of prokaryotes are active only as homodimers, and each dimer contains two active sites (Linder and Schultz, 2003). However, the $C_1$ and $C_2$ domains of mammalian cyclases are divergent (Sunahara *et al.*, 1996), and one of the two potential catalytic sites is degenerate. The diterpene forskolin, derived from the herb *Coleus forskohlii*, binds to this site (Tesmer *et al.*, 1997b) and, probably by promoting $C_1$–$C_2$ interactions, acts as an allosteric activator (Dessauer *et al.*, 1997; Seamon and Daly, 1981) (Fig. 5). Structural complimentarity between the $C_1$ and $C_2$ domains is essential to cyclase activity. The $C_2$ domain contributes residues that bind to the nucleoside moiety of ATP and confer specificity for the adenine base of the nucleotide. The $C_1$ domain provides the aspartyl residues that bind the two essential $Mg^{2+}$ (or $Mn^{2+}$) that are required for catalysis together with a helical P-loop sequence that enfolds the α and β phosphates of the nucleotide. As isolated domains, neither $C_1$ nor $C_2$ exhibits significant catalytic activity.

The $C_1$ and $C_2$ domains contain all of the biochemical determinants required for catalytic activity and regulation by forskolin, Gαs and Gαi (Dessauer and Gilman, 1996; Tang and Gilman, 1995). When expressed as separate proteins and mixed together, the $VC_1$ and the $IIC_2$ exhibit a turnover number of 0.6 μmol/min/mg. In the presence of Gαs•GTPγS, the turnover number increases 100-fold (Dessauer *et al.*, 1997). The affinity of $VC_1$ for $IIC_2$ also increases ~60-fold, so a simple explanation for the stimulatory affect of Gαs is that it promotes interactions between the catalytic domains. In addition, the Gαs-induced union of the two domains is accompanied by conformational remodeling that generates a catalytically functional catalytic site.

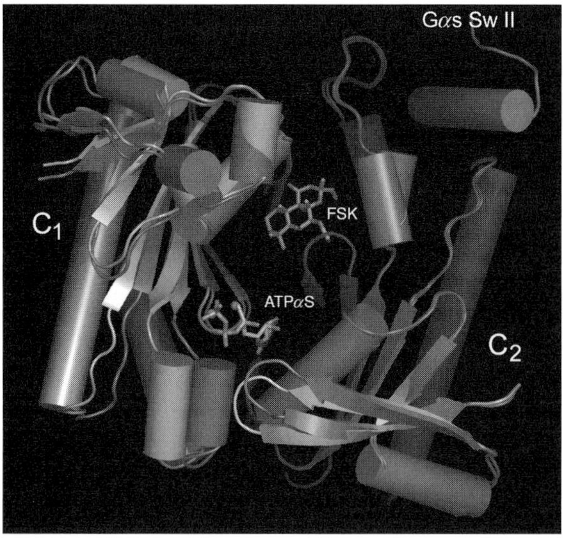


FIG. 5. The open (inactive) and closed (active) conformations of the adenylyl cyclase catalytic domains bound to Gα•GTPγS. Helices are shown as cylinders and β strands as ribbons. The $C_2$ domains of the two mAC molecules are superimposed. The open state of mAC, which was crystallized in the absence of an ATP analogue, is colored gray (PDB ID, 1AZS); the closed state, colored blue, was crystallized with ATPαS bound to the AC catalytic site, shown as a stick model (1CJT). Forskolin is present in both complexes. Switch II of Gαs is colored red. Note that binding of ATP is accompanied by segmental movement of residues in both domains toward the catalytic site. (See Color Insert.)

Structural and biochemical observations point to an allosteric role for Gαs in the activation of mAC. First, the $C_1$:$C_2$ interface and the catalytic site are remote from the Gαs-binding site (Figs. 3A and 5). The major contact between the mAC heterodimer and the G protein is formed by two adjacent helices at the periphery of the $C_2$ domain that interlock with the α2–α3-helices of Gαs. A second contact, also crucial to activation, involves a phenylalanine at the very N-terminus of the $C_1$ domain that stacks with the near-invariant α3 tryptophan of the Gαs effector-binding trough (Fig. 4A). If the $C_1$ phenylalanine is mutated, Gαs binds, but does not activate mAC (Yan *et al.*, 1997). Binding of Gαs to this peripheral site at the junction of $C_1$ and $C_2$ could affect the structure of the entire interface between the two CHDs. Gαs wedges the two helices of $C_2$ apart, while holding fast to the N-terminus of $C_1$; the distortion of $C_2$ in the context of a constrained $C_1$ domain could reconfigure the interface between the two CHDs, thereby increasing the activity of the interdomain catalytic site. Comparison of

G$\alpha$s-bound $C_1$:$C_2$ with the structure of the $C_2$ homodimer determined by Zhang *et al.* (1997) suggests that a 7° rotation of $C_1$ occurs with $C_2$ on binding of mAC to G$\alpha$s. However, the $C_2$ dimer may not be an accurate model of the G$\alpha$s-free low-activity state of $C_1$:$C_2$, the structure of which remains undetermined.

As an allosteric activator, G$\alpha$s increases the $V_{max}$ of V$C_1$:II$C_2$, but has only a modest affect on its affinity for ATP ($K_M$ for ATP decreases from 1 to 0.3 mM). Structures of G$\alpha$s/forskolin-bound V$C_1$:II$C_2$ suggest that the enzyme equilibrates between catalytically incompetent "open" and catalytically active "closed" states (Fig. 5). The structure determined in the absence of ligands or in the presence of ApCpp, a nonreactive substrate analog, reveals an "open" catalytic site, in which the $\alpha$1 helix of $C_1$—which forms the binding site for the phosphate groups of the substrate—is displaced from the purine-binding site in the $C_2$ domain. The substrate can be accommodated in the active site in this state, but cannot form productive interactions with catalytic residues or the $Mg^{2+}$ ions that are required for catalytic activity. In the presence of potent P-site inhibitors (Dessauer *et al.*, 1999) or substrate analogues, the $\alpha$1 helix and contiguous $\beta$ strands execute a hinge-like rotation toward $C_2$, in a motion that engenders compete complimentarity to the coupled ATP and $Mg^{2+}$-binding sites. This structural rearrangement is not accompanied by substantial changes in other parts of the G$\alpha$s:V$C_1$:II$C_2$ interface. G$\alpha$ may promote the closed state of mAC by inducing strain in the open state.

Synergy, rather than additivity with forskolin, is a third aspect of G$\alpha$s allosterism. In the presence of G$\alpha$s•GTP$\gamma$S, the apparent affinity of V$C_1$:II$C_2$ for forskolin increases 400-fold from a basal $K_d$ of 40 $\mu$M. The $V_{max}$ of the enzyme is two- to threefold higher in the presence of both activators than when either alone is present. Forskolin, in binding to the "degenerate" mAC catalytic site at the $C_1$:$C_2$ interface, indeed appears to act as molecular glue. On the other hand, G$\alpha$s appears more likely to promote conformational change. Yet, the two effects cannot be dissociated, and each of the two effectors contributes, through a distinct mechanism, to both modes of activation. Only regulation by G$\alpha$s is physiologically relevant, since there is no known naturally occurring counterpart to forskolin in mammalian systems. The apparent effects of G$\alpha$s on the conformation of the $C_1$:$C_2$ interface are probably the more critical, since interactions between the two domains are assured by virtue of their incorporation in the same polypeptide chain.

An intriguing but speculative role for G$\alpha$ subunits in the regulation of mAC is suggested by the observation that $C_1$ and $C_2$ domains readily form homodimers, while G$\alpha$s and G$\alpha$i have opposing effects on their conversion to $C_1$:$C_2$ heterodimers (Sunahara *et al.*, 1997a). The self-association constant for $IIC_2$ is in the submicromolar range, and the homodimer can be recovered by gel-filtration chromatography. Indeed, crystallographic analysis of the forskolin-bound $IIC_2$ homodimer provided the first clue to the structure of the cyclase catalytic core (Zhang *et al.*, 1997). Gαs•GTPγS displaces $C_2$ from the $(C_2)_2$ homodimer and forms a Gαs•GTP:$C_2$ complex that readily binds $C_1$, thereby generating a Gαs•GTPγS-activated $C_1$:$C_2$ catalytic complex. In contrast, Gαi1•GTPγS displaces $C_1$ from $(C_1)_2$ to generate a Gαi1•GTPγS:$C_1$ complex but does not support stable formation of a Gαi1•GTPγS:$C_1$:$C_2$ complex. Thus, Gαs favors but Gαi1 inhibits formation of productive interactions between the catalytic domains of mAC. Homodimeric association of $C_1$ or $C_2$ would naturally be inhibitory toward adenylyl cyclase activity, but would be biochemically feasible and relevant only if mAC holoenzymes form dimers or oligomers at the plasma membrane. In this context, homomeric association of $C_1$ or $C_2$ domains would sequester the catalytic domains of mAC oligomers into inactive complexes. Indeed, there is biochemical and mutagenic evidence for the formation of multimeric mAC species (Pfeuffer *et al.*, 1985; Tang *et al.*, 1995). mACs were shown to dimerize or oligomerize through their transmembrane domains (Gu *et al.*, 2002). Homomeric interactions between catalytic domains of dimerized AC molecules would result in low basal activity. Gαs•GTP could potentially activate mAC, first, by breaking up inhibitory interactions between the $C_2$ domains of dimerized molecules, and second, by allosteric potentiation of a productive $C_1$:$C_2$ interface.

The inhibitory action of Gαi toward type V and VI mAC is also exerted through an allosteric mechanism. Gαi weakens interactions between $C_1$ and $C_2$ and impairs binding of substrates. Accordingly, mutations that increase the affinity of $VC_1$ for $VC_2$ are more resistant to the inhibitory action of Gαi1 (Dessauer *et al.*, 2002). Gαi1•GTP does not compete for the binding of Gαs•GTP, but rather reduces the $V_{max}$ for Gαs-stimulated AC activity (Taussig *et al.*, 1993). In the presence of Gαi•GTP, a higher concentration of Gαs is required to achieve the same level of AC activity than is needed in its absence. Therefore, Gαi•GTP destabilizes the mAC:Gαs•GTP complex by interaction with the $C_1$ domain, possibly at a site that is structurally analogous to the Gαs-binding site of $C_2$. Mutations of residues within the α2- and α3-helices of $VC_1$ (which correspond to the helices of the $C_2$ domain that bind to Gαs) alter the sensitivity of type VAC to inhibition by Gαi1. Further investigation shows that residues C-terminal to the core of the $C_1$ domain (called $C_{1b}$) are also involved in the interaction (Dessauer *et al.*, 1998; Wittpoth *et al.*, 1999). The ternary complex containing weakly associated Gαi•GTP, Gαs•GTP, and mAC might possess the quasi-twofold symmetry of $C_1$:$C_2$, with the two Gα subunits attached to their respective binding sites on quasi-symmetry-related CHDs (Dessauer *et al.*, 1998). Depending on the relative concentration of Gαi•GTP and Gαs•GTP at the plasma membrane, this ternary complex would either dissociate to form active mAC:GαsGTP with $C_1$ and $C_2$ domains tightly engaged or inactive mAC:Gαi1•GTP with $C_1$ and $C_2$ domains dissociated. In the context of dimeric mAC, the inhibited state could be further reinforced by formation of homodimeric interactions between $C_2$ domains.

### 2. *G12/13 Stimulation of RhoGEFs*

Because p115RhoGEF family members are both effectors and GAPs, the mechanism by which they regulate Gα12, Gα13 or Gαq is complex. In this sense, the regulatory paradigm for p115RhoGEF is comparable to that for PLC*β* isoforms. p115RhoGEF, the founding member of this family of RhoGEFs (which we shall refer to as RGS-homology-RhoGEFs, or RH-RhoGEFs) contains an RGS-like domain near its N-terminus, linked to DH and PH domains that constitute the functional unit of GEF activity in most Rho-directed GEFs (Cerione and Zheng, 1996). In the two other known orthologues, LARG (Fukuhara *et al.*, 2000; Kourlas *et al.*, 2000) and PDZ-RhoGEF (Fukuhara *et al.*, 1999), an N-terminal PDZ domain precedes the RGS-like module. Dimerization or oligomerization mediated by C-terminally located coiled coil segments appears to inhibit the GEF activity of RH-RhoGEFs (Chikumi *et al.*, 2004). The three members of the RH-RhoGEF family differ with respect to their Gα specificity and GAP activity. p115RhoGEF is stimulated by activated Gα13; it binds but is not stimulated by Gα12 and has GAP activity toward both Gα12 and Gα13 (Hart *et al.*, 1998). LARG is activated by Gαq as well as Gα13 (Booden *et al.*, 2002), and if tyrosine-phosphorylated is also stimulated by Gα12 (Suzuki *et al.*, 2003). Of the three RH-RhoGEF orthologues, only PDZ-RhoGEF is without GAP activity (Wells *et al.*, 2002). A comprehensive discussion of RH-RhoGEFs is presented by Sternweis *et al.* in this volume. Here, we focus on mechanisms by which Gα13 might stimulate the GEF activity of p115RhoGEF. The discussion is largely informed by the structure of the complex between the rgRGS domain of p115RhoGEF with Gα13-i, a chimera of Gα13 and Gαi1 (Chen *et al.*, 2005) (Fig. 3C). The rgRGS fragment encompasses residues 1–239, which consists of the N-terminal polypeptide segment that is largely responsible for the GAP activity of p115RhoGEF (residues 1–42), followed by the RGS-like domain (43–239), which, in its mode of Gα binding, assumes the properties of an effector (in contrast to a GAP) domain. The mutual binding of both segments to Gα13 allows us to consider, in Section V.D , the mechanism by which the GAP and effector properties of p115RhoGEF and its homologues might be coordinated.

In general, RhoGEFs engage Rho-family GTPases directly through their DH domains; the participation of PH domains in the interaction differs among members of the RhoGEF family, and the nature of their contribution to GEF activity is not well understood (Rossman and Sondek, 2005). The DH-PH domains of P115RhoGEF, LARG, and PDZ-RhoGEF all stimulate nucleotide exchange on RhoA *in vitro*; the crystal structures of the DH-PH domains of the latter two RH-RhoGEFs bound to Rho have been determined (Derewenda *et al.*, 2004; Kristelly *et al.*, 2004). Gα13 has been reported to bind, but does not stimulate the GEF activity of the isolated DH-PH domain of p115RhoGEF (Wells *et al.*, 2001). Hence, it appears that Gα13 must also interact with the rgRGS domain of p115RhoGEF in order to stimulate GEF activity.

The contact surface between Gα13 and the rgRGS domain of p115RhoGEF is bipartite (Fig. 3C), such that the N-terminal residues that confer GAP activity occupy a distinct subsurface of Gα, comparable to that occupied by RGS GAPs (Section V.D). As described in Section III.A, the RGS-like domain of p115RhoGEF binds to the characteristic effector-binding site of Gα13. Mutations of residues within the N-terminal GAP segment of the rgRGS domain substantially reduce affinity for Gα13 and, consequently, GAP activity. In contrast, the same mutations have no effect on the ability of Gα13 to stimulate the GEF activity of p115RhoGEF (Chen *et al.*, 2003). Therefore, the GAP and effector elements (N-terminal and RGS-like domains) within the rgRGS domain are to some extent functionally independent, although both contribute to Gα13 binding.

Although the mechanism of p115RhoGEF GAP activity can be deduced from the structure of the rgRGS:Gα13-i complex (Section V.D), it is not apparent how Gα13 stimulates RhoGEF activity. The spatial relationship between the rgRGS domain and the DH-PH domain is not known for either the free or Gα13-bound p115RhoGEF. The Gα13-binding site within the DH-PH domain has not been identified, nor have the structural elements within Gα13 with which it interacts. It is remarkable that mutations in the β4 strand of Gα12, which is well away from the consensus effector-binding site, or that to which the rgRGS domain binds, abolished its ability to stimulate RhoA-mediated responses in cultured cells (Meigs *et al.*, 2005).

The structure of the DH-PH domains of LARG with RhoA may provide an indirect clue to the mechanism of Gα13 stimulation (Kristelly *et al.*, 2004). Preceding the DH-PH domains of lbc-family RhoGEFs, which include intersectin and the RH-RhoGEFs, is a characteristic and conserved block of residues that, in LARG, folds into an α–α hairpin. On binding to RhoA, the hairpin moves 2–3 Å and forms contacts with residues in Switch I of RhoA. On deletion of the hairpin, the truncated LARG DH-PH domain has only 1/5th the activity of the intact domain. Kristelly *et al.* (2004) suggest that this block of residues might serve as a point of regulation by Gα, possibly in conjunction with the domains of the exchange factor. The case of SOS provides a second paradigm for RhoGEF regulation, although the structural context is quite different from that of the RH-RhoGEFs (Margarit *et al.*, 2003). SOS contains both a DH-PH domain with exchange activity for Rac1 and a cdc25 domain, which is a GEF for Ras. The cdc25 and neighboring Rem domains contain a positive allosteric site for Ras•GTP, which is distinct from the site at which Ras GEF activity occurs. In the absence of Ras•GTP, the allosteric site is occluded by the DH domain of SOS, and Ras GEF activity is low. Ras•GTP displaces the DH domain, binds to the allosteric site, and increases Ras GEF activity, thereby acting as a feed-forward regulator. The Rac1-binding site of the DH domain is also occluded in apo-SOS, and is liberated in the complex with Ras•GTP, so that nucleotide exchange activities for both GTPases are coordinately upregulated. Release from autoinhibition is the general regulatory theme in this example, and may apply also to activation of RH-RhoGEFs by Gα. In the simplest model, the RGS-like domain inhibits the GEF activity of the DH-PH domains, and Gα13 relieves this inhibition by binding to the RGS-like domain. Additional interactions with Gα13 might directly activate the DH-PH domain. The ~110-residue linker segment that separates the DH-PH and RGS-like domains may play more than a passive role in the stimulatory action of Gα13, since certain deletions within this region profoundly affect both basal and Gα13-stimulated GEF activity (Wells *et al.*, 2001). Further insight into the mechanism by which Gα proteins stimulate the GEF activity of RH-RhoGEFs shall require comprehensive structural studies of the relevant protein–protein complexes.

### *3. Gαq Sequestration by GRK2*

GTP-dependent binding of G protein subunits to an effector may influence the direction and rate of regulatory events without changing the functional state of the effector itself. Simultaneous binding of Gβγ and Gq class Gα subunits to GRK2 provides an example. On phosphorylation by GRK2, the β2 adrenergic receptor becomes a target for binding and downregulation by arrestin (Pitcher *et al.*, 1998). The PH domain of GRK2 binds to Gβγ and consequently GRK2 is tethered to its site of action at the plasma membrane. However, this binding event does not induce major structural changes in the kinase domain of GRK2, which remains in an open, apparently inactive, conformation (Lodowski *et al.*, 2003, 2005). The RGS-like domain of GRK2 also engages activated Gq-class Gα subunits in an effector-like interaction (Fig. 3D). However, the site of Gαq interaction is far-removed from the kinase domain of GRK2, and, again, binding of Gαq does not appear to have structural consequences for the kinase domain of GRK2 (Tesmer *et al.*, 2005).

The Gαq-binding site on the RGS-like domain and the Gβγ-binding site near the PH domain do not interact directly, and Gαq and Gβγ are fully dissociated in the complex with GRK2. The structure of the complex thereby lends support for the classical notion that, on GTP activation of Gα, the heterotrimer is fully dissociated. The presumptive β2 adrenergic receptor-binding site of GRK2 remains unoccluded by interactions with Gαq•GTP and Gβγ (Lodowski *et al.*, 2003). Simply by tethering GRK2 to the plasma membrane, near sites of GPCR action, Gαq and Gβγ serve as effectors of receptor desensitization. At the same time, GRK2 sequesters Gβγ and the active form of Gαq. GRK2 thereby inhibits G protein-mediated signal transduction through a mechanism that is independent of its kinase activity (Carman *et al.*, 1999; Sallese *et al.*, 2000; Tesmer *et al.*, 2005; Usui *et al.*, 2000). Effectors such as mAC and PLCβ can be coregulated by Gα and Gβγ; GRK2 reverses this paradigm, by inhibiting the signaling activity of both. GRK2, which interacts through its RGS-like domain with the α2, α3 effector-binding site of Gαq, has little or no GAP activity, and therefore its complex with the G protein is potentially long-lived. However, the RGS-like domain of GRK2 does not block the putative binding site on Gαq•GTP for RGS GAPs (Tesmer *et al.*, 2005), and receptor-selective negative regulators such as RGS4 would have ready access to the activated G protein (Xu *et al.*, 1999; Zeng *et al.*, 1998). The emerging story of GRK2 illustrates the regulatory complexity of G protein signaling, where multiple and often synergistic interactions among preassembled signaling molecules are possible.

## IV. Signal Termination: The Mechanism of GTP Hydrolysis and Conformational Deactivation

GTP hydrolysis terminates the signaling state of Gα. The binding energy of GTP that is used to stabilize the Gα:effector complex is dissipated in this reaction. Gα subunits of the Gi, Gs, and Gq families catalyze GTP hydrolysis with single turnover rates in the range of 2–4 $min^{-1}$ (Gilman, 1987), a rate enhancement of about $10^5$–$10^6$ relative to the spontaneous rate of GTP hydrolysis in water, and about 50-fold faster than p21 Ras (Glennon *et al.*, 2000; Temeles *et al.*, 1985). Hence, the GTP complex of Gα has an intrinsic lifetime ($t_{1/2}$) of 10–50 s. With a $K_M$ for GTP of 0.3 $\mu$M, the catalytic efficiency of Gαs is about $1 \times 10^5\ s^{-1}\ M^{-1}$, near the low end of the range of $k_{cat}/K_M$ values for typical enzymes. The steady-state rate of GTP hydrolysis by G$\alpha$ is limited by product (GDP) release (Higashijima *et al.*, 1987a). The dissociation constant for the complex of G$\alpha$i1 with the GTP analogue GTP$\gamma$S in the presence of $Mg^{2+}$ is only 5 nM (Higashijima *et al.*, 1987c). The low catalytic activity of G$\alpha$ subunits reflects the remarkable stability of their complexes with GTP, and the energetic cost of perturbing them. In their ground state E•S complexes with GTP, G$\alpha$ proteins appear to be catalytically deficient, for example, in their ability to optimally stabilize the transition state or perhaps to orient water for nucleophilic attack on GTP. GTP hydrolysis ultimately results in a structural reorganization of the switch regions–in particular, an order $\rightarrow$ disorder transition in Switch II. The activation energy barrier to GTP hydrolysis could potentially arise from the energetic cost of a pretransition state enzyme reorganization or chemical steps involved in formation of the transition state itself. It is possible that the reorganization of the switch regions also encounters a kinetic barrier, coupled to product (Pi) release or an event preceding formation of the GDP•Pi complex.

To resolve the forgoing issues is to understand the kinetic and thermodynamic linkage between chemical events and conformational changes in the active sites of G$\alpha$ proteins. It is essential to know the chemical mechanism of GTP hydrolysis, and the order of events that take place in the active site. Structural reference points are provided by a series of complexes of G$\alpha$i1 and other G$\alpha$ proteins that we envision as mimics of structural states along the trajectory of G$\alpha$-catalyzed GTP hydrolysis (Coleman and Sprang, 1999a). These include structures of G$\alpha$i1•$Mg^{2+}$•GppNp, which mimics the GTP-bound E•S complex (protein databank code 1CIP) (Coleman and Sprang, 1999b); G$\alpha$i1•$Mg^{2+}$•GDP•$AlF_4^-$, an analogue of the activated state for catalysis (1GFI) (Coleman *et al.*, 1994); G$\alpha$i1•$Mg^{2+}$•GDP•$SO_4^{-2}$ (1BOF) (Coleman and Sprang, 1998) and G$\alpha$i1•GDP•Pi (Berghuis *et al.*, 1996; Raw *et al.*, 1997) (1GIT), which model ternary product complexes; and G$\alpha$i1•GDP, the stable binary product complex that is inactive in signaling (1GDD) (Mixon *et al.*, 1995). Although these structures provide insight into the mechanism of the G$\alpha$-catalyzed GTP reaction, they are not themselves reaction intermediates, but rather time-averaged equilibrium states crystallized under conditions far from those of the biochemical milieu in which they function. Nevertheless, these structures may approximate conformational states that occur along the reaction trajectory for GTP hydrolysis, and in the subsequent discussion, we take advantage of this assumption (Coleman and Sprang, 1999a). In this vein, we begin with a description of the ground state E•S complex for GTP hydrolysis before considering the events associated with GTP hydrolysis and signal termination.

### A. *Structure of the Ground State for GTP Hydrolysis*

The active sites of Gα isoforms are essentially identical (Coleman *et al.*, 1994; Kreutz *et al.*, 2006; Noel *et al.*, 1993; Sunahara *et al.*, 1997b; Tesmer *et al.*, 2005) (Fig. 6A). They share certain features common to all Ras superfamily members, as reviewed elsewhere (Kjeldgaard *et al.*, 1996; Sprang, 1997; Vetter and Wittinghofer, 2001). Here, we focus on the environment of the triphosphate moiety of the nucleotide where bond scission and conformational changes occur—discussions concerning the origin of guanine nucleotide specificity can be found in the reviews cited above. The ground state for the GTPase reaction is formed upon incorporation of $Mg^{2+}$ into the Gα•GTP complex. Magnesium ion binding is associated with a rapid increase in the intrinsic fluorescence at 340 nm that arises from changes in the environment of a conserved tryptophan residue in Switch II (residue 211 in Gαi1) (Faurobert *et al.*, 1993; Higashijima *et al.*, 1987a; Phillips and Cerione, 1986). The structure of the $Mg^{2+}$-free GTP complex of Gα is not known, but it is likely to be relatively less well ordered than that with $Mg^{2+}$•GTP, and possibly prone to functional inactivation (Zelent *et al.*, 2001). In structures of Gα proteins bound, with $Mg^{2+}$, to GTP analogues GppNp or GTPγS, the Switch II tryptophan is partially buried in a predominately hydrophobic environment, but is solvent-exposed in structures of Gα bound to GDP, GDP•Pi, or GDP•$Mg^{2+}$•$SO_4^{-2}$. Therefore, the $Mg^{2+}$-free Gα•GTP complex could resemble one of these product complexes.

In both the GDP- and GTP-bound states, the α and β phosphates are enfolded by the P-loop, and form hydrogen bonds with the backbone amides of five successive residues together with the amino group of the conserved P-loop lysine 46, which forms an ionic hydrogen bond with both β and γ phosphate oxygen atoms (Fig. 6A). The GTP γ phosphate is tethered by hydrogen bonds with the backbone amide groups of Switch I Thr 181 and Switch II Gly 203 of the Walker B DXXG motif (Gαi1 residue numbering). These highly conserved but mobile residues act as structural "sensors" of the γ phosphate. The magnesium ion is coordinated by β and γ phosphate oxygen atoms, together with the P-loop Ser 47, Switch I Thr 181 and, through a water molecule, the Switch II Asp 200 of the DXXG motif. Thus, $Mg^{2+}$ rigidifies the catalytic site through a network of interactions that link all of the structural elements involved in triphosphate binding and catalysis: the P-loop, Switch I, Switch II, and GTP. GTP hydrolysis and the ensuing conformational changes partly dismantle the $Mg^{2+}$-binding site.

In most GTP-bound Gα structures, the putative lytic water is positioned less than 4 Å from the γ phosphorus and, in Gαi1•GppNp, accepts hydrogen bonds from the main chain amide and side-chain amino group of

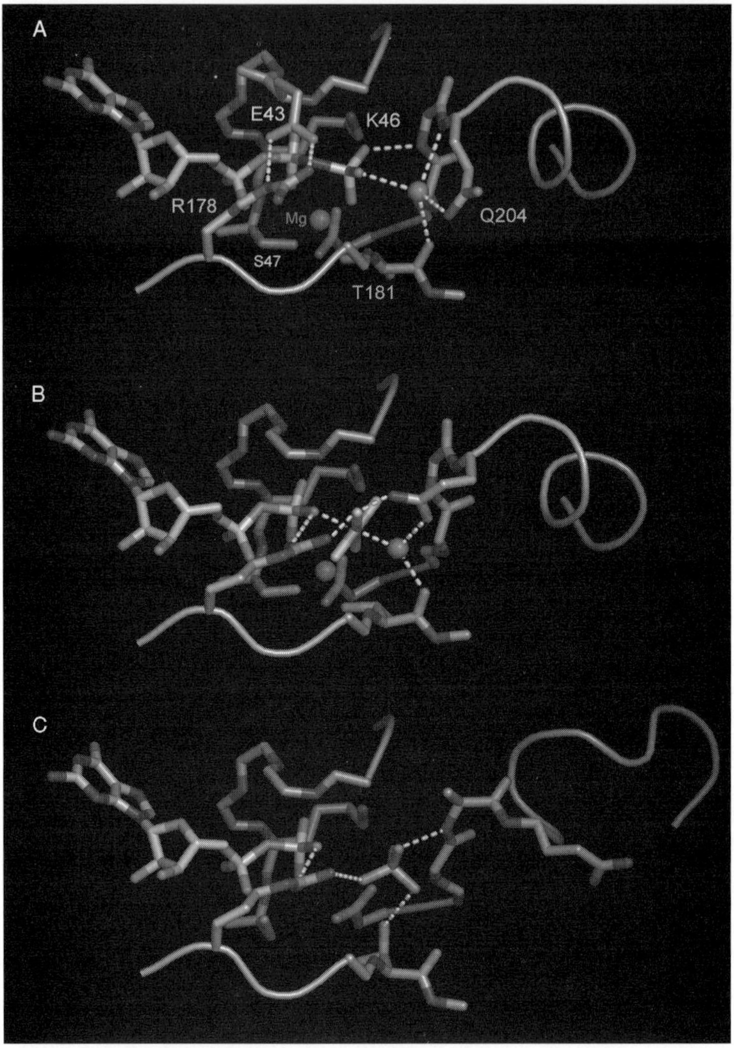


FIG. 6. Gαi1 complexes that represent possible states of the reaction trajectory for GTP hydrolysis; Switch I and Switch II are colored slate blue, and the P-loop tan; atom coloring follows the convention used in the preceding figures, except that carbon and phosphorus atoms of the nucleotide, and the aluminum atom of $AlF_4^-$ are colored gray; Fluorine atoms are colored light blue. The putative lytic water molecule is depicted as a red sphere. Hydrogen bonds involving Arg 178, the $\gamma$ phosphate of the nucleotide, $AlF_4^-$, Pi, and the putative lytic water molecule are indicated with yellow dashed lines. Panel A depicts the Gαi1·$Mg^{2+}$·GppNp complex (PDB ID 1CIP). Charged-hydrogen bond contacts between Arg 178 and Glu 43 sequesters Arg 178 from interaction with the nucleotide phosphate groups. The side chain hydroxyl groups of P-loop residue Ser 47 and Switch I Thr 181 are ligands of $Mg^{2+}$, as are a $\beta$ and a $\gamma$ phosphate oxygen atom of the

Gln 204 in Switch II and donates a hydrogen bond to a $\gamma$ phosphate oxygen and the main chain carbonyl oxygen of Thr 181 (the same residue that participates in $\gamma$ phosphate and $Mg^{2+}$ coordination) (Fig. 6A). The lytic water is less then 0.2 Å off-center of the path of nucleophilic attack. A similar arrangement is observed in the cryo-stabilized structure of Ras•GTP where the lytic water is further from the $\gamma$-phosphorus, but also in-line with the $\beta\gamma$ bridging oxygen of the leaving group (Scheidig *et al.*, 1999).

With respect to the disposition of catalytic residues, the ground state Gα•GTP active site appears to be catalytically disabled, relative to the catalytically activated state that is mimicked by the complex with GDP•$Mg^{2+}$•$AlF_4^-$ (Fig. 6B). One of these residues, Arg 178 in Switch I is conserved, and critical to catalytic activity in the Gα family of GTPases, but is not present in other members of the Ras superfamily (Sprang, 1997). This critical side chain stabilizes charge in the transition state and product complexes by forming ionic contacts with the $\beta$ and $\gamma$ phosphate moieties of GTP (Fig. 6B and C). However, in the crystal structures of Gαi•GTPγS, Gαi1•GppNp, and Gαs•GTPγS, Arg 178 exhibits high thermal mobility, and fails to interact with nucleotide phosphates. Rather, it is in ionic contact with the P-loop Glu 43 (Fig. 6A). Only in the structure of Gαt•GTPγS does the catalytic site arginine directly contact the nucleotide $\gamma$ phosphate (Noel *et al.*, 1993).

The catalytic glutamine at position 204 in Gαi1 is a conserved feature in all functional Gα proteins, and in most members of the Ras superfamily (residue 61 in Ras). Mutations of this residue in Gα and Ras abolish GTPase activity, are constitutively active, and contribute to cellular transformation (Barbacid, 1987; Bourne *et al.*, 1991; Gilman, 1987). Further, GTPase activity of the

---

nucleotide (bonds not shown) together with two water molecules (not shown). Note that the water nucleophile accepts hydrogen bonds from the side-chain amino group and the main chain amide of Gln 204, and donates hydrogen bonds to the carbonyl group of Thr 181 and the nucleotide $\gamma$ phosphate. Panel B shows the complex of Gαi1 with $Mg^{2+}$, $AlF_4^-$, and GDP (1GFI), which represents the catalytically active state of the enzyme that is able to stabilize the transition state of the enzyme. The GDP•$AlF_4^-$ complex, together with the lytic water molecule, is nearly isosteric with the putative pentacoordinate transition state or intermediate for GTP hydrolysis. The lytic water molecule and a $\beta$ phosphate oxygen atom of GDP are positioned ~2.0 Å from, and are axial ligands of, the aluminate. Arg 178 adopts a conformation that allows ionic contacts with the $\beta$-bridge oxygen of the GDP, and a fluoride ligand of , which mimic possible interactions with the pentacoordinate phosphate. Gln 204 adopts a conformation in which it accepts a hydrogen bond from the lytic water, and donates a hydrogen bond to a fluoride ligand of $AlF_4^-$. Panel C shows the structure of the G203A mutant of Gαi1 bound to GDP and Pi (1GIT), the ternary product complex of the GTPase reaction. $Mg^{2+}$ is not present in the complex. Arg 178 maintains ionic and hydrogen bond contact with a $\beta$ phosphate oxygen atom of GDP and of Pi. Switch II is kinked and the N-terminus of this helix is directed at Pi, which accepts a hydrogen bond from the main chain amide of Ala (Gly in the wild-type protein) 203. (See Color Insert.)

Q61L mutant of Ras is resistant to simulation by RasGAP (Adari *et al.*, 1988; Gideon *et al.*, 1992), as is the corresponding Q204L mutant of Gαi1 to RGS proteins (Berman *et al.*, 1996b). (Here and elsewhere, point mutants are described by the residue number corresponding to the mutation site, preceded by the one letter code for the residue in the wild-type protein, and followed by that for the residue in the mutant.) Structures of activated, GDP•$Mg^{2+}$•$AlF_x$-bound G proteins suggest that the glutamine binds the lytic water molecule that attacks the $\gamma$-phosphorus (or metaphosphate) of GTP. However, with the exception of Gαi1•GppNp, in none of the ground state complexes mentioned above does the catalytic glutamine directly contact the water nucleophile. In most complexes, the glutamine residue is less well ordered than other active site residues. As noted, Gln 204 is a hydrogen bond donor to the lytic water molecule in the Gαi1•GppNp complex. Although this interaction stabilizes the bound water nucleophile in the catalytic site, it would be anti-catalytic with respect to the role of the bound water as a nucleophile (Maegley *et al.*, 1996). Pasqualato *et al.* (2005) suggest that the hydrogen bonding network that supports the water nucleophile is anticatalytic in the ground state structure of Rab-11•GTP. Thus, while the active site of Gα binds tightly to GTP in the ground state, it is not well configured to catalyze GTP hydrolysis.

### B. *Reaction Trajectory for GTP Hydrolysis*

A minimal reaction trajectory, as suggested by states observed in crystal structures, is given by Scheme 1:

$$\mathrm{G \cdot GTP} \overset{1}{\rightleftharpoons} \mathrm{G \cdot GTP^{*}} \overset{2}{\rightleftharpoons} \mathrm{G \cdot GDP{\cdot}Pi} \overset{3}{\rightleftharpoons} \mathrm{G \cdot GDP + Pi} \qquad (1)$$

The transitions are: (1) active site reorganization, (2) formation of the ternary product complex, and (3) product (Pi) release. In this scheme, formation of the chemical transition state occurs at step 2, and the reorganization of the switch segments that lead to effector release occurs in steps 2 and 3. It is probable that the trajectory is more complex, but the scheme serves as a basis for discussion of the data at hand.

In principle, any of the steps in the reaction trajectory could be rate limiting, including active site reorganizations. Release of phosphate from the ternary complex is probably not rate limiting for Gαi1 (Raw *et al.*, 1997; Thomas *et al.*, 2004), although conflicting results have been reported for presteady-state GTP hydrolysis by Gαt in the presence of rhodopsin (Guy *et al.*, 1990; Ting and Ho, 1991).

### 1. *The Catalytically Active State of Gα*

Reorganization of the G protein catalytic site in advance of catalysis (step 1) results in an activated E•S complex similar in structure to that of Gαi1 and Gαt bound to $Mg^{2+}$•GDP•$AlF_4^-$ (Coleman *et al.*, 1994; Sondek *et al.*, 1994) (Fig. 6B). Rather than the "transition state" per se, the Gα•GDP•$Mg^{2+}$•$AlF_4^-$ complex is a catalytically activated form of the enzyme that is sterically and electrostatically complementary to the transition state in GTP hydrolysis. The complex exhibits the high intrinsic tryptophan fluorescence observed in GTP analogue-bound Gα (Higashijima *et al.*, 1987b; Phillips and Cerione, 1988), and therefore Switch II adopts a conformation similar to that of the ground state, as the crystallographic data indicate. The square-planar $AlF_4^-$ observed in these structures has been proposed to mimic the pentacoordinate $\gamma$ phosphate of which the attacking water nucleophile and the $\beta\gamma$ bridging oxygen are axial ligands. Corresponding complexes of Ras (Scheffzek *et al.*, 1997) and Cdc42 (Nassar *et al.*, 1998) contain $AlF_3$, which is isosteric to the trigonal $\gamma$ phosphoryl (or metaphosphate) in the transition state. Notable in all of these complexes is the reorientation of the catalytic glutamine such that it comes to accept a hydrogen bond from the water nucleophile, and donate to an equatorial fluorine atom of the aluminate (and by inference, an oxygen atom of the $\gamma$ phosphate). In this configuration, the glutamine is able to promote GTP hydrolysis by either increasing the nucleophilicity and basicity of the lytic water or aligning it for in-line attack. In the Gα proteins, the Switch I arginine reorders to form interactions with the aluminate ligands that mimic the equitorial $\gamma$ phosphate oxygen atoms, and is therefore functionally and structurally equivalent to the "arginine finger" provided by Ras- and Rho-GAPs (Brownbridge *et al.*, 1993; Markby *et al.*, 1993; Rittinger *et al.*, 1997; Scheffzek *et al.*, 1997). A small rearrangement in Switch I allows the main chain carbonyl of Thr 181 to maintain the hydrogen bond with the water nucleophile as the latter moves toward the inferred position of the GTP $\gamma$ phosphate. The structural changes proposed for the transition from the ground state to the catalytically active conformation of Gα are small, but could encounter a substantial activation energy barrier, as was proposed for Ras (Neal *et al.*, 1990). If we assume that the $10^3$-fold rate enhancement afforded by Gα is achieved solely by catalyzing the formation of the activated state, then the activation energy barrier at step 1 is in the neighborhood of 4 kcal/mol.

### 2. *Transition State Formation Is Rate Limiting for Gα-Catalyzed GTP Hydrolysis*

The progress of the GTPase reaction in Gα proteins can be monitored by first-order decay of the tryptophan fluorescence inherent to the ground state Gα•$Mg^{2+}$•GTP complex. The rate constant for the fluorescence transition is virtually identical to that for the production of free Pi (Higashijima *et al.*, 1987a). Yet, the events that correspond to these two experimental readouts are distinct. The former reflects the collapse (or reconfiguration) of Switch II, while the latter embodies a series of steps that culminate in the production of free phosphate. Since the two processes are kinetically indistinguishable, either could subsume a rate-limiting event. In any case, the conformational and chemical transitions are tightly coupled. In the presence of RGS4, which functions as a GAP for Gαi1, the rates of hydrolysis and tryptophan quenching are equally accelerated (Lan *et al.*, 2000). Thus, the kinetic coupling between these events remains intact on GAP stimulation.

Using Ras as a model system for G proteins in general, we sought to determine whether cleavage of the $\beta-\gamma$ bond of GTP constitutes the rate-limiting step in the conversion of Ras•GTP to Ras•GDP. The approach taken was to measure the $^{18}$O KIEs for the Ras-catalyzed GTPase reaction at steady state (Du *et al.*, 2004). KIE refers to the change in reaction rate when one or more atoms in the substrate are substituted with heavier isotopes, and is defined as a ratio of the reaction rate of the unlabeled substrate to that of the labeled substrate. Atoms in a molecule occupy discrete vibrational energy levels that depend on both their mass and the force constants of the chemical bonds by which they are restrained. Isotopic substitution lowers the energy levels of both ground state and transition state, but to different degrees because of the differences in chemical bonding in the two states. If the bond order of the atoms at or near the scissile bond decreases at the transition state, isotopic substitution reduces the energy levels of the transition state to a lesser extent than those of the ground state. Consequently, labeled substrate encounters a higher activation barrier, and the KIE is greater than one (such KIEs are considered "normal"). In the opposite case in which the bond order increases at the transition state, KIE is less than one and is considered "inverse." A KIE measured for a nonenzymatic reaction is a direct result of changes in bond order and geometry in the transition state. However, an enzymatic reaction is a multistep process that involves not only chemical steps but also substrate binding and release, product release, formation of intermediates, and the possibility of slow conformational changes. Consequently, the KIE measured in an enzymatic reaction is scaled down by a commitment factor, which is basically a measure of the extent to which the chemical step is rate limiting. Comparison of a KIE for an enzymatic reaction with that expected for the chemical steps alone would reveal whether conformational steps or chemistry is rate limiting.

We determined the $^{18}$O KIEs for the Ras-catalyzed GTPase reaction at steady state in the presence of the nucleotide exchange factor, cdc25 (Du *et al.*, 2004). The nucleotide exchange factor for Ras was included in the reaction mixtures to ensure that GTP release is facile and does not contribute to the commitment of the reaction (Lenzen *et al.*, 1998). In these experiments, $^{18}O$ KIE effects were indeed observed, with significant catalytic rate reduction with $^{18}O$ substituted at the $\beta$ phosphate oxygen atoms of the leaving group. Therefore, the rate-limiting step for Ras-catalyzed GTP corresponds to the activation energy barrier to formation of a chemical transition state (step 2 in Scheme 1). The same chemical step remains rate limiting even when the Ras-catalyzed reaction is accelerated by RasGAP, wherein substantial KIEs are also observed (X.D., unpublished data). While reorganization of the ground state structure of the Ras active site (step 1) may encounter a substantial activation energy barrier, it must be smaller than the energy difference between the ground and transition states for GTP hydrolysis.

### 3. *The Mechanism of GTP Hydrolysis*

Insights into the mechanism of G protein-catalyzed GTP hydrolysis are largely derived from biochemical, spectroscopic, structural, and theoretical studies of Ras GTPase (Wittinghofer, 2006). The structural similarity of the catalytic sites of Ras and Gα, particularly with respect to their complexes with GDP•$Mg^{2+}$•$AlF_x$, suggests that their catalytic mechanisms, and the structures of the transition states, are the same. Although many questions remain unresolved, views of catalysis have evolved, and some degree of consensus has emerged. Structures of GDP•$Mg^{2+}$•$AlF_x$-bound G proteins, in which the catalytic glutamine and arginine residues appear to provide charge stabilization to a pentacoordinate $\gamma$ phosphate, were strongly suggestive of an associative transition state. Yet, as Herschlag and colleagues have pointed out, phosphoryl transfer reactions in bulk solution are typically dissociative, in which case stabilization of charge at the $\gamma$ phosphate, or nucleophilic activation of the lytic water, would have little catalytic advantage (Maegley *et al.*, 1996). However, "associative" and "dissociative" represent mechanistic extremes, and it is possible that the reaction is concerted with a transition state characterized with some degree of bonding with both leaving and attacking groups. FTIR spectroscopic studies have demonstrated that Ras stabilizes charge on the $\beta$ phosphate oxygen atoms of GTP in the ground state (Allin and Gerwert, 2001; Cepus *et al.*, 1998; Du *et al.*, 2000). The same charge distribution is observed in the presence of RasGAP (Allin *et al.*, 2001). Hybrid MM–QM calculations are consistent with the spectroscopic observations and further suggest that much of the charge stabilization arises from the interactions with $Mg^{2+}$ and the tertiary amino group of the conserved P-loop lysine residue. Coordination by $Mg^{2+}$ and hydrogen bonding with the backbone amides are proposed to stretch the bond between the $\beta\gamma$ bridge oxygen and the $\gamma$-phosphorus (Klahn *et al.*, 2005). Hydrogen bonding between the $\beta$ nonbridge oxygen atoms and P-loop amides are also expected to contribute to charge localization at these atoms and weakening of the $\beta$–$\gamma$ bond. For a series of Ras mutants at the P-loop Gly 12, Du *et al.* (2000) found that the strength of binding to the GDP $\beta$ phosphate—as revealed by the wave number for the asymmetric stretch of $\beta O_3$—is positively correlated with enzyme activity, suggesting a GDP-like transition state.

The FTIR studies described above reveal the properties of GTP and its products in the ground states of the Ras-catalyzed reactions. However, transition state stabilization is an important component of enzyme catalysis, and in this context KIEs are highly informative as indicators of the electronic state of GTP in the active site of Ras at the transition state. KIE measured with $^{18}O$-labeled GTP substrates show that, in the rate-limiting transition state, charge is distributed between the two $\beta$ nonbridge oxygen atoms and the $\beta\gamma$ bridging oxygen atom. Further, the KIE observed at the $\gamma$ phosphate oxygen atoms is near unity (Du *et al.*, 2004). Therefore, the transition state has substantial dissociative character, in which charge is deposited on the leaving group while there is little change in the charge on $\gamma$ phosphate oxygen atoms. In the context of structural data, we can infer that the hydrogen bonding and electrostatic interactions between the $\beta$ phosphate group of GTP and the Ras active site residues, including $Mg^{2+}$, the P-loop lysine residue, and the P-loop amide hydrogen atoms, are conducive to catalysis. As charge increases at the $\beta\gamma$ bridge oxygen and $\beta$ nonbridge oxygens at the transition state, these interactions become stronger, affording preferential stabilization of the transition state.

Other aspects of the mechanism are less well resolved. First, it has not been established how, and at what point in the reaction trajectory, the lytic water is deprotonated. Second, the role of the catalytic site glutamine (residue 204 in G$\alpha$i1 and 61 in Ras) remains controversial. Third, there are hints from computational and experimental studies that a reaction intermediate may form in the GTPase reaction.

No evidence has emerged for the existence of an enzyme catalytic base to deprotonate the nucleophilic water (although this is not required to achieve a dissociative transition state). The catalytic glutamine is within hydrogen bonding distance, but is presumably to weak a base to abstract a proton from the lytic water (Langen *et al.*, 1992). Dependence of the GTPase reaction rate on the p$K_a$ of the GTP $\gamma$ phosphate provided evidence for a substrate-assisted mechanism in which the $\gamma$ phosphate extracts a proton from the attacking water molecule in a preequilibrium step (Schweins *et al.*, 1994, 1995, 1996). The high-resolution structure was determined of a catalytically defective mutant of the small GTPase Rab-11, in which the catalytic glutamate was substituted by leucine (Pasqualato and Cherfils, 2005). The reduced activity of this mutant and the conditions of crystallization allowed the GDP•$Mg^{2+}$•Pi ternary product complex to be trapped in the Rab-11 active site. An unusually short (2.4–2.5 Å), possibly LBHB was observed between oxygen atoms of the leaving group and the phosphate. It was proposed that the proton involved in this hydrogen bond could have been abstracted by the $\gamma$ phosphate from the lytic water in a pretransition state step, and subsequently transferred to the leaving group. The tightly bound proton was thought to stabilize negative charge at the $\beta$ and $\gamma$ phosphates in the transition state. However, protonation of the $\beta\gamma$ bridge oxygen would be expected to result in much lower KIE values than those we have observed for GTP containing $^{18}O$ in the $\beta\gamma$ bridge and the two $\beta$ nonbridge positions, suggesting that the bridge oxygen is not protonated in the transition state. A similar interaction between the $\beta$ phosphate group of GDP and an inorganic phosphate was observed in the Gαi1•GDP•Pi complex (Berghuis *et al.*, 1996; Raw *et al.*, 1997), where a normal hydrogen bond (~3.0 Å) between the leaving group and Pi is observed (Fig. 6C). This contrasts with the very short hydrogen bond observed in the Rab-11•GDP•$Mg^{2+}$•Pi complex; it is possible that the shorter hydrogen bond in the Rab-11 structure arises from coordination of the two phosphates by $Mg^{2+}$, which both sterically constrains the two groups and reduces charge repulsion between them. $Mg^{2+}$ is absent from the Gαi1•GDP•Pi complexes, as discussed in further detail below. The forgoing structures are consistent with formation of a monoanionic phosphate in the ternary product complex and by extension, the possibility that the $\gamma$ phosphate of GTP serves as a base. However, the structures provide no information about the origin of the presumptive phosphate monoanion (since it could be derived from the crystallization solution), or, if a direct product of hydrolysis, the points in the reaction trajectory at which putative proton transfer steps involved in its formation might have occurred.

The catalytic role of the active site glutamine remains something of a mystery. Structures of small G protein•GDP•$Mg^{2+}$•$AlF_x$ complexes bound to GAPs, and of free or GAP-bound Gα•GDP•$Mg^{2+}$•$AlF_4^-$, are similar to each other with respect to the coordination of the glutamine side chain. In each of these structures, the side chain carbonyl accepts a hydrogen bond from the axial ligand of the fluroaluminate (the presumptive lytic water), while the side chain amide donates a hydrogen bond to an equatorial fluorine substituent (which mimics the corresponding equatorial $\gamma$ phosphate oxygen in the pentacoordinate transition state or intermediate) (Fig. 6B). This geometry suggests a role for the glutamine in orienting or positioning the lytic water for nucleophilic attack. However, in Ras, this residue can be substituted by a nitroglutamine residue without loss of catalytic activity (Chung *et al.*, 1993). While this rules out the possibility that the glutamine acts as a general base, it does not rule out a role in aligning the lytic water. Nevertheless, empirical valence bond calculations of RasGAP-stimulated Ras GTPase indicate that the glutamine does not interact electrostatically or sterically with the transition state itself, but rather acts allosterically to stabilize the structure of the active site in the transition state; the loss of this stabilization is proposed to explain the GTPase-inactivating effect of the mutations of Gln 61 (Glennon *et al.*, 2000; Shurki and Warshel, 2004). The ground states (GTP-bound) of G$\alpha$i1 and Ras are not perturbed by mutation of the catalytic glutamine (Coleman *et al.*, 1994; Privé *et al.*, 1992). However, no structures are available for glutamine mutants in the $Mg^{2+}$•GDP•$AlF_x$ state (because they impair $Mg^{2+}$•GDP•$AlF_x$ binding). Therefore, the structural consequences due to the loss of the active site glutamine cannot be directly assessed. That subfamilies of GTPases (Rap, EF-Tu) use a threonine, serine, or histidine instead of glutamine, and yet are capable of weak intrinsic and GAP-stimulated activity, shows that glutamine is not absolutely required for GTPase activity (Vetter and Wittinghofer, 2001). Yet there is evidence that Rap1GAP provides such a glutamine (Daumke *et al.*, 2004). It was found that a member of the TBC domain family of RabGAPs provides both glutamine and arginine, in *trans*, to the catalytic site of Rab (the endogenous Rab glutamine is shunted to an interaction with the TBC domain), producing an active site configuration similar to that of an activated G$\alpha$ protein (Pan *et al.*, 2006).

Computational and experimental studies have suggested the possibility of an intermediate in the GTPase reaction catalyzed by Ras. Quantum mechanical simulations of the RasGAP-catalyzed reaction show a minimum energy reaction path for a two-stage transformation (Grigorenko *et al.*, 2005; Topol *et al.*, 2004). The first step results in the formation of a metaphosphate intermediate. Nucleophilic attack on the metaphosphate by water occurs in the second stage of the reaction. The water is proposed to be indirectly deprotonated by the $\gamma$ phosphate, using Gln 61 as a shuttle, in a mechanism similar to that suggested earlier by Sondek *et al.* (1994). This scenario is consistent with structural data in which the catalytic glutamine carbonyl appears as a hydrogen bond acceptor from the attacking water and a donor to the $\gamma$ phosphate (Fig. 6B). Grigorenko *et al.* (2005) propose that the imino tautomer formed in the course of this shuttle mechanism would not be unfavorable in the context of a metaphosphate as the ultimate proton acceptor. Real-time FTIR studies of GAP-stimulated GTP hydrolysis revealed the accumulation of a transient, enzyme-bound species that was identified with the cleaved and protein-bound phosphate (Allin *et al.*, 2001). In more recent time-resolved FTIR experiments in which the NF1-333 was employed as a GAP, the accumulating species was identified as $H_2PO_4^-$. The bound phosphate monoanion is less than 5-Å distant from the GDP $\beta$ phosphate and is therefore proposed to be hydrogen-bonded to the latter, as observed in the crystal structures of the GDP•Pi complexes of Rab-11 and Gαi1 as described above. These observations are in accord with kinetic studies of the NF1-333-catalyzed Ras GTPase reaction, in which NF1-333 and Pi are released with similar kinetics from the NF1-333:Ras•GDP•Pi complex, at a rate of threefold slower than bond cleavage. It is probable that NF1-333 dissociation is triggered by phosphate release (Phillips *et al.*, 2003). Isotope exchange experiments indicated that the rate of GTP regeneration from the GDP•Pi complex of Ras:RasGAP is 90-fold slower than that of its cleavage. Thus, for Ras GTPase, the rate of Pi release approaches that of bond cleavage in the presence of a GAP. Comparable mechanistic studies for Gα proteins have not been conducted, and so it is not possible to determine the generality of this finding to more divergent G protein superfamily members.

### *4. Coupling of GTP Hydrolysis to Switch Rearrangement*

The conformational changes, particularly in Switch II, that occur on GTP hydrolysis by Gα proteins are substantial. They are associated with the termination of signaling activity because they destructively perturb the effector-binding site. How are these conformational changes coordinated with GTP hydrolysis itself? Structural studies in our laboratory suggest that conformational collapse of Switch II may occur even before phosphate is released from the ternary complex. Although the ternary complex cannot be trapped in wild-type Gαi1, it is possible to crystallize the G42V and G203A mutants of Gαi1, as GTP•Pi complexes (Fig. 6C). Glycine 42 is a constituent of the P-loop GA*G*ES motif, and its substitution by valine produces a mimic of Gαz (for which the P-loop sequence is GT*S*NS) with respect to its substantially reduced GTPase activity (Raw *et al.*, 1997). Glycine 203 forms part of the Switch II DXXG motif. Mutation of this residue to alanine in Gαs does not affect GTPase activity but reduces affinity for GTPγS, $Mg^{2+}$, and impairs GPCR-mediated GTP-induced Gβγ release from Gs heterotrimers (Lee *et al.*, 1992; Miller *et al.*, 1988). Although they differ with respect to their guanine nucleotide binding and catalytic properties, both of these mutants could be induced to crystallize as complexes with GDP and Pi in the presence of high [Pi] at pH 5.5–6. The two GDP•Pi complexes of the two Gαi1 mutants adopt essentially identical conformations (Berghuis *et al.*, 1996; Raw *et al.*, 1997). Notably, $Mg^{2+}$ is not present in either of the Gαi1•GDP•Pi complexes, even though it is present in the crystallization buffer.

In the GDP•Pi state, Switch II is reorganized in part by rotation about glycine residues 202 and 203 (Fig. 6C). The Switch II glycine equivalent to Gly 203 in Gαi1 is the locus of guanine nucleotide-dependent conformational changes in Gα, Ras, and elongation factor 2 (Kjeldgaard *et al.*, 1993; Sprang, 1997). In the Gαi1•GDP•Pi complex, Switch II is withdrawn from the catalytic site and bent midway along the $\alpha 2$ helix, such that the helix dipole of the reorganized switch is directed at the Pi. This conformational adjustment allows the phosphate group to maintain hydrogen bond contact with Gly 203 in Switch II after the bond to the $\beta$ phosphate is cleaved. The structural change in Switch II and smaller adjustments in Switch I also relieve steric barriers that would block the displacement of Pi on cleavage from GTP. The structural changes relative to the GTP-bound state might be attributable to cleavage of the GTP $\beta$–$\gamma$ bond, but could also be due to the absence (or loss) of $Mg^{2+}$ from the catalytic site subsequent to bond cleavage.

It is possible that exclusion of $Mg^{2+}$ from crystals of G$\alpha$i•GDP•Pi is due to the low pH (5.5–6) at which the complexes were crystallized, wherein the phosphate is expected to be a monoanion. In the GTPase reaction, a phosphate monoanion is likely to be generated from deprotonation of the lytic water molecule. The separation of negative charges (localized on Pi and the GDP $\beta$ phosphate) in the ternary complex, in the context of the electropositive environment of the catalytic site (due to the P-loop backbone amides, Lys 46 and Arg 178), may favor dissociation of $Mg^{2+}$. Ras homologues, which do not possess a Switch I arginine, retain $Mg^{2+}$ with micromolar affinity in the GDP-bound state (John *et al.*, 1993). The collapse of the $Mg^{2+}$-binding site, with the concomitant loss of interactions between the nucleotide and the switches, is expected to destabilize the ground state conformation of Switch I and Switch II in G$\alpha$. Hence, the GDP•Pi complexes trapped in crystals of the two G$\alpha$i1 mutants may correspond to a *bona fide* product intermediate that is formed after $Mg^{2+}$ dissociates from the G$\alpha$•GDP•$Mg^{2+}$•Pi complex but before Pi is released. Thus, the first conformational step in G$\alpha$ signal termination may be due to dissociation of $Mg^{2+}$ from the ternary product complex.

Release of Pi from the GDP•Pi complex occasions a second conformational transition. Relative to both G$\alpha$i1•GTP and GDP•Pi states, Switch I moves further from the GDP-binding site and Switch II becomes almost fully disordered. This finding is based on GDP-bound structures of G$\alpha$i1 and G$\alpha$12 (Kreutz *et al.*, 2006; Mixon *et al.*, 1995). Neither complex contains $Mg^{2+}$, which, as noted above, has low affinity for G$\alpha$•GDP. Only in the crystal structure of G$\alpha$t•$Mg^{2+}$•GDP, which was crystallized in the presence of 200 mM $Mg^{2+}$, is Switch II well ordered. However, the GDP-bound conformation of G$\alpha$t precludes PDE$\gamma$ binding (Slep *et al.*, 2001). The mechanisms by which the switch segments of G$\alpha$t are differently configured in the GTP$\gamma$S and GDP-bound forms of G$\alpha$t have been discussed (Lambright *et al.*, 1994).

To understand how $Mg^{2+}$ might be accommodated in the catalytic site of G$\alpha$i1•GDP, we attempted a series of soaking and cocrystallization experiments with solutions containing up to 200-mM $Mg^{2+}$ (Coleman and Sprang, 1998).

Only in excess of ~5-mM $Mg^{2+}$ it was possible to observe significant occupancy of the $Mg^{2+}$ binding site, and this was accompanied by movement of Switch I toward the metal-binding site. The metal coordination sphere included the P-loop serine (residue 47), the Switch I Thr 181, a GDP $\beta$ phosphate oxygen, and three water molecules. In proportion to $Mg^{2+}$ occupancy, electron density for sulfate ion (present at >1 M in the crystallization solution) appeared near the Pi-binding locus, but unlike the $\gamma$ phosphate of GTP, did not participate in $Mg^{2+}$ coordination. Binding of $Mg^{2+}$ and $SO_4^-$ to Gαi1•GDP failed to induce extensive order in Switch II. This experimental result suggests that the conformational changes that occur on phosphate release are effectively irreversible. Accordingly, phosphate, even at concentrations in excess of 1 M, does not inhibit Gαi1-catalyzed GTP hydrolysis by Gαi1 (Raw *et al.*, 1997).

The structural integrity of Switch I, and particularly Switch II, in the active signaling state of Gα and other G proteins is largely engendered by their interactions with GTP and $Mg^{2+}$. The network of polar interactions that link GTP•$Mg^{2+}$ with the P-loop, Switch I, and Switch II is weakened even in the transition state for GTP hydrolysis, as electron density is shifted toward the GTP $\beta$ phosphate, and proceed successively toward collapse as the reaction is resolved. The amino acid sequences of Switch II segments are not strongly predicted to adopt helical structure and are only marginally stable in the absence of electrostatic interactions with the di-anionic $\gamma$ phosphate of GTP (Young *et al.*, 1999). The destabilization of Switch II, which forms a substantial portion of the binding surface for Gα effectors, confers an entropic penalty on effector binding by Gαs•GDP. Thus, the activation constant of Gαs•GDP toward adenylyl cyclase is about an order of magnitude less than that for Gαs•$Mg^{2+}$•GTP. Sequestration of Gαs•GDP by Gβγ subunits finally ensures complete signal termination (Sunahara *et al.*, 1997a).

The tight kinetic linkage between GTP hydrolysis and switch reorganization in Gα requires precise organization of active site residues, and perhaps, synchrony in the dynamic behavior of the enzyme itself. The two processes can be decoupled. Such is the unexpected consequence of substituting a proline for Lys 180 in Switch I of Gαi1 (Thomas *et al.*, 2004). Lysine at this site is not highly conserved in the G protein superfamily and a proline is present at the corresponding position in the sequences of Ras and Gαq. Like wild-type Gαi1, the K180P mutant shows elevated intrinsic tryptophan fluorescence on binding GTP analogues. The amplitude of excited fluorescence in the $Mg^{2+}$•GTP state and GDP•$Mg^{2+}$•$AlF_4^-$ complexes, and the extent of quenching on GTP hydrolysis are the same in the K180P mutant as in the wild-type protein. However, the rate of fluorescence quenching as catalysis proceeds is nearly tenfold higher than that of the wild-type protein. Ordinarily, this would suggest that the K180P mutant hydrolyzes GTP at a comparably greater rate. In fact, the rate of hydrolysis, as measured by

$^{32}$Pi release on acid/denaturant quench of the GTPase reaction, is sevenfold lower than that of wild-type G$\alpha$i1. Tryptophan fluorescence quenching and GTP hydrolysis both follow single exponential kinetics with no detectible burst phase. That tryptophan fluorescence is substantially lost well before a substantial fraction of GTP is hydrolyzed, and that no burst phase is observed, suggests that this G$\alpha$ mutant is catalytically active only after Switch II has undergone conformational rearrangement.

It is remarkable that the effects observed in the K180 mutant of G$\alpha$i1 were achieved by proline substitution; mutation of lysine 180 to alanine has no effect on catalysis or Switch II transition rates. In contrast to the other 19 naturally occurring amino acids, rotation about the $\phi$ (N–C$\alpha$) bond is constrained to approximately $-60^\circ$ in proline due to the pyrolidine ring formation at the N–C$\delta$ bond. Because the $\phi$ angle at position 180° is, in the wild-type protein, close to this value, the proline substitution can be structurally accommodated. However, proline does perturb the main chain conformation at the mutation site, and the perturbation is larger in the GDP•$Mg^{2+}$•$AlF_4^-$-bound complex than in the GppNp-bound state, particularly within the $Mg^{2+}$ coordination sphere. Further, we can expect that torsional fluctuations about the N–C$\alpha$ bond are hindered in the proline mutant. The difference between the $\phi$ and $\Psi$ angles of the activated and ground states are indeed smaller in the K180P mutant than in wild-type G$\alpha$i1. The kinetic decoupling of GTP hydrolysis from conformational change induced by the K180P mutation may arise both from subtle structural perturbations as well as changes in the global dynamic behavior of the enzyme. Molecular dynamic simulations indicate that the proline substitution alters the dynamic behavior of G$\alpha$i1 in the GTP-bound ground state, such that thermal motion appears decoupled from the ground state $\rightarrow$ activated state transition (A.E. Strand, unpublished data). The role of protein dynamics in catalysis is an emerging field of inquiry (Benkovic and Hammes-Schiffer, 2003), and it is intriguing to think that thermal fluctuations might influence probability of catalytic activation, an event so infrequent that GTP hydrolysis occurs in the timescale of seconds to minutes.

## V. Signal Termination Through GAPs and Effector GAP Domains

Most G$\alpha$-regulated physiological signaling mechanisms, for example photoreception and potassium channel regulation, require that G$\alpha$ proteins be deactivated in timescales $\sim 10^3$ faster than would be possible through their intrinsic GTPase activities (Arshavsky and Pugh, 1998; Ross and Wilkie, 2000; Zerangue and Jan, 1998). In these systems, GAPs or GAP domains within G$\alpha$ effectors or binding proteins stimulate GTPase activity. Thus, the high activation energy for GTP hydrolysis, subject to selective reduction by GAPs, confer specificity and efficiency to G protein signaling (Goody, 2003). In contrast to GAPs for Ras superfamily proteins, Gα GAPs have not been found to contribute catalytic residues to the G protein active site. Rather, they appear to act largely by promoting the configuration of residues in the active site that is optimal for their role in catalysis. In terms of Scheme 1, Gα GAPs promote step 1 and may also facilitate any subsequent enzyme reorganization steps in catalysis. GTPase-activating mechanisms for Gα proteins have been characterized at differing levels of physical detail for three systems. The structural mechanism of GAP activity appears to be different in each, but the chemical principles involved may be the same for all. First discovered was the GAP activity of PLCβ toward Gαq and Gα11 (Berstein *et al.*, 1992). Although the kinetic mechanism of GAP activity in PLCβ has been intensively studied, particularly in relationship to its tight coupling to GPCR activity (Biddlecome *et al.*, 1996; Mukhopadhyay and Ross, 1999), it remains poorly understood at the molecular level.

Within the last 10 years, there has been an explosion of information about members of the RGS family of proteins, many of which are GAPs for Gα proteins of the Gαi and Gαq families. Structures of Gαi1 bound to RGS4 (Tesmer *et al.*, 1997a), and of Gαt in complex with the RGS domain of RGS9 and PDEγ (Slep *et al.*, 2001), have provided mechanistic insight into the action of these domains. As discussed in Section III.B.2, it was discovered that the N-terminal segment of p115RhoGEF is a GAP for Gα13 (Kozasa *et al.*, 1998). The structure of a complex between the N-terminal GAP and RGS-like domains of p115RhoGEF bound to a Gα13-i chimera has been determined, revealing an unexpected mode of GAP activation (Chen *et al.*, 2005).

Structural and mutagenic studies of Gα:GAP complexes, as described in this section, show that the effector-binding and GAP-binding sites of Gα proteins are distinct and largely nonoverlapping. Nevertheless, the interactions of Gα with GAPs and effectors can be synergistic, and this synergism contributes to the spatial and temporal specificity of signal termination as well as the efficiency of steady-state signaling in the presence of GPCRs. The catalytic efficiency of GAP-stimulated, Gα-catalyzed GTP hydrolysis, when kinetically coupled to GPCR-catalyzed GDP release, is close to that expected for diffusion-limited enzyme catalysis ($>10^8\ M^{-1}\ s^{-1}$) (Ross and Wilkie, 2000).

### A. *Deactivation of Gαq by PLCβ*

When reconstituted into proteoliposomes with M1 muscarinic cholinergic receptor, PLCβ1 potently stimulates the GTPase activity of Gαq, achieving a steady-state rate 1000-fold greater than that in the absence of a GAP

(Mukhopadhyay and Ross, 1999). Gαq is also a potent activator of PLCβ1 (Smrcka *et al.*, 1991; Taylor *et al.*, 1991). Deletion mapping experiments demonstrated that Gαq binds to a region located C-terminal to the C2 domain of PLCβ1. The interaction of Gαq with this C-terminal region of PLCβ1 is required both for the stimulatory activity of Gαq and for the GAP activity of PLCβ1 (Lee *et al.*, 1993; Park *et al.*, 1993). Several C-terminal constructs derived from the C-terminus of PLCβ1 showed substantial GAP activity toward Gαq (Paulssen *et al.*, 1996). The most potent ($IC_{50} \sim 100$ nM, residues 902–1042) of these fragments was found to augment the GAP activity of PLCβ1 at low concentration, but was inhibitory at high concentration. This biphasic behavior suggested that PLCβ1 is active as a homodimer that is stabilized by intermolecular contacts within the C-terminal domain. Indeed, the C-terminus of turkey PLCβ (residues 878–1158, equivalent to human 890–1172) crystallizes as a dimer (Singer *et al.*, 2002). This fragment folds into an extended, parallel three-helix bundle, and forms an extensive antiparallel dimer interface along the long axis of the bundle. Although the PLCβ C-terminal fragment inhibits the GAP activity of mammalian PLCβ1, it has no GAP activity.

Using a computational protein-docking algorithm that exploits only molecular surface geometry (Katchalski-Katzir *et al.*, 1992), Singer *et al.* (2002) were able to generate a model of the PLCβ:Gαq complex. The model has several attractive features not imposed by the computational algorithm with which it was generated, among which are: exploitation of the positively charged surface of the PLCβ terminus to engage the negatively charged surface of Gαq, involvement of the Gαq switch regions at the interface with PLCβ1, participation of conserved PLCβ residues at the interaction site, and exploitation of the twofold symmetry PLCβ to allow docking of a second, symmetry-related Gαq. The results of scanning and targeted mutagenesis of the PLCβ1 C-terminal domain corroborate the model (Ilkaeva *et al.*, 2002; Kim *et al.*, 1996). Ross and colleagues found that mutations that affect PLCβ1 GAP activity can be clustered into two structural classes: the first consisting of basic residues at the dimer interface and the second forming a distributed patch of conserved residues that project from the positively charged concave surface of the C-terminal fragment. The latter falls within the predicted Gαq contact surface. Most of the mutations at both sites reduce both the affinity of Gαq as well as its efficacy as a stimulator of PLCβ1 catalytic activity, but increase only the $EC_{50}$ for GAP activity of PLCβ1 toward Gαq. Thus, GAP activity is largely determined by the affinity of the GAP toward the catalytically active state of Gα, a feature common to Gα GAPs. Determinants for GAP activity, on one hand, and susceptibility to activation by Gαq, on the other, are not easily separable. At least one PLCβ1 deletion mutant was found that retained sensitivity to the stimulatory activity of Gαq, but was not a GAP. Presumably, this mutant PLCβ1 binds to the GTP-bound form of Gα, but fails to stabilize the catalytically activated state.

The mechanism by which PLCβ1 is stimulated by or exerts GAP activity toward Gαq/11 is unknown, and is likely to remain so until the structure of the complex can be determined. The C2 domain of PLCβ1, which is located between the catalytic and C-terminal domains of the enzyme, was shown to bind to GTPγS-activated Gαq (Wang *et al.*, 1999). Thus, Gαq might act allosterically on the catalytic domains via contacts with C2 (Singer *et al.*, 2002), perhaps using the C-terminus as a scaffold.

### *B. RGS GAPs*

RGS proteins, defined by the presence of a 130-residue helical domain, referred to as the "RGS box" (Fig. 7A), constitute the largest and best characterized group of Gα GAPs. The discovery of these proteins and their function has been reviewed; for example, see Berman and Gilman (1998) and Koelle (1997). Classification schemes proposed by Zheng *et al.* (1999) and Ross and Wilkie (2000) are comparable. Both defined 5 classes of RGS domain, respectively A/RZ, B/R4, C/R7, D/R12, and E/RA, for the 40 or so RGS proteins that had been described. Zheng *et al.* define a sixth group that encompasses the RGS-like domains of the RH-RhoGEF family. Yet another family includes the RGS-like domains present in GRKs. Only RGS domains in the A/RZ-D/R12 family have demonstrated GAP activity. Proteins of the A/RZ and B/R4 groups have been described as "simple" in the sense that they contain no recognized domains other than the RGS domain itself (Hollinger and Hepler, 2002). However, they do possess N-terminal cysteine strings (A/RZ) or amphipathic helical regions (some A/RZ and all B/R4) that promote membrane localization or interactions with other signaling components in the cell, including GPCRs. Proteins of the remaining classes are "complex" in that they incorporate a variety of other interaction and signaling domains in addition to the RGS module. GAP activity of complex RGS proteins constitutes only one of several functions related to the coordination of effectors, G proteins, GPCRs, and molecules from other signaling pathways, as reviewed elsewhere (Abramow-Newerly *et al.*, 2006; Hollinger and Hepler, 2002; Ross and Wilkie, 2000; Wilkie and Kinch, 2005).

Key to understanding the mechanism of RGS GAPs, and GAPs in general, is the observation that they bind most tightly to the GDP•$Mg^{2+}$•$AlF_4^-$ complex of their Gα substrates (Berman *et al.*, 1996a; Chen *et al.*, 1996, 1997; Watson *et al.*, 1996) and thereby promote a conformation that is structurally most

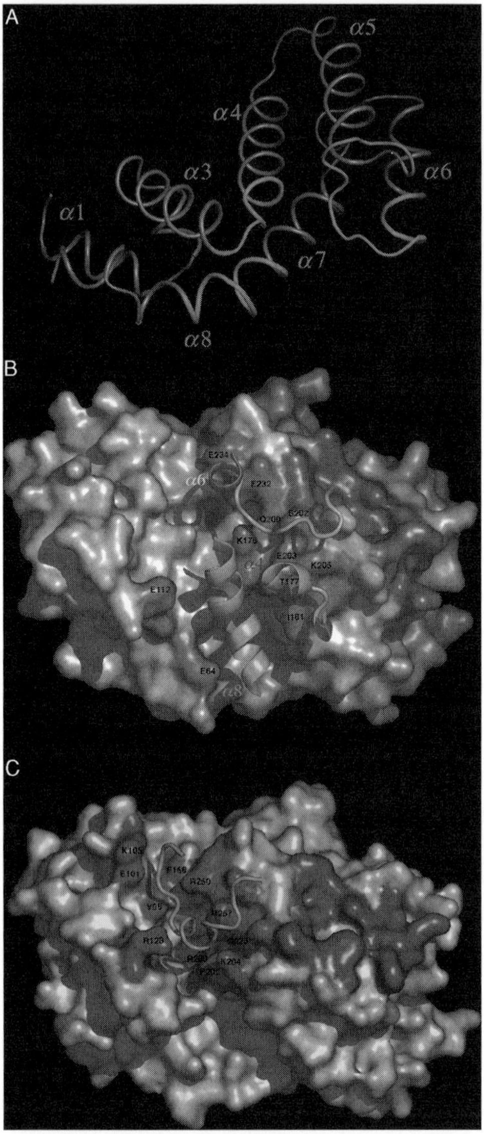


FIG. 7. Interaction of GAP domains with Gα. Panel A shows the Cα backbone trace of the RGS domain of RGS4. Loop segments that contact Gαi1 in the RGS4:Gαi1•GDP•$Mg^{2+}$•$AlF_4^-$ complex (1AGR) are colored green. Panel B shows the molecular surface of Gαt-i in the complex with RGS9 and PDEγ (1FQJ). Gαt-i residues within 4 Å of RGS9 are colored in green and labeled. Residues that form contacts with PDEγ are colored blue, and residues within 4 Å of both RGS9 and PDEγ are colored cyan. Backbone ribbons corresponding to the segments of RGS9 that contact Gαt-i are shown in orange.

complementary to the activated state for catalysis. The dissociation constants for binding of RGS4 to Giα1•GDP•$Mg^{2+}$•$AlF_4^-$, Giα1•$Mg^{2+}$•GTPγS, and Giα1•GDP were determined by isothermal titration calorimetry to be 0.1, 5, and 400 μM, respectively (Thomas *et al.*, 2004). However, this trend is not universal; Gαz, an exceptionally slow Gα GTPase, binds RGSZ1 with equal affinity in the GTPγS- and GDP•$AlF_4^-$ bound forms.

The RGS domains of RGS4 and RGS9 form similar interactions with Gαi1 and Gαt, respectively, despite the modest (~35%) sequence identity between them (Slep *et al.*, 2001; Tesmer *et al.*, 1997a). Eleven of seventeen Gα-contacting residues are conserved between RGS4 and RGS9 domains, and these are also well conserved in the A/RZ-D/R12 families of RGS domains (Ross and Wilkie, 2000). RGS domains that function as GAPs bind to a surface that is contiguous with, but distinct from, that occupied by effector domains (Figs. 2 and 7B). Three discrete segments of the RGS domain are involved in the contact with Gα: the acute connector between α3 and α4, an extended, platform-like loop between α5 and α6, and the convex surface of α7-α8 (Fig. 7A). Each of these RGS surfaces contacts a pair of structural elements on Gα (Fig. 7B). The α3-α4 hairpin binds Gα Switch I and Switch II; α5 and α6 contact Switch II and Switch III, and the extended α7-α8 surface links Switch I with the Gα helical domain. In this way the Gα switch segments are spatially constrained by overlapping interfaces with RGS. The highest density of RGS contacts is centered on Thr 182 of Switch I (Fig. 8A), and the RGS residues involved in this locus are the most highly conserved (in the ensuing discussion, residue numbers for RGS4 and Gαi1 are used, and RGS residue numbers are prefixed with "r").

RGS domains induce only subtle changes to the conformation of Gα•GDP•$AlF_4^-$. In both Gαi1 and Gαt, the side chain of Thr 182 rotates 120° as a consequence of its interaction with the RGS α4-residue Asp r88; this rotation allows the formation of a network of hydrogen bond interactions between Thr 182 and Switch II residues Glu 207 and Lys 210 that do not occur in the absence of RGS (Fig. 8A). On binding to RGS4, the helical and Ras-like domains of Gαi1 rotate together by 3°. Also observed is a twofold reduction in crystallographic temperature factors for the switch regions of Gαi1, indicative that these segments of polypeptide chain are

---

Panel C shows the molecular surface of Gα13-i•GDP•$Mg^{2+}$•$AlF_4^-$ in the complex with the rgRGS domain of p115RhoGEF (1SHZ), with the molecular surface of Gα shaded as in panel B, with αGAP-containing residues shaded green and RGS domain-contacting residues shaded blue. The Cα backbone ribbon of the αGAP segment that confers GAP activity is shown in orange. (See Color Insert.)

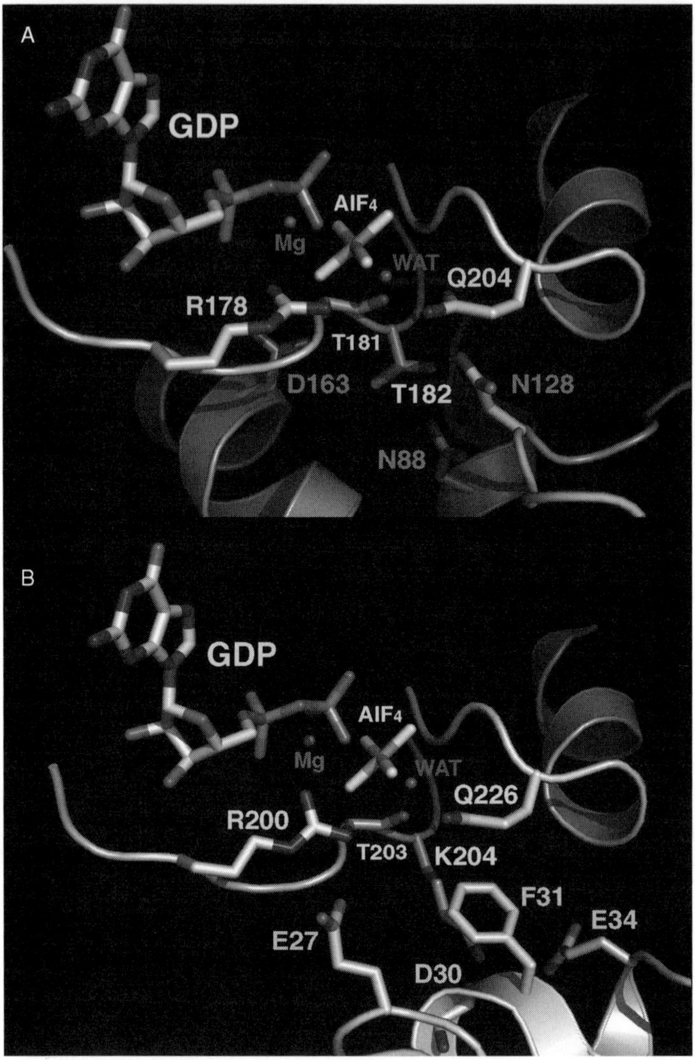


FIG. 8. Functionally analogous interactions of G$\alpha$ proteins with unrelated GAP domains. Panel A illustrates the contacts between RGS4 (tan) and the catalytic site of GDP•$Mg^{2+}$•$AlF_4^-$-bound G$\alpha$i1 (gray) (1AGR). Panel B shows the corresponding interface of the $\alpha$GAP segment of p115RhoGEF with the GDP•$Mg^{2+}$•$AlF_4^-$ complex of G$\alpha$13-i (1SHZ). Note that the $\alpha$GAP residues Glu 27 and Phe 31 adopt structural roles comparable to Asp 163 and Asn 128 of RGS4, respectively. (See Color Insert.)

being restrained from movement in the complex. A comparable change in the temperature factors of Gαt switch residues is not observed on binding to RGS9, however.

Interactions with RGS appear to stabilize the conformations of the switch segments in the catalytically activated state of Gα (Fig. 8A). Thus, Asp r163 in α6-α7 supports the conformation of Switch I by hydrogen bonding to the main chain of Thr 182, and the α3-α4 residue Asn 128 sterically traps the catalytic glutamine 204 of Gαi1, constraining it to interact with the lytic water and the GTP $\gamma$ phosphate (as mimicked by the fluorine atoms of $AlF_4^-$ (Coleman and Sprang, 1999b). In the RGS9:Gαt-i complex, the RGS asparagine and the Gα glutamine are hydrogen bonded. The RGS:Gα interface residues conserved in the A/R4–D/R12 class RGS domains are selective for Gαi and Gαq isoforms, and are incompatible with potential binding sites in Gαs and Gα12/13. The aspartate residue in Gαs that corresponds to the conserved serine 206 in Gαi/q/z isoforms cannot be sterically or electrostatically accommodated in the complex with RGS4. Substitution of this aspartate by serine renders Gαs a substrate for RGS4, while replacement of the serine with aspartate in Gαi abolishes RGS4 binding (Natochin and Artemyev, 1998a,b). Similarly, Thr 182 is replaced by a bulky lysine residue in Gα12/13, which precludes their binding to canonical RGS domains.

Certain point mutations in Gα proteins diminish RGS binding by indirect or allosteric effects. For example, replacement of the conserved Switch I glycine (position 183 in Gαi1) has no effect on intrinsic GTPase activity, but renders Gαo and yeast GPA1 insensitive to GAP activity of RGS4 and sst2, respectively (DiBello *et al.*, 1998; Lan *et al.*, 1998). Because Gly 183 adopts main chain conformation angles that are inaccessible to nonglycine residues, any substitution at this position is likely to disrupt the conformation of Switch I and thus interfere with RGS binding. The corresponding mutation does not alter the sensitivity of Gα13 to the GAP activity of p115RhoGEF (Grabocka and Wedegaertner, 2005), which does not exhibit GAP activity through its RGS-like domain. Natochin *et al.* found that replacement of Switch III Arg 238 with glutamine renders Gαt refractory to the GAP activity of RGS9, with only modest reduction in intrinsic GTPase activity. Substitution of Arg 238 presumably disrupts a network of largely solvent-inaccessible ionic interactions involving the P-loop Glu 38, and the catalytic Arg 178 in Switch I (Natochin and Artemyev, 2003).

To stabilize the catalytically active state of Gα, RGS domains distribute binding energy over the entire Gα interaction surface (Fig. 7B). Mutations at conserved residues in all three RGS4 contact zones impair GAP activity, and the effects of some of these mutations are thermodynamically additive (Chen *et al.*, 1997; Natochin *et al.*, 1998b; Posner *et al.*, 1999;

Srinivasa *et al.*, 1998). Almost in general, mutations that reduce the GAP activity of RGS4 also reduce affinity for Gα, consistent with a mechanism in which RGS functions as a GAP largely by binding specifically to the catalytically activated state of Gα. Accordingly, no specific catalytic function can be assigned to any single RGS residue. Asparagine r128, which is preserved in most RGS domains with GAP activity, is in polar contact with the catalytic glutamine of Gα in both RGS9 and RGS4 (Fig. 8A). It seemed possible that this RGS residue might stabilize the lytic water in the ground state Gαi1•GTP complex (Tesmer *et al.*, 1997a). However, mutations of r128 only impair GAP activity to the extent that they reduce affinity for Gα (Natochin *et al.*, 1998b; Posner *et al.*, 1999). Mutation of r128 to phenylalanine can be suppressed by a second mutation in the Switch I region of Gαi1. Thus, Asn 128 and its counterparts may serve a role in Gα specificity (as in GIAP, which is slightly selective toward Gαq, and where "r128" is a serine). Nonconservative mutations within the RGS domain of RGS16, but remote from the Gα-binding site, were found to abrogate pheromone signaling in a yeast transfection model, but did not block GDP•$AlF_4^-$-dependent binding to Gαi (Chen *et al.*, 1997). Thus, there may also be an allosteric component to the GAP activity of RGS.

Local structural differences between the Gαt:RGS9 and Gαi:RGS4 interfaces are largely confined to Switch III and helical-domain interaction sites, and are reflected in the concentration of nonconserved RGS residues in α7/α8. These structural and sequence variations provide mechanisms for selective interaction of certain RGS proteins with their respective Gα targets. Determinants for selectivity of RGS9 toward Gαt were shown to involve residues in the α helical domain of the G protein (Skiba *et al.*, 1999). In contrast to RGS4, which is active toward both Gαi and Gαq-class Gα proteins, RGS2 is attenuated in its activity toward Gαi, and is selective for Gαq (Heximer *et al.*, 1997; Ingi *et al.*, 1998). The origin of this selectivity can be traced to three nonconserved residues in α3/α4 and α7/α8 that are proposed to offer greater steric and electrostatic complementarity with Gαq than to Gαi (Heximer *et al.*, 1999). The differences between the interaction surfaces of the two Gα:RGS complexes may be subtle, and the consequences of these will be difficult to assess in the absence of experimentally derived structures.

### C. *Synergy Between RGS and Effector Domains*

The phototransduction apparatus in both rods and cones evolved an efficient mechanism to exert efficient GAP-mediated signal termination preferentially on effector-bound Gαt, as reviewed (Arshavsky and Pugh, 1998;

Chen, 2005). Early studies suggested that PDE$\gamma$ was a GAP for G$\alpha$t (Arshavsky and Bownds, 1992). However, it was later appreciated that PDE$\gamma$ has no GAP activity on its own, but rather stimulates a photoreceptor membrane-associated GAP activity three- to sixfold (Angleson and Wensel, 1993; Antonny *et al.*, 1993; Arshavsky *et al.*, 1994). The GAP was identified as RGS9, a member of the R7/C family of "complex" RGS proteins (Chen *et al.*, 2000; He *et al.*, 1998). RGS9 potentially resides at the plasma membrane as part of a multiprotein complex: it contains a G protein $\gamma$-like (GGL) domain that binds to G$\beta$5 (Makino *et al.*, 1999), and is tethered to a photoreceptor-specific plasma membrane anchoring protein R9AP through an N-terminal DEP domain (Hu and Wensel, 2002).

Synergy between PDE$\gamma$ and RGS9 arises from a network of hydrogen bond and van der Waals interactions that interconnect PDE$\gamma$, the RGS domain, and Switch II of G$\alpha$t (Fig. 9A). Packing contacts between Val 66 of PDE$\gamma$ and Trp 365 in the $\alpha$5/6 loop of RGS9 rigidify a scaffold that envelopes Switch II, and stabilizes a network of hydrogen bonds that orient the catalytic arginine and glutamine residues in the G$\alpha$t-i active site. The positive synergy is RGS9-specific. PDE$\gamma$ actually antagonizes the binding of RGS7—an R7 class RGS protein like RGS9—to G$\alpha$t (McEntaffer *et al.*, 1999; Nekrasova *et al.*, 1997). Evolutionary trace analysis suggests that sequence divergence within a cluster of residues that affect the $\alpha$5-$\alpha$6 loop accounts in part for positive or negative allosteric regulation of RGS domain GAP activity by PDE$\gamma$ (Sowa *et al.*, 2000). Mutations within this cluster affect the susceptibility to PDE$\gamma$ regulation and basal GAP activity (Sowa *et al.*, 2001). In this context, it is noteworthy that ERK1-dependent phosphorylation of Ser 151 in GAIP (equivalent to the $\alpha$5-$\alpha$6 loop residue Thr 124 in RGS4) causes mild stimulation of GAP activity (Ogier-Denis *et al.*, 2000). Thus, the susceptibility to specific effector regulation is an evolvable property of RGS domains.

### *D. Deactivation of G$\alpha$13 by the $\alpha$GAP Element of p115RhoGEF*

Biochemical studies of p115RhoGEF demonstrated that the GAP activity of an RGS-containing protein might arise from a structure other than the RGS domain itself. The N-terminal fragment of the protein, composed of an ~40-residue segment followed by an RGS-like domain (which together form the rgRGS domain), is both necessary and sufficient for GAP activity (Kozasa *et al.*, 1998). A finer dissection of the domain revealed that an element within the N-terminus, corresponding to residues 25–42 preceding the RGS-box, is absolutely required for GAP activity, and that deletion

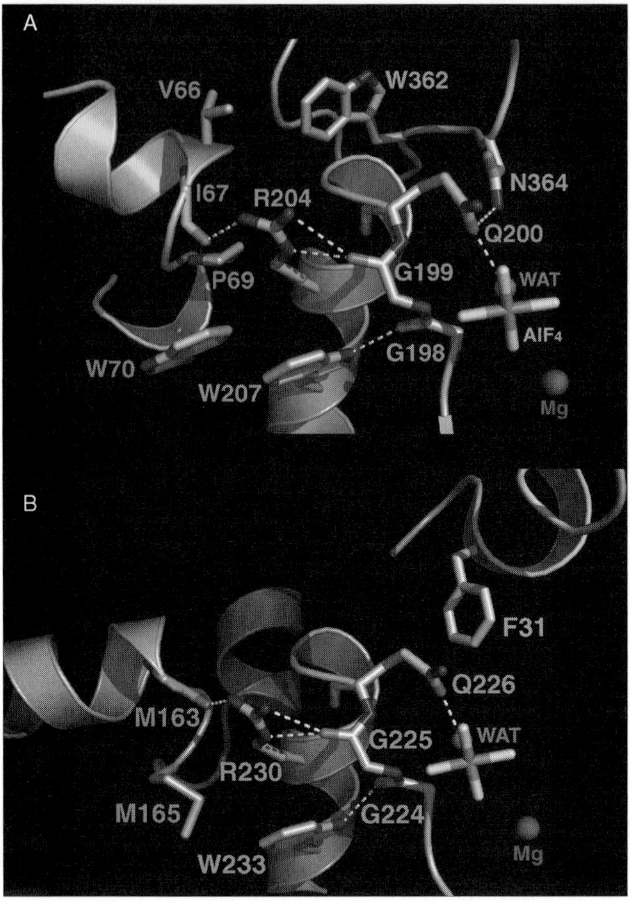


FIG. 9. Structural synergy between effector and GAP domains in their mutual interaction with G$\alpha$. Panels A and B illustrate the interaction sites of G$\alpha$t-i and G$\alpha$13-i, respectively, at which GAP and effector domains are in closest mutual contact (see text). Panel A depicts G$\alpha$t-i in the complex with RGS9 and PDE$\gamma$ (1FQJ). The RGS domain of RGS9 is colored brown and PDE$\gamma$ is green. Switch II of G$\alpha$t-i•GDP•$Mg^{2+}$•$AlF_4^-$ is gray. Atoms are colored according to the scheme used in preceding figures. Only $Mg^{2+}$•$AlF_4^-$ and the lytic water of the substrate complex is shown. Panel B shows the complex of the rgRGS domain of p115RhoGEF bound to the GDP•$Mg^{2+}$•$AlF_4^-$ complex of G$\alpha$13-i (1SHZ). The $\alpha$GAP domain of rgRGS is colored brown, the RGS green, and G$\alpha$13-i Switch II segment is gray. In both panels, hydrogen bonds are indicated by dotted yellow lines. (See Color Insert.)

of the N-terminal fragment substantially reduces the affinity of the rgRGS domain for G$\alpha$13 (Wells *et al.*, 2002). Approximately 20 residues that precede the RGS-like domain are enriched in acidic amino acids, and charge reversal mutations in three of these are sufficient to eliminate the

Gα13 binding and GAP activity of rgRGS. While mutation of several residues at the expected Gα-interaction surface of the RGS-like domain impaired GAP activity, only one of these mutations had drastic effects comparable with those of residues in the acidic cluster (Chen *et al.*, 2003). These experiments suggested that the N-terminal region of p115RhoGEF, and to a lesser extent the RGS-like domain, is crucial to GAP activity toward Gα13.

The structure of the Gα13-i:rgRGS complex shows that the core of the GAP is formed by a turn-helix motif (αGAP, residues 23–34) near the N-terminus of the rgRGS domain (Figs. 3C and 7C). This short segment encompasses the single hydrophobic and three acidic residues found to be essential for GAP activity. αGAP interacts with all three Gα13-i switch regions, together with residues in the helical domain that are unique to Gα12-class proteins. The extended αGAP peptide traverses the crevice between the Ras-like and helical domains of Gα13-i, and the succeeding helical segment emerges through a channel between the two domains (Fig. 7C). This protein–protein interaction is structurally unrelated to that between RGS domains and Gα proteins, but is functionally analogous (Fig. 8B). Glutamate 27 of αGAP forms an ion pair with the Gα13-i catalytic arginine in Switch I, and thereby appears to stabilize the interaction of the latter with the nucleotide $\beta$ and $\gamma$ phosphates. Replacement of the Glu 27 side chain with aspartate only reduces GAP activity, whereas replacement with lysine abolishes it. Phenylalanine 31 serves a function similar to that of the well-conserved Asn 128 in RGS4. van der Waals interactions with Phe 31 orient the backbone carbonyl of the Switch I Thr 203 (residue 181 in Gαi) as well as the carboxyamino group of Switch II Gln 226, promoting their hydrogen bonding interactions with the nucleophilic water. At the center of the interface, the lysine side chain of Switch I residue 204 is encompassed by a cavity formed by αGAP residues Glu 27, Phe 31, and Glu 34. Like the corresponding Thr 182 in Gαi1, Lys 204 is a critical determinant of specificity in the Gα13:αGAP interface. Mutation of this residue impairs binding of Gα13 to the rgRGS domains of both p115RhoGEF and LARG, and its sensitivity to GAP activity of these proteins. Loss of binding to rgRGS due to this mutation also renders Gα13 unable to stimulate the RhoGEF activity of p115RhoGEF and LARG (Nakamura *et al.*, 2004). Specificity for Gα12/13 is also mediated by ion pair interactions between αGAP and residues within an insertion into the helical domain that is present in Gα12/13, but not other members of the Gα family (Figs. 3C and 7C). Indeed, a charge reversal mutation of the αGAP Glu 29, one of two residues that interact with the Gα13 helical extension, abrogates binding of the rgRGS domain to Gα13. At a more global level, the highly electronegative surface of αGAP is electrostatically complementary to the positively charged surface at the junction between the Ras-like and helical domains of Gα12/13, which contrasts with a neutral or electronegative surface of proteins in the Gαi family (Chen *et al.*, 2003, 2005). Electrostatic complementarity may be generally exploited as a mechanism of RGS-domain/Gα specificity.

Binding of a GAP and effector to Gα, although they involve distinct sites, can be synergistic or antagonistic, as demonstrated by interaction of PDEγ and Gαt with RGS GAPs. The potential for synergy is amplified when GAP and effector functions are present in the same polypeptide, as is true for p115RhoGEF. In this case, much of the potential synergy arises from shared binding energy derived from the tethered GAP and effector domains. The rgRGS domain binds to Gα13 with five- to tenfold greater affinity than a truncation construct missing the first 42 residues. The contribution of the RGS-like domain to affinity for Gα13 is even more substantial: an αGAP peptide (residues 14–34) has only 0.1% of the rgRGS GAP activity toward Gα13. The RGS-like domain has no GAP activity itself, so it may act simply by increasing the effective concentration of αGAP at its Gα13-binding site. However, it is possible that the RGS-like domain potentiates the binding of αGAP by a mechanism similar to that by which PDEγ synergizes with RGS9. Although the αGAP segment and the RGS-like domain do not physically interact in the complex of rgRGS with Gα13-i, they participate in a network of interactions similar to that observed between PDEγ, RGS9, and Gαt (Fig. 9C). Because the αGAP segment is connected to the RGS-like domain by a disordered linker, the two elements are not as tightly buttressed against Switch II of Gα, as are PDEγ and RGS9 in the former complex. Nor does p115RhoGEF possess the potency as a GAP, at least *in vitro,* characteristic of RGS GAPs. It is possible that, as is true for PLCβ, the full GAP potency of p115RhoGEF and its homologues is realized only in the context of a complex with a GPCR and Gα12/13.

Generally, we may ask whether proteins that have dual effector and GAP activity interact differently with the GTP-bound and GDP•AlF-bound states of Gα. A suggestion that this might be true derives from the observation that GAP-deficient mutants of p115RhoGEF remain sensitive to the RhoGEF stimulatory activity of Gα13. This is in accordance with the detection of Gα-binding determinants in the DH-PH domains of p115RhoGEF. The interesting possibility arises that Gα, while bound to p115RhoGEF, and perhaps also to PLCβ, randomly samples different subsites of the effector/GAP-binding surface, or alternates between different conformational states or binding modes.

## VI. Conclusions

We end by restating the principal observations afforded by biochemical and structural studies of G protein effector activation and signal termination.

- Switch segments of Gα proteins serve as critical determinants for catalytic and effector-binding activity, thereby providing a mechanism for kinetic coupling of GTP binding to effector activity and GTP hydrolysis to signal termination.
- A structural motif, which is common to all Gα isoforms and composed of Switch II (α2) and α3 forms, in the GTP-bound state, the consensus binding site for Gα effector proteins. Limited sequence conservation within this region provides common interface elements; variation in sequence and conformation (particularly within α3 and the connecting loop to β5) confer Gα-effector specificity. Elements outside of the consensus binding site are involved in recognition processes for specific Gα–effector interactions. Effectors do not appear to produce GAP activity through interactions with the consensus effector-binding motif.
- Gα proteins exert their signaling activities either as passive binding partners or as allosteric ligands that modulate conformational equilibriums of effectors.
- The transition state for G protein-catalyzed GTP hydrolysis (with Ras as the model system) has dissociative character. Transition state formation, in contrast to enzyme reorganization is the rate-limiting step of the reaction, although in the presence of GAP, Pi release may become partially rate limiting. Even so, the activation energy for enzyme activation (attaining a state similar to that of the G protein•GDP•$Mg^{2+}$•$AlF_x$ complex) is substantial. Gα GAPs appear to act by reducing the activation energy barrier to this rearrangement. Ras (and presumably other members of the G protein superfamily) stabilizes charge at the β nonbridge oxygen atoms of GTP and the βγ bridge oxygen atom of the scissile bond. The P-loop backbone amide residues, the tertiary amino group of the conserved P-loop lysine, and the $Mg^{2+}$ ions are probably involved in the mechanism of charge stabilization. The endogenous Switch I arginine of Gα proteins and the exogenous "arginine finger" supplied by GAPs for Ras-family GTPases probably serve as a stereochemically specific ligand of the γ phosphate in the transition state (interacting also with the leaving group oxygen atom) and of Pi in the product complex. The role of the critical Switch II glutamine remains unresolved, but it is probably critical to the precise structural alignment of the lytic water and the

$\gamma$ phosphate of GTP of the activated enzyme. Proper orientation of the Switch I arginine and Switch II glutamine is achieved fully in the catalytically activated enzyme, but not in the ground state.

- The reorganization or collapse of the switch segments on GTP hydrolysis (signal termination) may occur before phosphate is released from the G$\alpha$•GDP•$Mg^{2+}$•Pi complex, and may be triggered by dissociation of $Mg^{2+}$ from the catalytic site or formation of a hypothetical catalytic intermediate (e.g., metaphosphate). GTP hydrolysis and switch reorganization can be mechanistically decoupled as a consequence of a point mutation that alters the dynamics of G$\alpha$ and perturbs the $Mg^{2+}$-binding site in the catalytically activated state.
- GAPs, whether exogenous proteins or effector-linked domains, bind to a discrete locus of G$\alpha$ that is composed of Switch I and the N-terminus of Switch II. This site is immediately adjacent to, but does not substantially overlap, the effector-binding site. Interactions of effectors and exogenous GAPs with G$\alpha$ proteins can be synergistic or antagonistic, mediated by allosteric interactions among the three molecules. The GAP–effector synergy, particularly where both functions are present in the same protein, may contribute substantially to binding of both domains (proteins) to G$\alpha$.

As a final note, it is worth remarking that the notion of "reserved" G$\alpha$-binding sites, exclusive to effectors and GAPs, may be extended to GEFs, such as GPCRs, which primarily engage the terminal $\alpha$ helix ($\alpha$4) of G$\alpha$ and surrounding structural elements to effect nucleotide release (Oldham and Hamm, this volume). The functional and structural modularity of G$\alpha$ proteins provides for independent, combinatorial, or synergistic regulation by effectors, GAPs, and exchange factors. Thus, G$\alpha$ and associated signaling molecules can exhibit emergent properties that are not apparent from their behavior as individual proteins, but which characterize the signal transduction complexes of which they are constituents.

## Acknowledgments

We are grateful to past and present members of our laboratory whose work we have cited in this chapter. We acknowledge the grants from the NIH, R01-DK046371 and R01-GM071442, the Robert Welch Foundation (I-1229), and past support from the Howard Hughes Medical Institute, to S.R.S.

## References

Abramow-Newerly, M., Roy, A. A., Nunn, C., and Chidiac, P. (2006). RGS proteins have a signalling complex: Interactions between RGS proteins and GPCRs, effectors, and auxiliary proteins. *Cell. Signal.* **18,** 579–591.

Adari, H., Lowy, D. R., Willumsen, B. M., Der, C. J., and McCormick, F. (1988). Guanosine triphosphatase activating protein (GAP) interacts with the p21 ras effector binding domain. *Science* **240,** 518–521.

Afshar, K., Willard, F. S., Colombo, K., Johnston, C. A., McCudden, C. R., Siderovski, D. P., and Gonczy, P. (2004). RIC-8 is required for GPR-1/2-dependent Galpha function during asymmetric division of *C. elegans* embryos. *Cell* **119,** 219–230.

Allin, C., and Gerwert, K. (2001). Ras catalyzes GTP hydrolysis by shifting negative charges from $\gamma$- to $\beta$-phosphate as revealed by time-resolved FTIR difference spectroscopy. *Biochemistry* **40,** 3037–3046.

Allin, C., Ahmadian, M. R., Wittinghofer, A., and Gerwert, K. (2001). Monitoring the GAP catalyzed H-Ras GTPase reaction at atomic resolution in real time. *Proc. Natl. Acad. Sci. USA* **98,** 7754–7759.

Ames, J. B., Porumb, T., Tanaka, T., Ikura, M., and Stryer, L. (1995). Amino-terminal myristoylation induces cooperative calcium binding to recoverin. *J. Biol. Chem.* **270,** 4526–4533.

Angleson, J. K., and Wensel, T. G. (1993). A GTPase-accelerating factor for transducin, distinct from its effector cGMP phosphodiesterase, in rod outer segment membranes. *Neuron* **11,** 939–949.

Antonny, B., Otto-Bruc, A., Chabre, M., and Vuong, T. M. (1993). GTP hydrolysis by purified $\alpha$-subunit of transducin and its complex with the cyclic GMP phosphodiesterase inhibitor. *Biochemistry* **32,** 8646–8653.

Arkinstall, S., Chabert, C., Maundrell, K., and Peitsch, M. (1995). Mapping regions of G alpha q interacting with PLC beta 1 using multiple overlapping synthetic peptides. *FEBS Lett.* **364,** 45–50.

Arshavsky, V. Y., and Bownds, M. D. (1992). Regulation of deactivation of photoreceptor G protein by its target enzyme and cGMP. *Nature* **357,** 416–417.

Arshavsky, V. Y., and Pugh, E. N., Jr. (1998). Lifetime regulation of G protein-effector complex: Emerging importance of RGS proteins. *Neuron* **20,** 11–14.

Arshavsky, V. Y., Dumke, C. L., Zhu, Y., Artemyev, N. O., Skiba, N. P., Hamm, H. E., and Bownds, M. D. (1994). Regulation of transducin GTPase activity in bovine rod outer segments. *J. Biol. Chem.* **269,** 19882–19887.

Barbacid, M. (1987). *ras* Genes. *Annu. Rev. Biochem.* **56,** 779–827.

Benkovic, S. J., and Hammes-Schiffer, S. (2003). A perspective on enzyme catalysis. *Science* **301,** 1196–1202.

Berghuis, A. M., Lee, E., Raw, A. S., Gilman, A. G., and Sprang, S. R. (1996). Structure of the GDP-Pi complex of Gly203 $\rightarrow$ Ala $G_{i\alpha1}$: A mimic of the ternary product complex of $G_\alpha$-catalyzed GTP hydrolysis. *Structure* **4,** 1277–1290.

Berlot, C. H., and Bourne, H. R. (1992). Identification of effector-activating residues of $G_{s\alpha}$. *Cell* **68,** 911–922.

Berman, D. M., and Gilman, A. G. (1998). Mammalian RGS proteins: Barbarians at the gate. *J. Biol. Chem.* **273,** 1269–1272.

Berman, D. M., Kozasa, T., and Gilman, A. G. (1996a). The GTPase-activating protein RGS4 stabilizes the transition state for nucleotide hydrolysis. *J. Biol. Chem.* **271,** 27209–27212.

Berman, D. M., Wilkie, T. M., and Gilman, A. G. (1996b). GAIP and RGS4 are GTPase-activating proteins (GAPs) for the $G_i$ subfamily of G protein $\alpha$ subunits. *Cell* **86,** 445–452.

Berstein, G., Blank, J. L., Jhon, D.-Y., Exton, J. H., Rhee, S. G., and Ross, E. M. (1992). Phospholipase C-$\beta$1 is a GTPase-activating protein for $G_{q/11}$, its physiologic regulator. *Cell* **70,** 411–418.

Biddlecome, G. H., Berstein, G., and Ross, E. M. (1996). Regulation of phospholipase C-$\beta$1 by $G_q$ and m1 muscarinic cholinergic receptor steady-state balance of receptor-mediated activation and GTPase-activating protein-promoted deactivation. *J. Biol. Chem.* **271,** 7999–8007.

Booden, M. A., Siderovski, D. P., and Der, C. J. (2002). Leukemia-associated Rho guanine nucleotide exchange factor promotes G$\alpha$q-coupled activation of RhoA. *Mol. Cell. Biol.* **22,** 4053–4061.

Bourne, H. R., Sanders, D. A., and McCormick, F. (1991). The GTPase superfamily: Conserved structure and molecular mechanism. *Nature* **349,** 117–127.

Brown, R. L. (1992). Functional regions of the inhibitory subunit of retinal rod cGMP phosphodiesterase identified by site-specific mutagenesis and fluorescence spectroscopy. *Biochemistry* **31,** 5918–5925.

Brownbridge, G. G., Lowe, P. N., Moore, K. J., Skinner, R. H., and Webb, M. R. (1993). Interaction of GTPase activating proteins (GAPs) with p21ras measured by a novel fluorescence anisotropy method. Essential role of ARG-903 of GAP in activation of GTP hydrolysis on p21ras. *J. Biol. Chem.* **268,** 10914–10919.

Bunemann, M., Frank, M., and Lohse, M. J. (2003). Gi protein activation in intact cells involves subunit rearrangement rather than dissociation. *Proc. Natl. Acad. Sci. USA* **100,** 16077–16082.

Carman, C. V., Parent, J. L., Day, P. W., Pronin, A. N., Sternweis, P. M., Wedegaertner, P. B., Gilman, A. G., Benovic, J. L., and Kozasa, T. (1999). Selective regulation of G$\alpha_{q/11}$ by an RGS domain in the G protein-coupled receptor kinase, GRK2. *J. Biol. Chem.* **274,** 34483–34492.

Casey, P. J. (1994). Lipid Modifications of G Proteins. *Curr. Opin. Cell Biol.* **6,** 219–225.

Cassel, D., and Selinger, Z. (1978). Mechanism of adenylate cyclase activation through the $\beta$-adrenergic receptor: Catecholamine-induced displacement of bound GDP by GTP. *Proc. Natl. Acad. Sci. USA* **75,** 4155–4159.

Cassel, D., Levkovitz, H., and Selinger, Z. (1977). The regulatory GTPase cycle of turkey erythrocyte adenylate cyclase. *J. Cyclic Nucleotide Res.* **3,** 393–406.

Cepus, V., Scheidig, A. J., Goody, R. S., and Gerwert, K. (1998). Time-resolved FTIR studies of the GTPase reaction of H-ras p21 reveal a key role for the beta-phosphate. *Biochemistry* **37,** 10263–10271.

Cerione, R. A., and Zheng, Y. (1996). The Dbl family of oncogenes. *Curr. Opin. Cell Biol.* **8,** 216–222.

Chen, C., Zheng, B., Han, J., and Lin, S. C. (1997). Characterization of a novel mammalian RGS protein that binds to G$\alpha$ proteins and inhibits pheromone signaling in yeast. *J. Biol. Chem.* **272,** 8679–8685.

Chen, C. K. (2005). The vertebrate phototransduction cascade: Amplification and termination mechanisms. *Rev. Physiol. Biochem. Pharmacol.* **154,** 101–121.

Chen, C.-K., Wieland, T., and Simon, M. I. (1996). RGS-r, a retinal specific RGS protein, binds an intermediate conformation of transducin and enhances recycling. *Proc. Natl. Acad. Sci. USA* **93,** 12885–12889.

Chen, C. K., Burns, M. E., He, W., Wensel, T. G., Baylor, D. A., and Simon, M. I. (2000). Slowed recovery of rod photoresponse in mice lacking the GTPase accelerating protein RGS9-1. *Nature* **403,** 557–560.

Chen, J., and Iyengar, R. (1993). Inhibition of cloned adenylyl cyclases by mutant-activated Gi-alpha and specific suppression of type 2 adenylyl cyclase inhibition by phorbol ester treatment. *J. Biol. Chem.* **268,** 12253–12256.

Chen, Z., Singer, W. D., Wells, C. D., Sprang, S. R., and Sternweis, P. C. (2003). Mapping the G$\alpha$13 binding interface of the rgRGS domain of p115RhoGEF. *J. Biol. Chem.* **278,** 9912–9919.

Chen, Z., Singer, W. D., Sternweis, P. C., and Sprang, S. R. (2005). Structure of the p115RhoGEF rgRGS domain-Gα13/i1 chimera complex suggests convergent evolution of a GTPase activator. *Nat. Struct. Mol. Biol.* **12,** 191–197.

Chikumi, H., Barac, A., Behbahani, B., Gao, Y., Teramoto, H., Zheng, Y., and Gutkind, J. S. (2004). Homo- and hetero-oligomerization of PDZ-RhoGEF, LARG and p115RhoGEF by their C-terminal region regulates their *in vivo* Rho GEF activity and transforming potential. *Oncogene* **23,** 233–240.

Chung, H.-H., Benson, D. R., and Schultz, P. G. (1993). Probing the structure and mechanism of ras protein with an expanded genetic code. *Science* **259,** 806–809.

Coleman, D. E., and Sprang, S. R. (1998). Crystal structures of the G-protein Gia1 complexed with GDP and $Mg^{2+}$: A crystallographic titration experiment. *Biochemistry* **37,** 14376–14385.

Coleman, D. E., and Sprang, S. R. (1999a). Reaction dynamics of G-protein catalyzed hydrolysis of GTP as viewed by X-ray crystallographic snapshots of $Gi_{\alpha 1}$. *In* ''Methods in Enzymology: Enzyme Kinetics and Mechanism, Part E: Energetics of Enzyme Catalysis'' (V. L. Schramm and D. L. Puvich, Eds.), Vol. 308, pp. 70–92. Academic Press, New York.

Coleman, D. E., and Sprang, S. R. (1999b). Structure of $G_{i\alpha 1}$•GppNHp, Autoinhibition in a $G_{\alpha}$ protein-substrate complex. *J. Biol. Chem.* **274,** 16669–16672.

Coleman, D. E., Berghuis, A. M., Lee, E., Linder, M. E., Gilman, A. G., and Sprang, S. R. (1994). Structures of active conformations of $G_{i\alpha 1}$ and the mechanism of GTP hydrolysis. *Science* **265,** 1405–1412.

Cooper, D. M. (2003). Regulation and organization of adenylyl cyclases and cAMP. *Biochem. J.* **375,** 517–529.

Daumke, O., Weyand, M., Chakrabarti, P. P., Vetter, I. R., and Wittinghofer, A. (2004). The GTPase-activating protein Rap1GAP uses a catalytic asparagine. *Nature* **429,** 197–201.

Day, P. W., Tesmer, J. J., Sterne-Marr, R., Freeman, L. C., Benovic, J. L., and Wedegaertner, P. B. (2004). Characterization of the GRK2 binding site of Gαq. *J. Biol. Chem.* **279,** 53643–53652.

De Waard, M., Hering, J., Weiss, N., and Feltz, A. (2005). How do G proteins directly control neuronal $Ca^{2+}$ channel function? *Trends Pharmacol. Sci.* **26,** 427–436.

Derewenda, U., Oleksy, A., Stevenson, A. S., Korczynska, J., Dauter, Z., Somlyo, A. P., Otlewski, J., Somlyo, A. V., and Derewenda, Z. S. (2004). The crystal structure of RhoA in complex with the DH/PH fragment of PDZRhoGEF, an activator of the Ca(2+) sensitization pathway in smooth muscle. *Structure* **12,** 1955–1965.

Dessauer, C. W., and Gilman, A. G. (1996). Purification and characterization of a soluble form of mammalian adenylyl cyclase. *J. Biol. Chem.* **271,** 16967–16974.

Dessauer, C. W., Scully, T. T., and Gilman, A. G. (1997). Interactions of forskolin and ATP with the cytosolic domains of mammalian adenylyl cyclase. *J. Biol. Chem.* **272,** 22272–22277.

Dessauer, C. W., Tesmer, J. J., Sprang, S. R., and Gilman, A. G. (1998). Identification of a $G_{i\alpha}$ binding site on type V adenylyl cyclase. *J. Biol. Chem.* **273,** 25831–25839.

Dessauer, C. W., Tesmer, J. J., Sprang, S. R., and Gilman, A. G. (1999). The interactions of adenylate cyclases with P-site inhibitors. *Trends Pharmacol. Sci.* **20,** 205–210.

Dessauer, C. W., Chen-Goodspeed, M., and Chen, J. (2002). Mechanism of Gαi-mediated inhibition of type V adenylyl cyclase. *J. Biol. Chem.* **277,** 28823–28829.

DiBello, P. R., Garrison, T. R., Apanovitch, D. M., Hoffman, G., Shuey, D. J., Mason, K., Cockett, M. I., and Dohlman, H. G. (1998). Selective uncoupling of RGS action by a single point mutation in the G protein α-subunit. *J. Biol. Chem.* **273,** 5780–5781.

Dowal, L., Elliott, J., Popov, S., Wilkie, T. M., and Scarlata, S. (2001). Determination of the contact energies between a regulator of G protein signaling and G protein subunits and phospholipase C beta 1. *Biochemistry* **40,** 414–421.

Dowal, L., Provitera, P., and Scarlata, S. (2006). Stable association between G{alpha}q and phospholipase C$\beta$1 in living cells. *J. Biol. Chem.* **281,** 23999–24014.

Du, X., Frei, H., and Kim, S. H. (2000). The mechanism of GTP hydrolysis by Ras probed by Fourier transform infrared spectroscopy. *J. Biol. Chem.* **275,** 8492–8500.

Du, X., Black, G. E., Lecchi, P., Abramson, F. P., and Sprang, S. R. (2004). Kinetic isotope effects in Ras-catalyzed GTP hydrolysis: Evidence for a loose transition state. *Proc. Natl. Acad. Sci. USA* **101,** 8858–8863.

Faurobert, E., Otto-Bruc, A., Chardin, P., and Chabre, M. (1993). Tryptophan W207 in transducin T$\alpha$ is the fluorescence sensor of the G protein activation switch and is involved in the effector binding. *J. Eur. Mol. Biol. Org.* **12,** 4191–4198.

Frank, M., Thumer, L., Lohse, M. J., and Bunemann, M. (2005). G protein activation without subunit dissociation depends on a G$\alpha_i$-specific region. *J. Biol. Chem.* **280,** 24584–24590.

Fukuhara, S., Murga, C., Zohar, M., Igishi, T., and Gutkind, J. S. (1999). A novel PDZ domain containing guanine nucleotide exchange factor links heterotrimeric G proteins to Rho. *J. Biol. Chem.* **274,** 5868–5879.

Fukuhara, S., Chikumi, H., and Gutkind, J. S. (2000). Leukemia-associated Rho guanine nucleotide exchange factor (LARG) links heterotrimeric G proteins of the G (12) family to Rho. *FEBS Lett.* **485,** 183–188.

Gideon, P., John, J., Frech, M., Lautwein, A., Clark, R., Scheffler, J. E., and Wittinghofer, A. (1992). Mutational and kinetic analyses of the GTPase-activating protein (GAP)-p21 interaction: The C-terminal domain of GAP is not sufficient for full activity. *Mol. Cell. Biol.* **12,** 2050–2056.

Gilman, A. G. (1987). G proteins: Transducers of receptor-generated signals. *Annu. Rev. Biochem.* **56,** 615–649.

Glennon, T. M., Villa, J., and Warshel, A. (2000). How does GAP catalyze the GTPase reaction of Ras? A computer simulation study. *Biochemistry* **39,** 9641–9651.

Goody, R. S. (2003). The significance of the free energy of hydrolysis of GTP for signal-transducing and regulatory GTPases. *Biophys. Chem.* **100,** 535–544.

Grabocka, E., and Wedegaertner, P. B. (2005). Functional consequences of G$\alpha$13 mutations that disrupt interaction with p115RhoGEF. *Oncogene* **24,** 2155–2165.

Grigorenko, B. L., Nemukhin, A. V., Topol, I. A., Cachau, R. E., and Burt, S. K. (2005). QM/MM modeling the Ras-GAP catalyzed hydrolysis of guanosine triphosphate. *Proteins* **60,** 495–503.

Grishina, G., and Berlot, C. H. (1997). Identification of common and distinct residues involved in the interaction of $\alpha_{i2}$ and $\alpha_s$ with adenylyl cyclase. *J. Biol. Chem.* **272,** 20619–20626.

Gu, C., Cali, J. J., and Cooper, D. M. (2002). Dimerization of mammalian adenylate cyclases. *Eur. J. Biochem.* **269,** 413–421.

Guy, P. M., Koland, J. G., and Cerione, R. A. (1990). Rhodopsin-stimulated activation-deactivation cycle of transducin: Kinetics of the intrinsic fluorescence response of the alpha subunit. *Biochemistry* **29,** 6954–6964.

Hamm, H. E. (1998). The many faces of G protein signaling. *J. Biol. Chem.* **273,** 669–672.

Hanoune, J., and Defer, N. (2001). Regulation and role of adenylyl cyclase isoforms. *Annu. Rev. Pharmacol. Toxicol.* **41,** 145–174.

Hart, M. J., Jiang, X., Kosaza, T., Roscoe, W., Singer, W. D., Gilman, A. G., Sternweis, P. C., and Bollag, G. (1998). Direct stimulation of the guanine nucleotide exchange activity of p115 RhoGEF by $G\alpha_{13}$. *Science* **280,** 2112–2114.

He, W., Cowan, C. W., and Wensel, T. G. (1998). RGS9, a GTPase accelerator for phototransduction. *Neuron* **20,** 95–102.

Hess, H. A., Roper, J. C., Grill, S. W., and Koelle, M. R. (2004). RGS-7 Completes a receptor-independent heterotrimeric G protein cycle to asymmetrically regulate mitotic spindle positioning in *C. elegans*. *Cell* **119,** 209–218.

Heximer, S. P., Watson, N., Linder, M. E., Blumer, K. J., and Hepler, J. R. (1997). RGS2/G0S8 is a selective inhibitor of Gqalpha function. *Proc. Natl. Acad. Sci. USA* **94,** 14389–14393.

Heximer, S. P., Srinivasa, S. P., Bernstein, L. S., Bernard, J. L., Linder, M. E., Hepler, J. R., and Blumer, K. J. (1999). G protein selectivity is a determinant of RGS2 function. *J. Biol. Chem.* **274,** 34253–34259.

Higashijima, T., Ferguson, K. M., Smigel, M. D., and Gilman, A. G. (1987a). The effect of GTP and $Mg^{2+}$ on the GTPase activity and the fluorescent properties of Go. *J. Biol. Chem.* **262,** 757–761.

Higashijima, T., Ferguson, K. M., Sternweis, P. C., Ross, E. M., Smigel, M. D., and Gilman, A. G. (1987b). The effect of activating ligands on the intrinsic fluorescence of guanine nucleotide-binding regulatory proteins. *J. Biol. Chem.* **262,** 752–756.

Higashijima, T., Ferguson, K. M., Sternweis, P. C., Smigel, M. D., and Gilman, A. G. (1987c). Effects of $Mg^{2+}$ and the beta gamma-subunit complex on the interactions of guanine nucleotides with G proteins. *J. Biol. Chem.* **262,** 762–766.

Hollinger, S., and Hepler, J. R. (2002). Cellular regulation of RGS proteins: Modulators and integrators of G protein signaling. *Pharmacol. Rev.* **54,** 527–559.

Hu, G., and Wensel, T. G. (2002). R9AP, a membrane anchor for the photoreceptor GTPase accelerating protein, RGS9-1. *Proc. Natl. Acad. Sci. USA* **99,** 9755–9760.

Hubbard, K. B., and Hepler, J. R. (2006). Cell signalling diversity of the Gqalpha family of heterotrimeric G proteins. *Cell. Signal.* **18,** 135–150.

Hurley, J. H. (1999). Structure, mechanism, and regulation of mammalian adenylyl cyclase. *J. Biol. Chem.* **274,** 7599–7602.

Ilkaeva, O., Kinch, L. N., Paulssen, R. H., and Ross, E. M. (2002). Mutations in the carboxyl-terminal domain of phospholipase C-beta 1 delineate the dimer interface and a potential Gαq interaction site. *J. Biol. Chem.* **277,** 4294–4300.

Ingi, T., Krumins, A. M., Chidiac, P., Brothers, G. M., Chung, S., Snow, B. E., Barnes, C. A., Lanahan, A. A., Siderovski, D. P., Ross, E. M., Gilman, A. G., and Worley, P. F. (1998). Dynamic regulation of RGS2 suggests a novel mechanism in G-protein signaling and neuronal plasticity. *J. Neurosci.* **18,** 7178–7188.

John, J., Rensland, H., Schlichting, I., Vetter, I., Borasio, G. D., Goody, R. S., and Wittinghofer, A. (1993). Kinetic and structural analysis of the Mg(2+)-binding site of the guanine nucleotide-binding protein p21H-ras. *J. Biol. Chem.* **268,** 923–929.

Katchalski-Katzir, E., Shariv, I., Eisenstein, M., Friesem, A. A., Aflalo, C., and Vakser, I. A. (1992). Molecular surface recognition: Determination of geometric fit between proteins and their ligands by correlation techniques. *Proc. Natl. Acad. Sci. USA* **89,** 2195–2199.

Kim, C. G., Park, D., and Rhee, S. G. (1996). The role of carboxyl-terminal basic amino acids in $G_q^{\alpha}$-dependent activation, particulate association, and nuclear localization of phospholipase C-$\beta$1. *J. Biol. Chem.* **271,** 21187–21192.

Kimple, R. J., Kimple, M. E., Betts, L., Sondek, J., and Siderovski, D. P. (2002). Structural determinants for GoLoco-induced inhibition of nucleotide release by Galpha subunits. *Nature* **416,** 878–881.

Kjeldgaard, M., Nissen, P., Thirup, S., and Nyborg, J. (1993). The crystal structure of elongation factor EF-Tu from *thermus aquaticus* in the GTP conformation. *Structure* **1,** 35–50.

Kjeldgaard, M., Nyborg, J., and Clark, B. F. C. (1996). The GTP binding motif: Variations on a theme. *FASEB J.* **10,** 1347–1368.

Klahn, M., Schlitter, J., and Gerwert, K. (2005). Theoretical IR spectroscopy based on QM/MM calculations provides changes in charge distribution, bond lengths, and bond angles of the GTP ligand induced by the Ras-protein. *Biophys. J.* **88,** 3829–3844.

Koelle, M. R. (1997). A new family of G-protein regulators—the RGS proteins. *Curr. Opin. Struct. Biol.* **9,** 143–147.

Kourlas, P. J., Strout, M. P., Becknell, B., Veronese, M. L., Croce, C. M., Theil, K. S., Krahe, T., Ruutu, T., Knuutila, S., Bloomfield, C. D., and Caligiuri, M. A. (2000). Identification of a gene at 11q23 encoding a guanine nucleotide exchange factor: Evidence for its fusion with MLL in acute myeloid leukemia. *Proc. Natl. Acad. Sci. USA* **97,** 2145–2150.

Kozasa, T., Jiang, X., Hart, M. J., Sternweis, P. M., Singer, W. D., Gilman, A. G., Bollag, G., and Sternweis, P. C. (1998). p115 RhoGEF, a GTPase activating protein for $G\alpha_{12}$ and $G\alpha_{13}$. *Science* **280,** 2109–2111.

Kreutz, B., Yau, D. M., Nance, M. R., Tanabe, S., Tesmer, J. J., and Kozasa, T. (2006). A new approach to producing functional Gα subunits yields the activated and deactivated structures of G alpha(12/13) proteins. *Biochemistry* **45,** 167–174.

Kristelly, R., Gao, G., and Tesmer, J. J. (2004). Structural determinants of RhoA binding and nucleotide exchange in leukemia-associated RhoGEF. *J. Biol. Chem.* **279,** 47362–47392.

Krupnick, J. G., and Benovic, J. L. (1998). The role of receptor kinases and arrestins in G protein-coupled receptor regulation. *Annu. Rev. Pharmacol. Toxicol.* **38,** 289–319.

Lambright, D. G., Noel, J. P., Hamm, H. E., and Sigler, P. B. (1994). Structural determinants for activation of the α-subunit of a heterotrimeric G protein. *Nature* **369,** 621–628.

Lambright, D. G., Sondek, J., Bohm, A., Skiba, N. P., Hamm, H., and Sigler, P. B. (1996). The 2. 0Å crystal structure of a heterotrimeric G protein. *Nature* **379,** 311–319.

Lan, K. L., Sarvazyan, N. A., Taussig, R., Mackenzie, R. G., DiBello, P. R., Dohlman, H. G., and Neubig, R. R. (1998). A point mutation in Galphao and Galphai1 blocks interaction with regulator of G protein signaling proteins. *J. Biol. Chem.* **273,** 12794–12797.

Lan, K. L., Zhong, H., Nanamori, M., and Neubig, R. R. (2000). Rapid kinetics of regulator of G-protein signaling (RGS)-mediated Gαi and Gαo deactivation. Galpha specificity of RGS4 and RGS7. *J. Biol. Chem.* **275,** 33497–33503.

Langen, R., Schweins, T., and Warshel, A. (1992). On the mechanism of guanosine triphosphate hydrolysis in *ras* p21 proteins. *Biochemistry* **31,** 8691–8696.

Lanier, S. M. (2004). AGS proteins, GPR motifs and the signals processed by heterotrimeric G proteins. *Biol. Cell* **96,** 369–372.

Lee, E., Taussig, R., and Gilman, A. (1992). The G226A mutant of Gs alpha highlights the requirement for dissociation of G protein subunits. *J. Biol. Chem.* **267,** 1212–1218.

Lee, S., Shin, S., Hepler, J., Gilman, A., and Rhee, S. (1993). Activation of phospholipase C-$\beta$2 mutants by G protein alpha q and beta gamma subunits. *J. Biol. Chem.* **268,** 25952–25957.

Lenzen, C., Cool, R. H., Prinz, H., Kuhlmann, J., and Wittinghofer, A. (1998). Kinetic analysis by fluorescence of the interaction between Ras and the catalytic domain of the guanine nucleotide exchange factor Cdc25Mm. *Biochemistry* **37,** 7420–7430.

Linder, J. U., and Schultz, J. E. (2003). The class III adenylyl cyclases: Multi-purpose signalling modules. *Cell. Signal.* **15,** 1081–1089.

Linder, M., Pang, I., Duronio, R., Gordon, J., Sternweis, P., and Gilman, A. (1991). Lipid modifications of G protein subunits. Myristoylation of Go$\alpha$ increases its affinity for $\beta\gamma$. *J. Biol. Chem.* **266,** 4654–4659.

Lipkin, V. M., Dumler, I. L., Muradov, K. G., Artemyev, N. O., and Etingof, R. N. (1988). Active sites of the cyclic GMP phosphodiesterase $\gamma$-subunit of retinal rod outer segments. *FEBS Lett.* **234,** 287–290.

Lodowski, D. T., Pitcher, J. A., Capel, W. D., Lefkowitz, R. J., and Tesmer, J. J. (2003). Keeping G proteins at bay: A complex between G protein-coupled receptor kinase 2 and G$\beta\gamma$. *Science* **300,** 1256–1262.

Lodowski, D. T., Barnhill, J. F., Pyskadlo, R. M., Ghirlando, R., Sterne-Marr, R., and Tesmer, J. J. (2005). The role of G$\beta\gamma$ and domain interfaces in the activation of G protein-coupled receptor kinase 2. *Biochemistry* **44,** 6958–6970.

Maegley, K. A., Admiraal, S. J., and Herschlag, D. (1996). Ras-catalyzed hydrolysis of GTP: A new perspective from model studies. *Proc. Natl. Acad. Sci. USA* **93,** 8160–8166.

Makino, E. R., Handy, J. W., Li, T., and Arshavsky, V. Y. (1999). The GTPase activating factor for transducin in rod photoreceptors is the complex between RGS9 and type 5 G protein $\beta$ subunit. *Proc. Natl. Acad. Sci. USA* **96,** 1947–1952.

Margarit, S. M., Sondermann, H., Hall, B. E., Nagar, B., Hoelz, A., Pirruccello, M., Bar-Sagi, D., and Kuriyan, J. (2003). Structural evidence for feedback activation by Ras. GTP of the Ras-specific nucleotide exchange factor SOS. *Cell* **112,** 685–695.

Markby, D. W., Onrust, R., and Bourne, H. R. (1993). Separate GTP binding and GTPase activating domains of a G$\alpha$ subunit. *Science* **262,** 1895–1901.

McEntaffer, R. L., Natochin, M., and Artemyev, N. O. (1999). Modulation of transducin GTPase activity by chimeric RGS16 and RGS9 regulators of G protein signaling and the effector molecule. *Biochemistry* **38,** 4931–4937.

Meigs, T. E., Juneja, J., DeMarco, C. T., Stemmle, L. N., Kaplan, D. D., and Casey, P. J. (2005). Selective uncoupling of G$\alpha$12 from Rho-mediated signaling. *J. Biol. Chem.* **280,** 18049–18055.

Miller, R. T., Masters, S. B., Sullivan, K. A., Beiderman, B., and Bourne, H. R. (1988). A mutation that prevents GTP-dependent activation of the $\alpha$ chain of Gs. *Nature* **334,** 712–715.

Mixon, M. B., Lee, E., Coleman, D. E., Berghuis, A. M., Gilman, A. G., and Sprang, S. R. (1995). Tertiary and quaternary structural changes in $G_{i\alpha1}$ induced by GTP hydrolysis. *Science* **270,** 954–960.

Morris, A. J., and Malbon, C. C. (1999). Physiological regulation of G protein-linked signaling. *Physiol. Rev.* **79,** 1373–1430.

Mou, T. C., Gille, A., Fancy, D. A., Seifert, R., and Sprang, S. R. (2005). Structural basis for the inhibition of mammalian membrane adenylyl cyclase by 2 (3′)-O-(N-Methylanthraniloyl)-guanosine 5′-triphosphate. *J. Biol. Chem.* **280,** 7253–7261.

Mukhopadhyay, S., and Ross, E. M. (1999). Rapid GTP binding and hydrolysis by G(q) promoted by receptor and GTPase-activating proteins. *Proc. Natl. Acad. Sci. USA* **96,** 9539–9544.

Nakamura, S., Kreutz, B., Tanabe, S., Suzuki, N., and Kozasa, T. (2004). Critical role of lysine 204 in switch I region of Gα13 for regulation of p115RhoGEF and leukemia-associated RhoGEF. *Mol. Pharmacol.* **66,** 1029–1034.

Nassar, N., Hoffman, G. R., Manor, D., Clardy, J. C., and Cerione, R. A. (1998). Structures of Cdc42 bound to the active and catalytically compromised forms of Cdc42GAP. *Nat. Struct. Biol.* **5,** 1047–1052.

Natochin, M., and Artemyev, N. O. (1998a). A single mutation Asp229 → Ser confers upon Gsα the ability to interact with regulators of G protein signaling. *Biochemistry* **37,** 13776–13780.

Natochin, M., and Artemyev, N. O. (1998b). Substitution of transducin $Ser^{202}$ by Asp abolishes G-protein/RGS interaction. *J. Biol. Chem.* **273,** 4300–4303.

Natochin, M., and Artemyev, N. O. (2003). A point mutation uncouples transducin-alpha from the photoreceptor RGS and effector proteins. *J. Neurochem.* **87,** 1262–1271.

Natochin, M., Granovsky, A. E., and Artemyev, N. O. (1998a). Identification of effector residues on photoreceptor G protein, transducin. *J. Biol. Chem.* **273,** 21808–21815.

Natochin, M., McEntaffer, R. L., and Artemyev, N. O. (1998b). Mutational analysis of the Asn residue essential for RGS protein binding to G proteins. *J. Biol. Chem.* **273,** 6731–6735.

Neal, S. E., Eccleston, J. F., and Webb, M. R. (1990). Hydrolysis of GTP by $p21^{NRAS}$, the *NRAS* protooncogene product, is accompanied by a conformational change in the wild-type protein: Use of a single fluorescent probe at the catalytic site. *Proc. Natl. Acad. Sci. USA* **87,** 3562–3565.

Nekrasova, E., Berman, D., Rustandi, R., Hamm, H., Gilman, A., and Arshavsky, V. (1997). Activation of transducin guanosine triphosphatase by two proteins of the RGS family. *Biochemistry* **36,** 7638–7643.

Noel, J. P., Hamm, H. E., and Sigler, P. B. (1993). The 2.2 A crystal structure of transducin-α complexed with GTPγS. *Nature* **366,** 654–663.

Ogier-Denis, E., Pattingre, S., El Benna, J., and Codogno, P. (2000). Erk1/2-dependent phosphorylation of Gα-interacting protein stimulates its GTPase accelerating activity and autophagy in human colon cancer cells. *J. Biol. Chem.* **275,** 39090–39095.

Otto-Bruc, A., Antonny, B., Vuong, T. M., Chardin, P., and Chabre, M. (1993). Interaction between the retinal cyclic GMP phosphodiesterase inhibitor and transducin. Kinetics and affinity studies. *Biochemistry* **32,** 8636–8645.

Pan, X., Eathiraj, S., Munson, M., and Lambright, D. G. (2006). TBC-domain GAPs for Rab GTPases accelerate GTP hydrolysis by a dual-finger mechanism. *Nature* **442,** 303–306.

Park, D., Jhon, D.-Y., Lee, C.-W., Ryu, S. H., and Rhee, S. G. (1993). Removal of the carboxyl-terminal region of phospholipase C-β1 by calapin abolishes activation by Gαq. *J. Biol. Chem.* **268,** 3710–3714.

Pasqualato, S., and Cherfils, J. (2005). Crystallographic evidence for substrate-assisted GTP hydrolysis by a small GTP binding protein. *Structure* **13,** 533–540.

Paulssen, R. H., Woodson, J., Liu, Z., and Ross, E. M. (1996). Carboxyl-terminal fragments of phospholipase C-β1 with intrinsic $G_q$ GTPase-activating protein (GAP) activity. *J. Biol. Chem.* **271,** 26622–26629.

Peleg, S., Varon, D., Ivanina, T., Dessauer, C. W., and Dascal, N. (2002). $G\alpha_i$ controls the gating of the G protein-activated K(+) channel, GIRK. *Neuron* **33,** 87–99.

Pfeuffer, E., Mollner, S., and Pfeuffer, T. (1985). Adenylate cyclase from bovine brain cortex: Purification and characterization of the catalytic unit. *EMBO J.* **4,** 3675–3679.

Phillips, R. A., Hunter, J. L., Eccleston, J. F., and Webb, M. R. (2003). The mechanism of Ras GTPase activation by neurofibromin. *Biochemistry* **42,** 3956–3965.

Phillips, W. J., and Cerione, R. A. (1986). The intrinsic fluorescence of the α subunit of transducin. *J. Biol. Chem.* **263,** 15498–15505.

Phillips, W. J., and Cerione, R. A. (1988). The intrinsic fluorescence of the α subunit of transducin. Measurement of receptor-dependent guanine nucleotide exchange. *J. Biol. Chem.* **263,** 15498–15505.

Pitcher, J. A., Freedman, N. J., and Lefkowitz, R. J. (1998). G protein-coupled receptor kinases. *Annu. Rev. Biochem.* **67,** 653–692.

Posner, B. A., Mukhopadhyay, S., Tesmer, J. J., Gilman, A. G., and Ross, E. M. (1999). Modulation of the affinity and selectivity of RGS protein interaction with Gα subunits by a conserved asparagine/serine residue. *Biochemistry* **38,** 7773–7779.

Privé, G. G., Milburn, M. V., Tong, L., deVos, A. M., Yamaizumi, Z., Nishimura, S., and Kim, S.-H. (1992). X-ray crystal structures of transforming p21 Ras mutants suggest a transition-state stabilization mechanism for GTP hydrolysis. *Proc. Natl. Acad. Sci. USA* **89,** 3649–3653.

Raw, A. S., Coleman, D. E., Gilman, A. G., and Sprang, S. R. (1997). Structural and biochemical characterization of the GTPγS-, GDP.$P_i^-$, and GDP-bound forms of a GTPase deficient $Gly^{42} \rightarrow$ Val mutant of $G_{i\alpha 1}$. *Biochemistry* **36,** 15660–15669.

Rhee, S. G. (2001). Regulation of phosphoinositide-specific phospholipase C. *Annu. Rev. Biochem.* **70,** 281–312.

Rittinger, K., Walker, P. A., Eccleston, J. F., Smerdon, S. J., and Gamblin, S. J. (1997). Structure at 1. 65 A of RhoA and its GTPase-activating protein in complex with a transition-state analogue. *Nature* **389,** 758–762.

Ross, E. M., and Wilkie, T. M. (2000). GTPase-activating proteins for heterotrimeric G proteins: Regulators of G protein signaling (RGS) and RGS-like proteins. *Annu. Rev. Biochem.* **69,** 795–827.

Rossman, K. L., and Sondek, J. (2005). Larger than Dbl: New structural insights into RhoA activation. *Trends Biochem. Sci.* **30,** 163–165.

Sadja, R., Alagem, N., and Reuveny, E. (2003). Gating of GIRK channels: Details of an intricate, membrane-delimited signaling complex. *Neuron* **39,** 9–12.

Sallese, M., Mariggio, S., D'Urbano, E., Iacovelli, L., and De Blasi, A. (2000). Selective regulation of Gq signaling by G protein-coupled receptor kinase 2: Direct interaction of kinase N terminus with activated Gαq. *Mol. Pharmacol.* **57,** 826–831.

Scheffzek, K., Ahmadian, M. R., Kabsch, W., Wiesmüller, L., Lautwein, A., Schmitz, F., and Wittinghofer, A. (1997). The Ras-RasGAP complex: Structural basis for GTPase activation and its loss in oncogenic Ras mutants. *Science* **277,** 333–338.

Scheidig, A. J., Burmester, C., and Goody, R. S. (1999). The pre-hydrolysis state of p21 (ras) in complex with GTP: New insights into the role of water molecules in the GTP hydrolysis reaction of ras-like proteins. *Structure* **7,** 1311–1324.

Schild, D., and Restrepo, D. (1998). Transduction mechanisms in vertebrate olfactory receptor cells. *Physiol. Rev.* **78,** 429–466.

Schweins, T., Langen, R., and Warshel, A. (1994). Why have mutagenesis studies not located the general base in ras p21. *Nat. Struct. Biol.* **1,** 476–484.

Schweins, T., Geyer, M., Scheffzek, K., Warshel, A., Kalbitzer, H. R., and Wittinghofer, A. (1995). Substrate-assisted catalysis as a mechanism for GTP hydrolysis of $p21^{ras}$ and other GTP-binding proteins. *Nat. Struct. Biol.* **2,** 36–44.

Schweins, T., Geyer, M., Kalbitzer, H. R., Wittinghofer, A., and Warshel, A. (1996). Linear free energy relationships in the intrinsic and GTPase activating protein-stimulated guanosine 5′-triphosphate hydrolysis of p21ras. *Biochemistry* **35,** 14225–14231.

Seamon, K., and Daly, J. (1981). Forskolin: A unique diterpene activator of cyclic AMP-generating systems. *J. Cyclic Nucleotide Res.* **7,** 201–224.

Shurki, A., and Warshel, A. (2004). Why does the Ras switch ''break'' by oncogenic mutations? *Proteins* **55,** 1–10.

Simonds, W. F. (1999). G protein regulation of adenylate cyclase. *Trends Pharmacol. Sci.* **20,** 66–73.

Singer, A. U., Waldo, G. L., Harden, T. K., and Sondek, J. (2002). A unique fold of phospholipase C-beta mediates dimerization and interaction with G$\alpha$q. *Nat. Struct. Biol.* **9,** 32–36.

Singer, W. D., Brown, H. A., and Sternweis, P. C. (1997). Regulation of eukaryotic phosphatidylinositol-specific phospholipase C and phospholipase D. *Annu. Rev. Biochem.* **66,** 475–509.

Sinha, S. C., and Sprang, S. R. (2006). Structures, mechanism, regulation and evolution of class III nucleotidyl cyclases. *Rev. Physiol. Biochem. Pharmacol.* **157,** 105–140.

Skiba, N. P., Bae, H., and Hamm, H. E. (1996). Mapping of effector binding sites of transducin $\alpha$-subunit using $G\alpha_t/G\alpha_{i1}$ chimeras. *J. Biol. Chem.* **271,** 413–424.

Skiba, N. P., Yang, C. S., Huang, T., Bae, H., and Hamm, H. E. (1999). The alpha-helical domain of G$\alpha$t determines specific interaction with regulator of G protein signaling 9. *J. Biol. Chem.* **274,** 8770–8778.

Slep, K. C., Kercher, M. A., He, W., Cowan, C. W., Wensel, T. G., and Sigler, P. B. (2001). Structural determinants for regulation of phosphodiesterase by a G protein at 2. 0 A. *Nature* **409,** 1071–1077.

Slepak, V. Z., Artemyev, N. O., Zhu, Y., Dumke, C. L., Sabacan, L., Sondek, J., Hamm, H. E., Bownds, M. D., and Arshavsky, V. Y. (1995). An effector site that stimulates G-protein GTPase in photoreceptors. *J. Biol. Chem.* **270,** 14319–14324.

Smrcka, A. V., and Sternweis, P. C. (1993). Regulation of purified subtypes of phosphatidylinositol-specific phospholipase C$\beta$ by G protein $\alpha$ and $\beta$g subunits. *J. Biol. Chem.* **268,** 9667–9674.

Smrcka, A. V., Hepler, J. R., Brown, K. O., and Sternweis, P. C. (1991). Regulation of polyphosphoinositide-specific phospholipase C activity by purified Gq. *Science* **251,** 804–807.

Sondek, J., Lambright, D. G., Noel, J. P., Hamm, H. E., and Sigler, P. B. (1994). GTPase mechanism of G proteins from the 1. 7 Å crystal structure of transducin $\alpha$-GDP-$AlF_4^-$. *Nature* **372,** 276–279.

Sowa, M. E., He, W., Wensel, T. G., and Lichtarge, O. (2000). A regulator of G protein signaling interaction surface linked to effector specificity. *Proc. Natl. Acad. Sci. USA* **97,** 1483–1488.

Sowa, M. E., He, W., Slep, K. C., Kercher, M. A., Lichtarge, O., and Wensel, T. G. (2001). Prediction and confirmation of a site critical for effector regulation of RGS domain activity. *Nat. Struct. Biol.* **8,** 234–237.

Sprang, S. R. (1997). G protein mechanisms: Insights from structural analysis. *Annu. Rev. Biochem.* **66,** 639–678.

Srinivasa, S. P., Bernstein, L. S., Blumer, K. J., and Linder, M. E. (1998). Plasma membrane localization is required for RGS4 function in *Saccharomyces cerevisiae*. *Proc. Natl. Acad. Sci. USA* **95,** 5584–5589.

Sterne-Marr, R., Tesmer, J. J., Day, P. W., Stracquatanio, R. P., Cilente, J. A., O'Connor, A. N., Pronin, A. N., Benovic, J. L., and Wedegaertner, P. B. (2003). G protein-coupled receptor Kinase2/$G\alpha_{q/11}$ interaction. A novel surface on a regulator of G protein signaling homology domain for binding G alpha subunits. *J. Biol. Chem.* **278,** 6050–6058.

Sunahara, R. K., and Taussig, R. (2002). Isoforms of mammalian adenylyl cyclase: Multiplicities of signaling. *Mol. Interv.* **2,** 168–184.

Sunahara, R. K., Dessauer, C. W., and Gilman, A. G. (1996). Complexity and diversity of mammalian adenylyl cyclases. *Annu. Rev. Pharmacol. Toxicol.* **36,** 461–480.

Sunahara, R. K., Dessauer, C. W., Whisnant, R. E., Kleuss, C., and Gilman, A. G. (1997a). Interaction of $G_{s\alpha}$ with the cytosolic domains of mammalian adenylyl cyclase. *J. Biol. Chem.* **272,** 22265–22271.

Sunahara, R. K., Tesmer, J. J. G., Gilman, A. G., and Sprang, S. R. (1997b). Crystal structure of the adenylyl cyclase activator Gsα. *Science* **278,** 1943–1947.

Suzuki, N., Nakamura, S., Mano, H., and Kozasa, T. (2003). Gα12 activates Rho GTPase through tyrosine-phosphorylated leukemia-associated RhoGEF. *Proc. Natl. Acad. Sci. USA* **100,** 733–738.

Tall, G. G., Krumins, A. M., and Gilman, A. G. (2003). Mammalian Ric-8A (synembryn) is a heterotrimeric Gα protein guanine nucleotide exchange factor. *J. Biol. Chem.* **278,** 8356–8362.

Tanaka, T., Ames, J. B., Harvey, T. S., Stryer, L., and Ikura, M. (1995). Sequestration of the membrane-targeting myristoyl group of recoverin in the calcium-free state. *Nature* **376,** 444–447.

Tang, W.-J., and Gilman, A. G. (1995). Construction of a soluble adenylyl cyclase activated by Gsα and forskolin. *Science* **268,** 1769–1772.

Tang, W. J., Stanzel, M., and Gilman, A. G. (1995). Truncation and alanine-scanning mutants of type I adenylyl cyclase. *Biochemistry* **34,** 14563–14572.

Taussig, R., Iniguez-Lluhi, J. A., and Gilman, A. G. (1993). Inhibition of adenylyl cyclase by Giα. *Science* **261,** 218–221.

Taylor, S. J., Chae, H. Z., Rhee, S. G., and Exton, J. H. (1991). Activation of the beta 1 isozyme of phospholipase C by α subunits of the Gq class of G proteins. *Nature* **350,** 516–518.

Temeles, G. L., Gibbs, J. B., D'Alonzo, J. S., Sigal, I. S., and Scolnick, E. M. (1985). Yeast and mammalian ras proteins have conserved biochemical properties. *Nature* **313,** 700–703.

Tesmer, J. J., and Sprang, S. R. (1998). The structure, catalytic mechanism and regulation of adenylyl cyclase. *Curr. Opin. Struct. Biol.* **8,** 713–719.

Tesmer, J. J. G., Berman, D. M., Gilman, A. G., and Sprang, S. R. (1997a). Structure of RGS4 bound to $AlF_4^-$-activated $G_{i\alpha 1}$: Stabilization of the transition state for GTP hydrolysis. *Cell* **89,** 251–261.

Tesmer, J. J. G., Sunahara, R. K., Gilman, A. G., and Sprang, S. R. (1997b). Crystal structure of the catalytic domains of adenylyl cyclase in a complex with $G_{s\alpha}$•GTPγS. *Science* **278,** 1907–1916.

Tesmer, J. J. G., Sunahara, R. K., Johnson, R. A., Gilman, A. G., and Sprang, S. R. (1999). Two metal ion catalysis in adenylyl cyclase. *Science* **285,** 756–760.

Tesmer, J. J., Dessauer, C. W., Sunahara, R. K., Murray, L. D., Johnson, R. A., Gilman, A. G., and Sprang, S. R. (2000). Molecular basis for P-site inhibition of adenylyl cyclase. *Biochemistry* **39,** 14464–14471.

Tesmer, V. M., Kawano, T., Shankaranarayanan, A., Kozasa, T., and Tesmer, J. J. (2005). Snapshot of activated G proteins at the membrane: The Gαq-GRK2-Gβγ complex. *Science* **310,** 1686–1690.

Thomas, C. J., Du, X., Li, P., Wang, Y., Ross, E. M., and Sprang, S. R. (2004). Uncoupling conformational change from GTP hydrolysis in a heterotrimeric G protein α-subunit. *Proc. Natl. Acad. Sci. USA* **101,** 7560–7565.

Ting, T. D., and Ho, Y.-K. (1991). Molecular mechanism of GTP hydrolysis by bovine transducin: Pre-steady-state kinetic analyses. *Biochemistry* **30,** 8996–9007.

Topol, I. A., Cachau, R. E., Nemukhin, A. V., Grigorenko, B. L., and Burt, S. K. (2004). Quantum chemical modeling of the GTP hydrolysis by the RAS-GAP protein complex. *Biochim. Biophys. Acta* **1700,** 125–136.

Tu, Y., Wang, J., and Ross, E. M. (1997). Inhibition of brain Gz GAP and other RGS proteins by palmitoylation of G protein $\alpha$ subunits. *Science* **278,** 1132–1135.

Usui, H., Nishiyama, M., Moroi, K., Shibasaki, T., Zhou, J., Ishida, J., Fukamizu, A., Haga, T., Sekiya, S., and Kimura, S. (2000). RGS domain in the amino-terminus of G protein-coupled receptor kinase 2 inhibits Gq-mediated signaling. *Int. J. Mol. Med.* **5,** 335–340.

Venkatakrishnan, G., and Exton, J. H. (1996). Identification of determinants in the $\alpha$-subunits of $G_q$ required for phospholipase C activation. *J. Biol. Chem.* **271,** 5066–5072.

Vetter, I. R., and Wittinghofer, A. (2001). The guanine nucleotide-binding switch in three dimensions. *Science* **294,** 1299–1304.

Walker, J. E., Saraste, M., Runswick, M. J., and Gay, N. J. (1982). Distantly related sequences in the $\alpha$- and $\beta$-subunits of ATP synthase, myosin, kinases and other ATP-requiring enzymes and a common nucleotide binding fold. *EMBO J.* **1,** 945–951.

Wall, M. A., Coleman, D. E., Lee, E., Iñiguez-Lluhi, J. A., Posner, B. A., Gilman, A. G., and Sprang, S. R. (1995). The structure of the G protein heterotrimer $G_{i\alpha1}\beta_1\gamma_2$. *Cell* **80,** 1047–1058.

Wall, M. A., Posner, B. A., and Sprang, S. R. (1998). Structural basis of activity and subunit recognition in G protein heterotrimers. *Structure* **6,** 1169–1183.

Wang, J., Ducret, A., Tu, Y., Kozasa, T., Aebersold, R., and Ross, E. M. (1998). RGSZ1, a Gz-selective RGS protein in brain. Structure, membrane association, regulation by Galphaz phosphorylation, and relationship to a Gz GTPase-activating protein subfamily. *J. Biol. Chem.* **273,** 26014–26025.

Wang, T., Pentyala, S., Elliott, J. T., Dowal, L., Gupta, E., Rebecchi, M. J., and Scarlata, S. (1999). Selective interaction of the C2 domains of phospholipase C-$\beta_1$ and -$\beta_2$ with activated G$\alpha$q subunits: An alternative function for C2-signaling modules. *Proc. Natl. Acad. Sci. USA* **96,** 7843–7846.

Watson, N., Linder, M. E., Druey, K. M., Kehrl, J. H., and Blumer, K. J. (1996). RGS family members: GTPase-activating proteins for heterotrimeric G-protein $\alpha$-subunits. *Nature* **383,** 172–175.

Wedegaertner, P. B., Wilson, P. T., and Bourne, H. R. (1995). Lipid modifications of trimeric G proteins. *J. Biol. Chem.* **270,** 503–506.

Wells, C. D., Gutowski, S., Bollag, G., and Sternweis, P. C. (2001). Identification of potential mechanisms for regulation of p115 RhoGEF through analysis of endogenous and mutant forms of the exchange factor. *J. Biol. Chem.* **276,** 28897–28905.

Wells, C. D., Liu, M. Y., Jackson, M., Gutowski, S., Sternweis, P. M., Rothstein, J. D., Kozasa, T., and Sternweis, P. C. (2002). Mechanisms for reversible regulation between G13 and Rho exchange factors. *J. Biol. Chem.* **277,** 1174–1181.

Wilkie, T. M., and Kinch, L. (2005). New roles for G$\alpha$ and RGS proteins: Communication continues despite pulling sisters apart. *Curr. Biol.* **15,** R843–R854.

Wilkie, T. M., and Yokoyama, S. (1994). Evolution of the G protein $\alpha$ subunit multigene family. *Soc. Gen. Physiol. Ser.* **49,** 249–270.

Willard, F. S., Kimple, R. J., and Siderovski, D. P. (2004). Return of the GDI: The GoLoco motif in cell division. *Annu. Rev. Biochem.* **73,** 925–951.

Wittinghofer, A. (2006). Phosphoryl transfer in Ras proteins, conclusive or elusive? *Trends Biochem. Sci.* **31,** 20–23.

Wittpoth, C., Scholich, K., Yigzaw, Y., Stringfield, T. M., and Patel, T. B. (1999). Regions on adenylyl cyclase that are necessary for inhibition of activity by $\beta\gamma$ and $G_{i\alpha}$ subunits of heterotrimeric G proteins. *Proc. Natl. Acad. Sci. USA* **96,** 9551–9556.

Xu, X., Zeng, W., Popov, S., Berman, D. M., Davignon, I., Yu, K., Yowe, D., Offermanns, S., Muallem, S., and Wilkie, T. M. (1999). RGS proteins determine signaling specificity of Gq-coupled receptors. *J. Biol. Chem.* **274,** 3549–3556.

Yan, S. Z., Huang, Z. H., Rao, V. D., Hurley, J. H., and Tang, W. J. (1997). Three discrete regions of mammalian adenylyl cyclase form a site for Gα activation. *J. Biol. Chem.* **272,** 18849–18854.

Young, M., Kirshenbaum, K., Dill, K. A., and Highsmith, S. (1999). Predicting conformational switches in proteins. *Protein Sci.* **8,** 1752–1764.

Zelent, B., Veklich, Y., Murray, J., Parkes, J. H., Gibson, S., and Liebman, P. A. (2001). Rapid irreversible G protein α subunit misfolding due to intramolecular kinetic bottleneck that precedes $Mg^{2+}$ "lock" after GTP/GDP exchange. *Biochemistry* **40,** 9647–9656.

Zeng, W., Xu, X., Popov, S., Mukhopadhyay, S., Chidiac, P., Swistok, J., Danho, W., Yagaloff, K. A., Fisher, S. L., Ross, E. M., Muallem, S., and Wilkie, T. M. (1998). The N-terminal domain of RGS4 confers receptor-selective inhibition of G protein signaling. *J. Biol. Chem.* **273,** 34687–34690.

Zerangue, N., and Jan, L. Y. (1998). G protein signaling: Fine-tuning signaling kinetics. *Curr. Biol.* **8,** R313–R316.

Zhang, G., Liu, Y., Ruoho, A. E., and Hurley, J. H. (1997). Structure of the adenylyl cyclase catalytic core. *Nature* **386,** 247–253.

Zheng, B., De Vries, L., and Gist Farquhar, M. (1999). Divergence of RGS proteins: Evidence for the existence of six mammalian RGS subfamilies. *Trends Biochem. Sci.* **24,** 411–414.

Zheng, J., Knighton, D. R., Xuong, N.-H., Taylor, S. S., Sowadski, J. M., and Ten Eyck, L. F. (1993). Crystal structures of the myristylated catalytic subunit of cAMP-dependent protein kinase reveal open and closed conformations. *Prot. Sci.* **2,** 1559–1573.

Zhong, H., Wade, S. M., Woolf, P. J., Linderman, J. J., Traynor, J. R., and Neubig, R. R. (2003). A spatial focusing model for G protein signals. Regulator of G protein signaling (RGS) protien-mediated kinetic scaffolding. *J. Biol. Chem.* **278,** 7278–7284.

# HOW DO RECEPTORS ACTIVATE G PROTEINS?

By WILLIAM M. OLDHAM AND HEIDI E. HAMM

Department of Pharmacology, Vanderbilt University School of Medicine, Nashville, Tennessee 37232



## Abstract

Heterotrimeric G proteins couple the activation of heptahelical receptors at the cell surface to the intracellular signaling cascades that mediate the physiological responses to extracellular stimuli. G proteins are molecular switches that are activated by receptor-catalyzed GTP for GDP exchange on the G protein $\alpha$ subunit, which is the rate-limiting step in the activation of all downstream signaling. Despite the important biological role of the receptor–G protein interaction, relatively little is known about the structure of the complex and how it leads to nucleotide exchange. This chapter will describe what is known about receptor and G protein structure and outline a strategy for assembling the current data into improved models for the receptor–G protein complex that will hopefully answer the question as to how receptors flip the G protein switch.

## I. Introduction

Heptahelical receptors are the single most diverse class of cell surface receptors, which bind a variety of hormones, neurotransmitters, chemokines, sensory stimuli, and greater than 50% of all prescribed drugs. Heterotrimeric guanine nucleotide-binding proteins (G proteins; subunits $\alpha$, $\beta$, and $\gamma$) couple the activation of these receptors to the intracellular signaling pathways that underlie the physiological responses of tissues and organisms. As the primary intermediary between receptors and effectors,


DOI: 10.1016/S0065-3233(07)74002-0


0065-3233/07 $35.00

G proteins play a critical role in determining the specificity and temporal characteristics of the cellular responses to a diverse array of extracellular stimuli (Cabrera-Vera *et al.*, 2003).

G proteins are inactive in the heterotrimeric conformation where G$\alpha$ binds GDP and the constitutive G$\beta\gamma$ dimer. On binding an extraceullar stimulus, heptahelical receptors undergo a conformational change that permits G protein binding and catalyzes GDP release from the G protein $\alpha$ subunit (G$\alpha$). Once GDP is released, a stable, high-affinity complex between the activated receptor (R*) and G protein is formed. Binding of GTP to G$\alpha$ destabilizes this complex, allowing both subunits, G$\alpha$(GTP) and G$\beta\gamma$, to interact with downstream effector proteins. The cellular response is terminated when G$\alpha$ hydrolyzes GTP to GDP and reassociates with G$\beta\gamma$, thus completing the cycle (Fig. 1).

A great deal of structural information about G proteins is known from x-ray crystallographic studies, providing insight into GTP-mediated conformational changes in G$\alpha$, subunit interactions with effector proteins, and the mechanism of GTP hydrolysis. By contrast, relatively little structural information is known about the interaction between the receptor and G protein and how this interaction leads to GDP release. After an overview of the structure of heptahelical receptors and heterotrimeric G proteins, this chapter will discuss the current models of the receptor–G protein complex and proposed mechanisms for receptor-catalyzed nucleotide exchange.

### A. *Structure of Heptahelical Receptors*

Receptors that couple to heterotrimeric G proteins have a common body plan with seven transmembrane-spanning $\alpha$ helices with an extracellular N-terminus, intracellular C-terminus, and three interhelical loops on each side of the membrane (Fig. 2). The C-terminal tail contains an eighth $\alpha$ helix and palmitoylation sites that create a fourth intracellular loop (IC 4). Although the overall sequence homology among receptors is low, there are several highly conserved microdomains, including helix VIII and the palmitoylation sites, as well as the D(E)RY (aspartate/glutamate-arginine-tyrosine) and NPxxYx$_{5/6}$F (asparagine-proline-x-x-tyrosine-x$_{5/6}$-phenylalanine) motifs (reviewed in Kristiansen, 2004). Currently, high-resolution structural information is only available for the dim light photoreceptor, rhodopsin, in its inactive conformation bound to the inverse agonist, 11-*cis*-retinal. Although these structures provide little direct insight into the mechanism of G protein coupling, they have provided a useful context for the interpretation of numerous biophysical studies on the conformational changes associated with receptor activation (reviewed in Hubbell

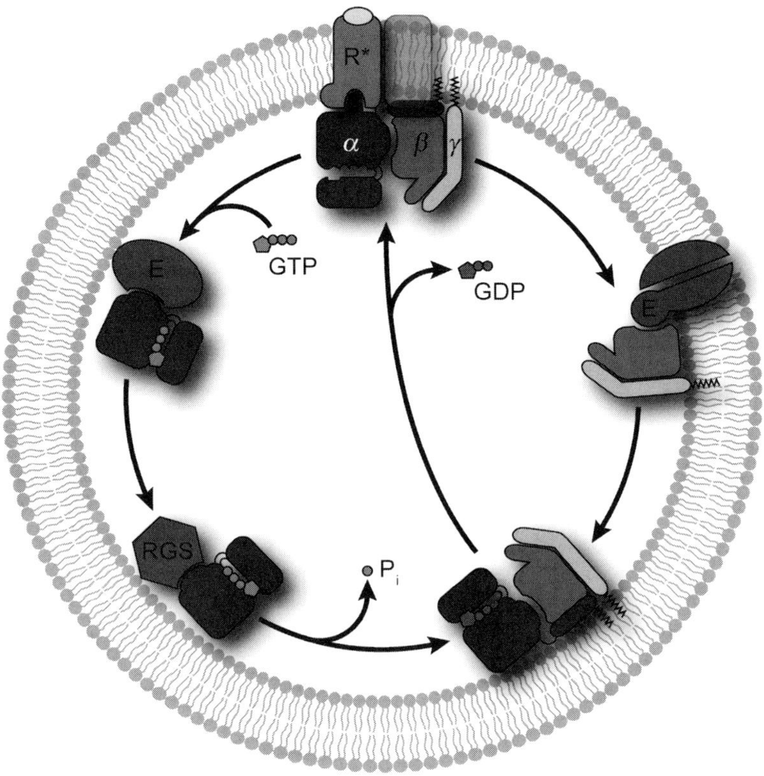


FIG. 1. Overview of the G protein cycle. In the resting state, G proteins are heterotrimers of GDP-bound $\alpha$ (*blue*), $\beta$ (*green*), and $\gamma$ (*gold*) subunits. Agonist binding to heptahelical receptors (*pink*) in the cell membrane results in a conformational change leading to G protein binding and, subsequently, GDP release. The high-affinity, nucleotide-free complex is stable until GTP binding causes dissociation of R*, G$\alpha$(GTP), and G$\beta\gamma$. The G protein subunits go on to activate a variety of downstream effector proteins (E, *purple*). The signal is terminated on the hydrolysis of GTP to GDP by G$\alpha$, which may be catalyzed by RGS proteins (*dark red*). (See Color Insert.)

*et al.*, 2003). Site-directed spin-labeling (SDSL) experiments have shown that photoactivation of rhodopsin primarily results in an outward movement of helix VI, thereby opening a crevice in the intracellular face of the receptor (Fig. 2; Farrens *et al.*, 1996). This conformational change appears to be essential for G protein activation as disulfide and metal-ion cross-links between helices III and VI prevent G protein activation (Farrens *et al.*, 1996; Sheikh *et al.*, 1996). Evidence for a similar conformational change has been reported for other receptors, suggesting that this is a conserved

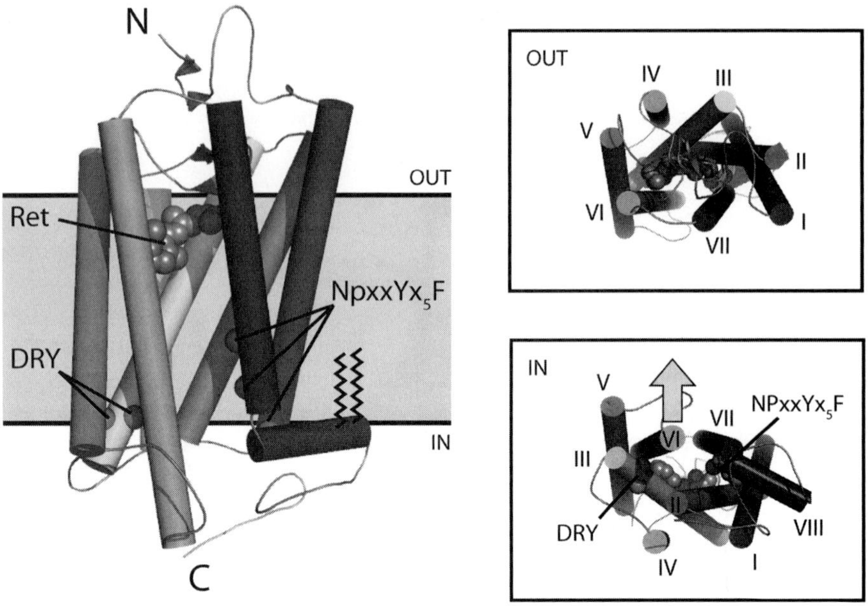


FIG. 2. Structure of a heptahelical receptor. Cartoon model of dark (inactive) bovine rhodopsin (1U19), showing the seven transmembrane-spanning $\alpha$ helices (*red* to *blue*) and 11-*cis*-retinal (*gray spheres*). Conserved residues important for receptor and G protein activation are shown (*magenta spheres*), including the DRY motif on helix III (*yellow*) and $NpxxYx_5F$ motif on helices VII and VIII (*blue* and *purple*). The extracellular and intracellular faces of rhodopsin are shown. Receptor activation results in an outward movement of helix VI (*yellow arrow*), which opens a gap in the cytoplasmic face of the receptor, exposing residues critical for G protein activation, such as the DRY motif on helix III (*yellow*). (See Color Insert.)

feature of receptor activation necessary for G protein binding (reviewed in Gether, 2000; Gether *et al.*, 2002).

### B. *Heterotrimeric G Protein Structure*

Although nearly 900 different heptahelical receptors have been identified in the human genome (Fredriksson and Schiöth, 2005), they activate relatively few types of G proteins, composed of $\alpha$, $\beta$, and $\gamma$ subunits. At last count, there are 21 G$\alpha$ subunits encoded by 16 genes, 6 G$\beta$ subunits encoded by 5 genes, and 12 G$\gamma$ subunits (Downes and Gautam, 1999). The heterotrimeric G proteins are typically classified into four families based on the primary sequence similarity of the G$\alpha$ subunit: G$\alpha_s$, G$\alpha_i$, G$\alpha_q$, and $G\alpha_{12}$ (Simon *et al.*, 1991). Many crystal structures of these proteins in a variety of conformations have been solved, revealing a conserved protein fold composed of two domains, a GTPase domain and a helical domain (Fig. 3). The GTPase domain is conserved in all members of the G protein superfamily, including small G proteins and elongation factors. This domain hydrolyzes GTP and contains the sites for binding to the $G\beta\gamma$ dimer, receptors, and effector proteins. Additionally, this domain contains three flexible loops, termed Switches I, II, and III, where the only crystallographic differences between the GDP-bound (Lambright *et al.*, 1994; Mixon *et al.*, 1995) and GTP$\gamma$S-bound (Coleman *et al.*, 1994; Noel *et al.*, 1993) conformations of $G\alpha$ are found (Fig. 3). The helical domain is unique to $G\alpha$ and is composed of a bundle of six $\alpha$ helices that form a lid over the nucleotide-binding pocket, burying it in the core of the protein. All $G\alpha$ subunits have fatty acid modifications at the N-terminus that regulate G protein localization and protein–protein interactions.

The $G\beta$ subunit has a seven-bladed $\beta$-propeller structure composed of seven WD-40 sequence repeats. The N-terminus of $G\beta$ adopts an $\alpha$ helical conformation that forms a coiled-coil with the N-terminus of $G\gamma$, while the C-terminus of $G\gamma$ binds to blades five and six (Fig. 3; Sondek *et al.*, 1996). All of the $G\gamma$ subunits undergo post-translational isoprenylation at their C-termini with a 15-carbon farnesyl ($G\gamma_1$, $G\gamma_8$, and $G\gamma_{11}$) or 20-carbon geranylgeranyl (all others) moiety. The G protein $\beta$ and $\gamma$ subunits form a functional unit that is not dissociable except by denaturation (Schmidt *et al.*, 1992).

Two crystal structures of G protein heterotrimers, $G\alpha_{t/i}\beta_1\gamma_1$ and $G\alpha_{i1}\beta_1\gamma_2$, show two sites of interaction between $G\alpha$ and the constitutive

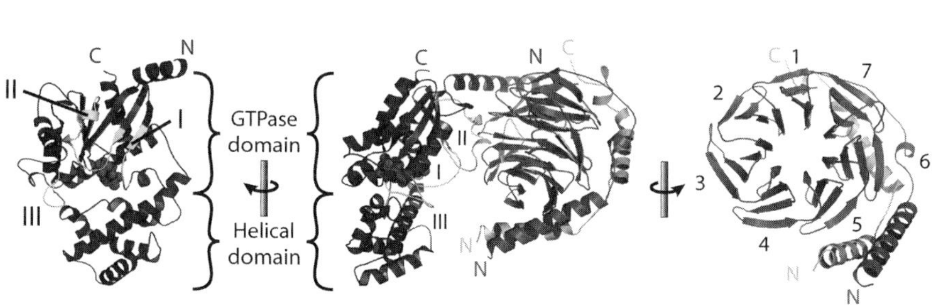


FIG. 3. Structural features of heterotrimeric G proteins. Ribbon model of $G\alpha_{t/i}$(GDP)$\beta_1\gamma_1$ heterotrimer (1GOT), where $\alpha$ is *blue*, $\beta$ is *green*, and $\gamma$ is *gold*. The three Switch regions in $G\alpha$ are highlighted in *yellow*, and GDP (*red*) is buried between the GTPase and helical domains of $G\alpha$. The subunits have been rotated outward to show the intersubunit interface, which is primarily composed of the $\alpha$N helix (truncated in this view) and Switch II on $G\alpha$ and blades 1–3 on $G\beta$. (See Color Insert.)

$G\beta\gamma$ dimer (Lambright *et al.*, 1996; Wall *et al.*, 1995). The primary interaction surface is between a hydrophobic pocket formed by Switches I and II on $G\alpha$(GDP) and five of seven blades of $G\beta$. A smaller interaction surface is formed between the N-terminus of $G\alpha$ and blade one of $G\beta$. On formation of the heterotrimer, $G\beta\gamma$ significantly alters the conformation of Switches I and II, thereby dismantling the binding sites for $Mg^{2+}$ and the GTP $\gamma$ phosphate, explaining the observation that GTP and $G\beta\gamma$ binding to $G\alpha$ are negatively cooperative (Higashijima *et al.*, 1987b). In contrast to the structural rearrangements observed in $G\alpha$ on heterotrimer formation, significant changes are not observed in $G\beta\gamma$.

## II. Toward a Model of the Receptor–G Protein Complex

### A. *Structural Determinants of Receptor–G Protein Specificity*

Relatively few types of G proteins transduce the signals from a vast number of receptors. Each member of the heterotrimeric G protein family must, therefore, be capable of interacting with many different receptors. Additionally, many receptors are capable of activating multiple different G proteins. Although many interfacial contacts between these proteins have been described, the connections that define the coupling between receptors and specific G proteins remain unclear.

Much of the difficulty in understanding coupling specificity between receptors and G proteins arises from the poor sequence homology of the intracellular loops that comprise the G protein–binding site (reviewed in Wess, 1998). Even closely related receptors that activate the same G protein can have dissimilar intracellular loops, making coupling determinations based on primary structure impossible (Hedin *et al.*, 1993). Extensive studies have shown that IC 2 and 3 are the most common selectivity determinants in receptors, while occasionally, IC 1 and 4 modulate receptor–G protein interactions (reviewed in Wess, 1997). Frequently, mutations in other regions of the receptor distant from the G protein–binding site can affect coupling, suggesting that the global conformation of the receptor or changes in its dynamics may be just as important as specific side chain interactions in determining receptor–G protein selectivity (Gilchrist *et al.*, 1996; Perez *et al.*, 1996).

Indeed, the observation that different agonists can affect which G proteins are activated by a given receptor supports a model where specific receptor conformations may be more or less favorable for coupling to specific G proteins (reviewed in Kenakin, 2003; Perez and Karnik, 2005). In one recent example of ligand-directed signaling, structurally different ligands of the $CB_1$-cannabinoid receptor differentially regulated the binding of the three homologous $G\alpha_i$ proteins to the receptor. While one ligand was shown to behave as an agonist for $G_{i1}$ and $G_{i2}$ and as an inverse agonist for $G_{i3}$, another ligand was shown to behave exactly opposite, as an inverse agonist of $G_{i1}$ and $G_{i2}$ and as an agonist for $G_{i3}$ (Mukhopadhyay and Howlett, 2005). Similarly, the functional responses of protease-activated receptor 1 (PAR1) activation differ whether the receptor is activated by thrombin or an agonist peptide. Mathematical modeling of the downstream signaling pathways indicates that the peptide favors $G\alpha_q$ activation over $G\alpha_{12/13}$ activation by 800-fold, while thrombin-mediated activation of PAR1 does not discriminate between the G proteins (McLaughlin *et al.*, 2005). Clearly, specific receptor–G protein coupling results from a complex relationship between specific side chain interactions between these proteins as well as the agonist-induced tertiary structure of the G protein–binding site.

The most well-characterized receptor contact site in G proteins is the extreme C-terminus of Gα (Fig. 4). Structural information for this region is lacking as the C-terminus is either absent or disordered in most

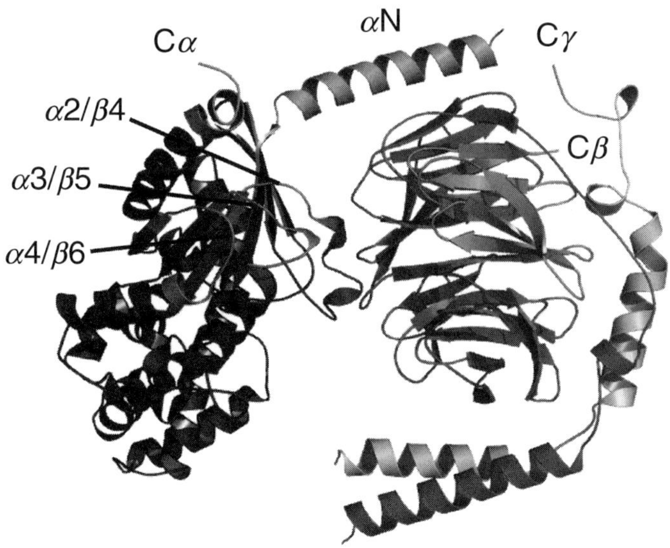


FIG. 4. Receptor contact sites on the G protein. Ribbon diagram of the heterotrimeric G protein (1GOT) with the C-terminal tail of $G\alpha_t$ attached (1AQG), which has been rotated toward the viewer from the putative receptor-binding orientation to visualize the receptor contact sites (*magenta*). (See Color Insert.)

G protein crystal structures. However, a peptide corresponding to the 11 C-terminal residues of G$\alpha_t$ adopts an $\alpha_L$-type C-cap motif on binding to light-activated rhodopsin as determined by nuclear magnetic resonance (NMR) spectroscopy (Dratz *et al.*, 1993; Kisselev *et al.*, 1998; Koenig *et al.*, 2002). The angle between the helical axis of the C-terminus and the membrane normal was determined to be ~40° with the combination of transfer nuclear Overhauser effect and residual dipolar coupling NMR data (Koenig *et al.*, 2002). This C-terminal undecapeptide stabilizes the metarhodopsin II (Meta II)-signaling conformation of rhodopsin and will compete with the transducin heterotrimer for binding to this receptor (Aris *et al.*, 2001; Hamm *et al.*, 1988). Screening a combinatorial library for peptides that could stabilize Meta II demonstrated that the C-terminal seven amino acids of G$\alpha_t$ are the most important for this interaction (Martin *et al.*, 1996). Several mutations in this region have been shown to uncouple G proteins from their receptors, including an arginine-to-proline mutation in G$\alpha_s$ first identified in the *unc* clone of S49 murine lymphoma cells that has since been associated with Albright's hereditary osteodystrophy (Osawa and Weiss, 1995; Schwindinger *et al.*, 1994; Sullivan *et al.*, 1987). Similarly, the pertussis toxin-catalyzed ADP-ribosylation of the C-terminal cysteine in G$\alpha_i$ family members also uncouples signaling through G$_i$-coupled receptors (West *et al.*, 1985). This cysteine can only be substituted with other hydrophobic residues (Aris *et al.*, 2001), evidence that it is binding to a hydrophobic pocket in the activated receptor. Indeed, a hydrophobic patch is exposed on the inner face of TM VI on rhodopsin activation that was shown to bind the C-terminal peptide of G$\alpha_t$ by fluorescence quenching (Janz and Farrens, 2004). In addition to providing a critical contact surface on the G protein, the C-terminus has also been shown to be an important determinant of coupling specificity. Frequently, C-terminal chimeras of G$\alpha$ are used to switch receptor–effector coupling (Blahos *et al.*, 1998; Conklin *et al.*, 1993, 1996; Kostenis *et al.*, 1997b; Natochin *et al.*, 2000). For example, replacing the C-terminus of G$\alpha_q$ with that of G$\alpha_i$ results in a G protein that is activated by G$_i$-coupled receptors, but stimulates phospholipase C (PLC), a G$\alpha_q$ effector (Conklin *et al.*, 1993).

In addition to the C-terminus, several other regions of G$\alpha$ have been implicated in receptor binding by a variety of experimental approaches (Fig. 4). Residues in the $\alpha 4/\beta 6$ loop contribute to the binding surface as demonstrated by alanine-scanning mutagenesis (Onrust *et al.*, 1997), studies with chimeric G$\alpha$ proteins (Bae *et al.*, 1997, 1999), sequence analysis of conserved amino acids in G$\alpha$ subclasses (Lichtarge *et al.*, 1996), chemical cross-linking (Cai *et al.*, 2001), and protection from tryptic proteolysis (Mazzoni and Hamm, 1996). The C-terminal 11 amino acids of G$\alpha_t$ and

$G\alpha_i$ are nearly identical, yet the $5HT_{1B}$-serotonin receptor activates only $G_i$ through a specific interaction with two amino acids in the $\alpha4$ helix of $G\alpha_i$ (Bae *et al.*, 1999). Five residues in the $\alpha3/\beta5$ loop of $G\alpha_s$, when replaced with the homologous amino acids in $G\alpha_{i2}$, decrease receptor-mediated GTP binding while increasing receptor affinity (Grishina and Berlot, 2000). The $\alpha$N helix also interacts with the receptor, as shown by peptide competition (Hamm *et al.*, 1988), mutagenesis (Onrust *et al.*, 1997), chemical cross-linking (Itoh *et al.*, 2001; Taylor *et al.*, 1994), and $G\alpha$ chimeras (Ho and Wong, 2000). A conserved 6 amino acid extension in the $\alpha$N helix of $G\alpha_{11}$ constrains the selectivity of this protein to $G_q$-coupled receptors (Kostenis *et al.*, 1997a, 1998), and two residues in the $\alpha N/\beta1$ loop of $G\alpha_{16}$ were identified as one sequence used by the metabotropic glutamate receptor 8 ($mGluR_8$) to discriminate between $G\alpha_{16}$ and its murine homologue $G\alpha_{15}$ (Blahos *et al.*, 2001). In a similar study, substitution of the $\alpha2/\beta4$ loop of $G\alpha_{11}$ with the homologous region of $G\alpha_{16}$ was important for conferring C5a receptor-induced activation (Lee *et al.*, 1995). Using SDSL, we have observed striking decreases in side chain mobility for labeled residues at the C-terminus, the $\alpha4/\beta6$ loop, and the $\alpha2/\beta4$ loop of $G\alpha$ on binding to R*, which has enabled us to directly trace the receptor-binding surface (Oldham *et al.*, 2006).

Interestingly, mutations in the $\alpha2/\beta4$ and $\alpha4/\beta6$ loops in $G\alpha_s$ do not affect binding to the $\beta_2$-adrenergic receptor, while mutations in the $\alpha3/\beta5$ loop increase receptor binding (Grishina and Berlot, 2000). This may suggest differential determinants of G protein coupling in different G protein families. A mutation in a highly conserved glycine in Linker 1 of $G\alpha$ was shown to increase the activation of $G\alpha_q$ by $G_i$- and $G_s$-coupled receptors, demonstrating that sites distant from the receptor–G protein interface may also be important for regulating the specificity of this interaction (Heydorn *et al.*, 2004). Further, coupling specificity may involve cooperative interactions between Linker 1 and the C-terminus (Kostenis *et al.*, 2005). Indeed, studies suggest that coupling selectivity involves subtle and cooperative interactions among all of these various domains involved in receptor recognition (Slessareva *et al.*, 2003).

Both $G\beta$ and $G\gamma$ also contribute to the receptor interacting surface on G proteins (Fig. 4). The C-terminal 60 amino acids of $G\beta$ can be cross-linked to a peptide corresponding to IC 3 of the $\alpha_2$-adrenergic receptor (Taylor *et al.*, 1994, 1996). Alanine-scanning mutagenesis in $G\beta$ identified several mutants with normal heterotrimer formation, but defective receptor-mediated GTP$\gamma$S uptake, indicating a role for $G\beta\gamma$ in this process, either directly or indirectly through stabilization of the $G\alpha$ subunit (Ford *et al.*, 1998). A farnesylated C-terminal 12 amino acid peptide of $G\gamma_1$ stabilizes Meta II (Kisselev *et al.*, 1994, 1995, 1999). This $G\gamma$ peptide forms an amphipathic helix on interaction with activated rhodopsin (Kisselev and Downs, 2003). A similar peptide corresponding to the geranylgeranylated C-terminus of G$\gamma_5$ has also been shown to inhibit G protein activation by the $M_2$-muscarinic receptor (Azpiazu *et al.*, 1999).

Specific isoforms of G$\beta$ and G$\gamma$ have been shown to preferentially interact with specific receptors. For example, both G$\beta_1\gamma_2$ and G$\beta_5\gamma_2$ can couple G$\alpha_q$ to the endothelin B and $M_1$-muscarinic receptors; however, only G$\beta_1\gamma_2$ promotes endothelin B receptor binding to G$\alpha_i$ (Lindorfer *et al.*, 1998). Additional work showed that G$\beta_5\gamma_2$ specifically couples G$\alpha_q$ to receptors (Fletcher *et al.*, 1998). Similarly, the $M_2$-muscarinic receptor interacts with G$\alpha_o$-heterotrimers containing G$\beta_4\gamma_2$, but not G$\beta_1\gamma_2$ (Hou *et al.*, 2001). The $A_1$-adenosine receptor couples equally well to G$\alpha_{i1}$ heterotrimers containing G$\beta_1\gamma_2$, G$\beta_1\gamma_3$, G$\beta_2\gamma_2$, and G$\beta_2\gamma_3$ (Figler *et al.*, 1996); however, G$\beta_1$ dimers with G$\gamma_2$ and G$\gamma_3$ had a much faster receptor-catalyzed nucleotide exchange rate as compared to dimers with G$\gamma_1$ (Figler *et al.*, 1997). A similar study measuring the ability of each of the five G$\beta\gamma_2$ isoforms to couple to the $\beta_1$-adrenergic and $A_{2a}$-adenosine receptors found that G$\beta_4$ coupled the best while G$\beta_5$ coupled the worst. Interestingly, although G$\beta_1$ has a similar potency to G$\beta_4$ in coupling to the $\beta_1$-adrenergic receptor, its potency was 20-fold lower when coupling to the adenosine receptor (McIntire *et al.*, 2001).

The G$\gamma$ subunit also plays an important role as a determinant of receptor–G protein coupling specificity. Coupling of G$\alpha_{i1}$ to the $\alpha_{2a}$-adrenergic receptor was significantly affected by which one of eight G$\gamma$ subunits participated in the interaction (Richardson and Robishaw, 1999). Similarly, the $M_2$-muscarinic receptor does not interact with G$\alpha_o$ heterotrimers with G$\beta_1\gamma_2$, although G$\beta_1\gamma_5$ and G$\beta_1\gamma_7$ could mediate binding to this receptor (Hou *et al.*, 2000), and several other studies offer additional examples of G$\gamma$-directed coupling specificity in other receptor systems (Butkerait *et al.*, 1995; Jian *et al.*, 1999; Kisselev and Gautam, 1993; Lim *et al.*, 2001). The principal determinant of this specificity is likely the primary sequence of the C-terminal third of G$\gamma$, independent of the type of lipid modification, as shown by mutagenesis and studies with chimeric G$\gamma$ subunits (Jian *et al.*, 2001; Kisselev *et al.*, 1994; Myung *et al.*, 2005). However, the type of lipid modification also has an effect on receptor coupling, as a geranylgeranyl moiety on G$\gamma_1$ or G$\gamma_2$ increases the affinity of the G protein for rhodopsin (Jian *et al.*, 2001) and the $A_1$-adenosine receptor (Yasuda *et al.*, 1996), respectively. Since many of these coupling studies were performed *in vitro*, additional studies are necessary to address the physiological significance of different heterotrimer composition on receptor coupling and effector activation *in vivo*.

### *B. Point-to-Point Interactions Between Receptors and G Proteins*

Despite the extensive data describing the receptor-binding surface on G proteins, much less is known about the specific point-to-point interactions between these proteins. Generally, the membrane-proximal sections of IC 2, 3, and 4 form the primary interaction surface on the receptor. Additional contact sites in the transmembrane helices may also be formed as the G$\alpha$ C-terminus enters deeply into the pocket formed in the cytoplasmic face of the receptor on activation (Janz and Farrens, 2004). Peptides corresponding to IC 2, 3, and 4 block the stabilization of Meta II by G$\alpha_t$ (Konig *et al.*, 1989; Marin *et al.*, 2000). A reverse substitution approach, where each intracellular loop was replaced with polyalanine sequences and then individual alanine residues were mutated back to the native amino acid, demonstrated that residues adjacent to the transmembrane helices in these loops were the most important for G protein activation (Natochin *et al.*, 2003).

Chemical cross-linking experiments between rhodopsin and G$\alpha_t$ indicate that residue 240 in IC 3 of rhodopsin is near the N- and C-termini of G$\alpha$, as well as the $\alpha 4/\beta 6$ loop (Fig. 5; Cai *et al.*, 2001; Itoh *et al.*, 2001). Five residues at the C-terminus of G$\alpha$ were shown to specifically interact with four noncontiguous residues in IC 3 (Liu *et al.*, 1995). Peptides corresponding to IC 3 also cross-link to the N-terminus of G$\alpha$ and the C-terminus of G$\beta$ (Taylor *et al.*, 1994, 1996). As mentioned previously, the binding site for the C-terminal peptide of G$\alpha_t$ was mapped to the inner face of helix VI (Janz and Farrens, 2004), and both the G$\alpha$ and G$\gamma$ C-termini of G$_t$ interact with IC 4 of rhodopsin (Ernst *et al.*, 2000). These data provide important information about specific contacts between receptors and G proteins; however, more constraints are needed to improve our current model of the receptor–G protein interface.

### *C. Current Approaches to Modeling the Receptor–G Protein Complex*

Computational techniques have allowed investigators to use this extensive amount of biochemical and biophysical information to develop the available structural information, including the inactive structure of rhodopsin (Teller *et al.*, 2001), the GDP-bound G$_t$ heterotrimer (Lambright *et al.*, 1994), and NMR structures of the C-termini of G$\alpha_t$ and G$\gamma_1$ (Kisselev and Downs, 2003; Kisselev *et al.*, 1998), into molecular models of the receptor–G protein complex (Ciarkowski *et al.*, 2005; Fotiadis *et al.*, 2004; Slusarz and Ciarkowski, 2004). Since the C-terminus of G$\alpha$, a critical receptor-binding site, is absent in both structures of G protein heterotrimers (Lambright

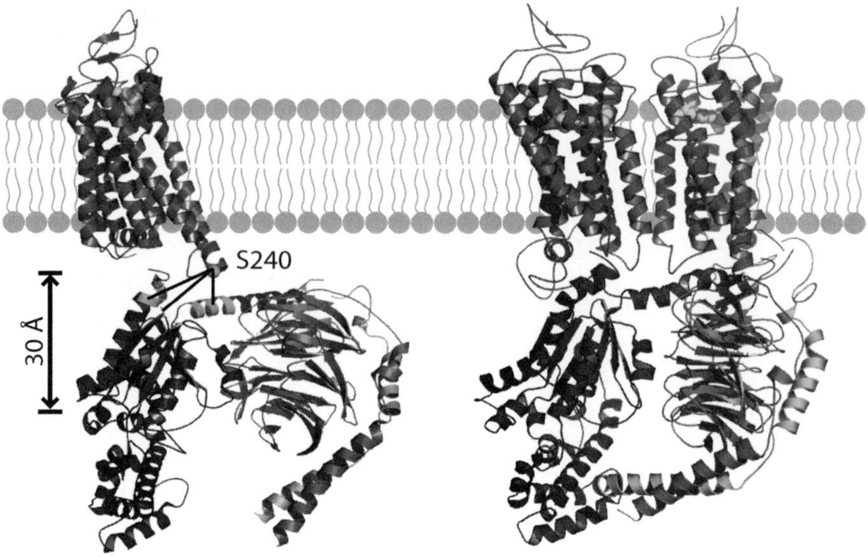


FIG. 5. Models of the receptor–G protein complex. Two representations of receptor–G protein complexes are shown. The R*·Gα(O)βγ complex created by manually docking the G protein onto an activated receptor model based on the rhodopsin crystal structure (1GZM) (*left panel*). Sites on Gα that cross-link to residue S240 (*cyan sphere*) on the receptor are highlighted (*cyan*). This sort of data will be critical for improving models of the receptor–G protein complex, as it provides constraints for the location of IC 3 relatively to Gα. In this model, the nucleotide-binding pocket is some 30 Å away from the receptor-binding surface. The model of an $R_2$·Gα(O)βγ-complex is based on coordinates generously provided by K. Palczewski (published in Fotiadis *et al.*, 2004). (See Color Insert.)

*et al.*, 1996; Wall *et al.*, 1995), it is typically appended to the heterotrimer structure using the NMR coordinates of the C-terminal peptide bound to activated rhodopsin (Kisselev *et al.*, 1998).

Assembling this complex *in silico* initially involves modeling the conformational changes associated with receptor activation to generate the binding site for the G protein C-terminus. Two different techniques have been described to develop models of activated rhodopsin guided by the distance constraints provided by SDSL studies of the protein (Nikiforovich and Marshall, 2003; Slusarz and Ciarkowski, 2004). Both of these methods resulted in reasonable working models for activated rhodopsin. In the first, simplified energy calculations were used to find sterically and energetically reasonable arrangements of the transmembrane helices around the agonist, all-*trans* retinal. From all of the possible arrangements, the model that agreed the most with experimentally derived distance constraints was selected and the nontransmembrane structural elements were reconstructed (Nikiforovich and Marshall, 2003). The second technique used target-driven molecular dynamics simulations to forcefully modify the cytoplasmic side of rhodopsin to fit the distances from SDSL (Slusarz and Ciarkowski, 2004). The rhodopsin model generated from this approach was subsequently used to examine G protein binding.

The G protein was manually docked onto the activated rhodopsin model. The topology and electrostatic potentials of the binding sites seem to limit the possible orientations between the receptor and G protein. This particular complex highlights three clusters of conserved interactions between the G$\alpha$ C-terminus and the receptor, suggesting that the conserved residues in these proteins form a core interaction surface that may be shared among different receptor–G protein complexes (Slusarz and Ciarkowski, 2004). According to this model, the cytoplasmic surface of the receptor is not large enough to accommodate both the G$\alpha$ C-terminus in its binding pocket and the G$\gamma$ C-terminus (Fig. 5; Slusarz and Ciarkowski, 2004).

One possible resolution to this dilemma is the realization that receptors function as dimers, and increasing experimental evidence suggests this to be the case (reviewed in Javitch, 2004; Milligan, 2004), including the demonstration by neutron scattering of an $R_2 \cdot G\alpha\beta\gamma$ assembly between a leukotriene $B_4$ receptor (BLT1) dimer and $G_i$ (Baneres and Parello, 2003). Rhodopsin appears to be arranged in paracrystalline rows of dimers in native disk membranes by atomic force microscopy (AFM) (Fotiadis *et al.*, 2003, 2004; Liang *et al.*, 2003). By packing the rhodopsin crystal structure into the unit cell constraints determined by AFM, a model of the rhodopsin dimer was constructed (Fotiadis *et al.*, 2004; Liang *et al.*, 2003). The interface between the two rhodopsin monomers is between helices IV and V in this model as other orientations were excluded due to geometrical or energetic limitations. After modeling the activated receptor using a 90° outward rotation of helix VI in one monomer and adding the missing termini to the G$\alpha$ and G$\gamma$ subunits, the G protein was docked onto a rhodopsin dimer. In the resulting structure, the G$\alpha$ C-terminus is inserted into the pocket formed in the cytoplasmic face of the activated receptor, the N-terminal helix of G$\alpha$ occupies a groove that runs between the two monomers and the farnesyl group at the C-terminus of G$\gamma$ interacts with the inactive receptor in the dimer (Fig. 5; Fotiadis *et al.*, 2004). An alternative model based on a different orientation of rhodopsin monomers within the rows of dimers has also been proposed (Ciarkowski *et al.*, 2005). However, the existence and physiological significance of rhodopsin-like receptor dimers remains a contentious issue (Chabre and le Maire, 2005).

## III. Molecular Basis for G Protein Activation

Activated receptors catalyze GDP release from G protein heterotrimers. This is the rate-limiting step in G protein activation and, consequently, the activation of downstream signaling proteins (Higashijima *et al.*, 1987a). Mapping the lipid modifications and the biochemically defined receptor contact sites onto the G protein crystal structures demonstrates that the nucleotide-binding pocket is some 30 Å away from the nearest receptor contact site (Fig. 5; Lambright *et al.*, 1996; Wall *et al.*, 1995). Thus, the receptor must induce some conformational change in the G protein to cause GDP release over this distance (Bourne, 1997; Hamm, 1998). Several potential mechanisms for receptor-catalyzed GDP release have been proposed, mainly based on the effects of various mutations on the basal and receptor-catalyzed nucleotide exchange rates of the G protein. To date, there is little direct evidence of specific receptor-mediated conformational changes in G proteins, and the molecular mechanisms for receptor-catalyzed nucleotide exchange remain one of the most important unanswered questions in the field of G protein signaling.

Inspection of the G protein crystal structures highlights several potential routes from known receptor contact sites to the nucleotide-binding pocket. Since these structural elements may form the allosteric connection between receptor binding and GDP release, they have been the primary targets in the study of receptor-catalyzed G protein activation (Fig. 6). Perhaps the most apparent connection is the $\alpha5$ helix, which links the C-terminus of $G\alpha$ to the nucleotide-binding pocket at the $\beta6/\alpha5$ loop. Alanine-scanning mutagenesis of this helix in $G\alpha_t$ identified three mutations on the inward face of the helix with substantially increased rates of basal and receptor-catalyzed nucleotide exchange, as well as several mutations on the outward face of the helix that decreased the receptor-catalyzed exchange rate (Marin *et al.*, 2001b). A subsequent study based on proline-scanning mutagenesis generally resulted in mutants with increased rates of basal and decreased rates of catalyzed exchange (Marin *et al.*, 2002). An isoleucine-to-proline mutation at the C-terminus of the $\alpha5$ helix has been shown to reduce the stimulatory activity of a $D_2$-dopamine receptor mimetic peptide on nucleotide exchange in $G_i$ (Nanoff *et al.*, 2006). These studies provide evidence that perturbations of the $\alpha5$ helix are coupled to nucleotide exchange and modulate receptor coupling.

When a flexible five-glycine linker is inserted between the C-terminus and the $\alpha5$ helix, receptor binding is only slightly decreased while receptor-catalyzed nucleotide exchange is severely impaired. Alternatively, doubling the length of the $\alpha5$ helix by repeating its sequence has only a

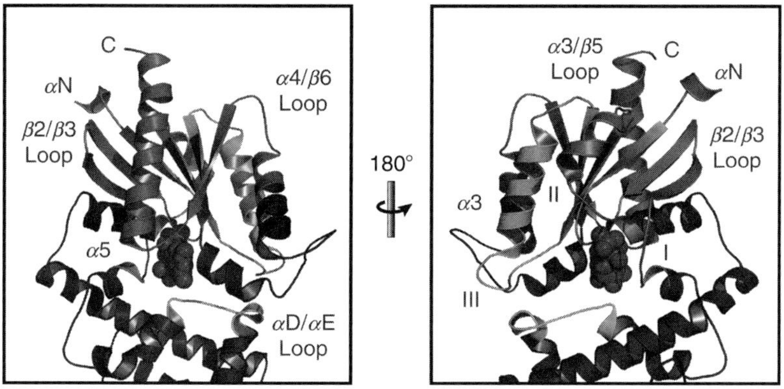


FIG. 6. Structural elements connecting receptor contact sites to the nucleotide-binding pocket. The α5 helix and β6 strand connect the C-terminus and α4/β6 loop to the nucleotide-binding pocket at the β6/α5 loop (*pink*). The α3/β5 loop is connected to Switch III by the α3 helix (*lime*). Perturbations in Switch III may be communicated across the interdomain interface through its interactions with the αD/αE loop (*lime*). The αN helix is connected to the P-loop and Switches I and II through its interactions with β strands 1–3 (*teal*). Additionally, these β strands also interact with the α5 helix (*pink*) and may participate in front-to-back communication between these elements. (See Color Insert.)

modest effect on receptor coupling. These data suggest that preserving the rigid connection between the C-terminus and GDP-binding pocket is important for nucleotide exchange. Furthermore, the α5 helix may be sufficient for coupling as extending the helix increases the distance between the receptor and the protein core of Gα, thereby disrupting other potential receptor contacts (Natochin *et al.*, 2001). Using SDSL, we have been able to observe receptor activation-dependent conformational changes in the α5 helix at sites distant from the receptor contact surface, and the five-glycine linker uncouples these conformational changes from receptor binding (Oldham *et al.*, 2006).

These conformational changes may be coupled to GDP release through the β6/α5 loop, which contains the conserved guanine ring-binding TCAT (threonine-cysteine-alanine-threonine) motif. Mutations in this loop greatly increase spontaneous GDP release from Gα, resulting in high basal rates of GTP binding and G protein activation (Iiri *et al.*, 1994; Posner *et al.*, 1998; Thomas *et al.*, 1993), including the activating mutation in $G\alpha_s$, A366S, that causes pseudohypoparathyroidism and gonadotropin-independent precocious puberty in male patients (Iiri *et al.*, 1994). A homologous mutation in $G\alpha_{i1}$ also greatly enhances GDP release (Posner *et al.*, 1998). No significant structural differences were observed in the crystal structure of the $G\alpha_{i1}$-A326S heterotrimer compared to the wild-type protein, but the GDP-binding site was only partially occupied, suggesting that GDP release can occur without large conformational changes in $G\alpha$ (Posner *et al.*, 1998).

Additional evidence supports a role in receptor-catalyzed nucleotide exchange for the structural elements extending from other receptor contact sites to the nucleotide-binding pocket (Fig. 6). Running parallel to the $\alpha5$ helix, the $\beta6$ strand may connect receptor binding of the $\alpha4/\beta6$ loop to the binding pocket at the $\beta6/\alpha5$ loop in a similar manner. Several alanine mutations in this region demonstrated reduced receptor-catalyzed activation, despite normal binding to the receptor (Onrust *et al.*, 1997). The $\alpha3/\beta5$ loop in $G\alpha_s$ appears to be a receptor contact site (Grishina and Berlot, 2000) that may transmit a conformational change along the $\alpha3$ helix to Switch III. Interactions between Switch III and the $\alpha D/\alpha E$ loop at the domain interface have been implicated in receptor-mediated activation of $G\alpha_s$ (Grishina and Berlot, 1998). Disruption of these interactions may open the gap between the helical and GTPase domains to release GDP. Interestingly, an arginine-to-glutamate mutation in Switch III has been shown to stabilize the nucleotide-free conformation of $G\alpha_t$, and excitingly, this mutant blocks the rhodopsin-catalyzed activation of wild-type $G\alpha_t$ (Pereira and Cerione, 2005). If applicable in other systems, this dominant-negative mutant may be valuable for the investigation of G protein coupling specificity *in vivo.*

Several intramolecular contacts with the N-terminus have also been shown to influence G protein activation (Fig. 6). C-terminal truncations in $G\alpha_o$ and $G\alpha_{i2}$ resulted in reduced affinity for GDP because of impaired interactions with the N-terminus (Denker *et al.*, 1992, 1995). A hydrophobic interaction between the N-terminus and $\alpha5$ helix in $G\alpha_t$ stabilizes the GDP-bound state (Natochin *et al.*, 2000), and a salt bridge between the N-terminus and $\beta3$ strand in $G\alpha_t$ has also been implicated in maintaining such a low rate of basal nucleotide exchange in this $G\alpha_i$ family member (Thomas *et al.*, 2001). This interaction connects the N-terminus to Switch II via the $\beta3$-strand. Movement of the $\alpha N$ helix could also affect the $\beta1$ and $\beta2$ strands, and in so doing, perturb Switches I, II, and the phosphate-binding (P-) loop, all of which form important interactions with GDP (Fig. 6). Additionally, receptor-induced perturbations in the $\alpha5$ helix may be transmitted to the GDP-binding pocket via the $\alpha1$ helix and $\beta2$ and $\beta3$ strands, as suggested by mutagenesis and computational studies (Ceruso *et al.*, 2004; Marin *et al.*, 2001b). Thus, conformational changes at the N-terminus may also work in concert with other receptor-mediated changes to cause GDP release.

Bound GDP is buried between the GTPase and helical domains of $G\alpha$. Some of the contacts between these two domains may need to be broken in order for GDP to be released. Several studies have identified interdomain interactions that affect nucleotide exchange; however, the results seem to differ depending on the G$\alpha$ subunit involved. In G$\alpha_s$ and G$\alpha_{i1}$, disrupting these interactions resulted in increased rates of basal or decreased rates of receptor-catalyzed exchange (Grishina and Berlot, 1998; Warner and Weinstein, 1999; Warner *et al.*, 1998). Increased rates of basal exchange indicate that an opening of the interdomain cleft reduces GDP affinity. Reduced rates of catalyzed exchange suggest that the receptor may transmit a conformational change across the domain interface via these interactions. However, these connections do not seem to be important for regulating the basal and catalyzed exchange rates in G$\alpha_t$ (Marin *et al.*, 2001a). Thus, the role of these interdomain interactions remains unclear, but inspection of the crystal structure seems to require a movement of the GTPase and helical domains relative to each other in order to release GDP. Indeed, an interdomain reorientation has been suggested by molecular dynamics simulations of G$\alpha_t$ (Ceruso *et al.*, 2004) and glycine-to-proline mutations in each of the interdomain linkers enhance the basal nucleotide exchange rate of G$\alpha$ (Majumdar *et al.*, 2004).

Outside of the G$\alpha$ subunit, two models for receptor-mediated GDP release involve an active role for the G$\beta\gamma$ subunit (Fig. 7). These models are based on the hypothesis that G$\beta\gamma$ functions in heterotrimeric G proteins as a structural homologue to the guanine nucleotide exchange factors for

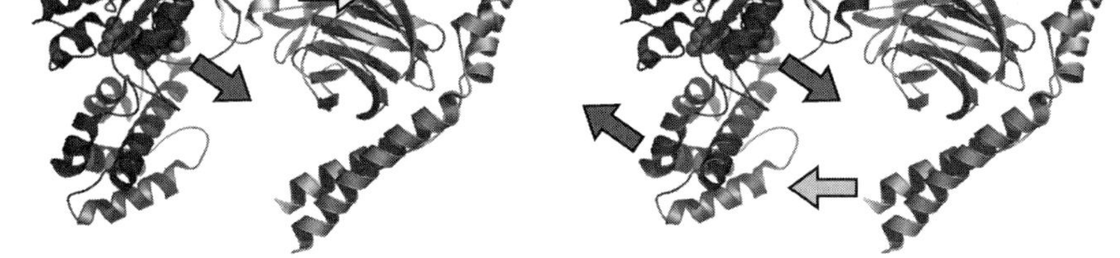


FIG. 7. Two models also propose that R* may use G$\beta\gamma$ to catalyze GDP release. In the lever-arm model, the receptor uses the $\alpha$N helix as a lever to pull G$\beta\gamma$ away from G$\alpha$, thereby prying Switch II away from the nucleotide-binding pocket and causing GDP release. The alternative, gearshift model requires that the receptor pushes G$\beta\gamma$ closer to G$\alpha$. This allows the N-terminus of G$\gamma$ to engage the helical domain, thus forcing the binding pocket due to the reorientation of the two domains. (See Color Insert.)

monomeric G proteins. In the first model, the receptor uses the N-terminal helix of $G\alpha$ as a lever-arm to pull $G\beta\gamma$ away from $G\alpha$. This would pry Switches I and II away from the nucleotide-binding pocket, resulting in GDP release (Iiri *et al.*, 1998). In support of this hypothesis, deletion of four amino acids at the C-terminus of the $\alpha$N helix of $G\alpha$ to mimic the putative orientation induced by the receptor allows $G\beta\gamma$ to increase basal nucleotide exchange in $G\alpha_s$ (Rondard *et al.*, 2001). The second model proposes that the receptor engages a train of three gears composed of (1) the receptor's interaction with the N- and C-termini of $G\alpha$ and the C-terminus of $G\gamma$, (2) $G\beta$'s interaction with the nucleotide-binding pocket at Switches I and II, and (3) the N-terminus of $G\gamma$ binding to the helical domain of $G\alpha$. In the gearshift model, the receptor uses the $\alpha$N helix to force $G\beta\gamma$ into $G\alpha$, allowing the N-terminus of $G\gamma$ to engage the helical domain of $G\alpha$. This interaction may cause the interdomain gap between the helical and GTPase domains of $G\alpha$ to open, causing GDP release (Cherfils and Chabre, 2003).

Regardless of the mechanism, the interaction between $G\alpha$ and $G\beta\gamma$ plays a critical role in G protein activation, as heterotrimer formation is required for efficient coupling to receptors. Indeed, several alanine mutations at the Switch interface of $G\beta$ are defective in receptor-catalyzed nucleotide exchange despite normal heterotrimer formation (Ford *et al.*, 1998). Furthermore, mutations in the C-terminus of $G\gamma$ have been shown to increase receptor-catalyzed nucleotide exchange (Azpiazu and Gautam, 2001). A small peptide derived from phage display has also recently been shown to increase spontaneous nucleotide exchange in $G\alpha_{i1}$. Structural studies indicate that this peptide dramatically displaces Switch II away from the nucleotide-binding pocket, presumably leading to GDP release, although the nucleotide remains bound in the crystal structure of the complex (Johnston *et al.*, 2005). This structure supports the lever-arm model of G protein activation; however, much work remains in order to define the contributions of $G\beta\gamma$ to the GDP release mechanism.

## IV. Summary and Conclusions

Despite the fundamental importance of receptor-catalyzed G protein activation in cellular signaling, relatively little is known about this process as compared to the other regulatory events in the G protein cycle. Crystal structures of inactive rhodopsin and G protein heterotrimers provide the structural context for the meaningful interpretation of the results of the numerous biochemical and biophysical studies described in the preceding sections. Enough data has been accumulated to begin to develop rudimentary models of the receptor–G protein complex that meet some of the important experimentally derived structural constraints. However, the currently available information about point-to-point interactions between the proteins remains sparse and the structural resolution of these models remains too low to answer questions about determinants of receptor–G protein specificity or the mechanism of activation. Many experiments remain to be done in order to refine our current models of this complex.

Particularly important information will come from additional distance constraints between the receptor and G protein, such as novel cross-links or interprobe measurements from fluorescence or electron paramagnetic resonance spectroscopy. Initially, relatively few long-range distance constraints would be necessary to define the relative orientation of the receptor and G protein, which is a fundamental but unanswered question about the complex. Once the proper orientation has been established experimentally, the model will suggest additional distance measurements that will be necessary to "pin down" the receptor's intracellular loops onto the surface of the G protein. Proceeding in such an iterative fashion should provide the experimental evidence that is critically important in refining the current models of the receptor–G protein complex.

Additionally, any improvement in the high-resolution structural information available for the complex will greatly enhance our models of it. The current models of the complex were assembled from models of an active receptor based on the inactive structure and biophysical data and G protein with C-termini of G$\alpha$ and G$\gamma$ based on NMR structures of the corresponding peptides. Furthermore, the modeling process does not yet take into account conformational changes in the G protein, which are necessary to cause GDP release from G$\alpha$, because little is known about specific receptor-mediated structural rearrangements in this protein. Although each of the conformational changes described above is supported by mutagenesis and biochemical data, additional information is needed to understand the structural basis of receptor-mediated G protein activation. Certainly, a high-resolution crystal structure would answer many questions about the orientation of these two proteins, the sites of interaction between them, and the activating conformational changes that lead to complex formation and trigger GDP release. Unfortunately, this complex has proven exceedingly difficult to crystallize, and a high-resolution structure may not be solved in the near future. In the meantime, alternative experimental approaches must be employed to refine current molecular models of the receptor–G protein complex and to characterize the conformational changes leading to GDP release.

Once these experimental approaches provide some knowledge about the nature of the receptor-mediated conformational changes in G proteins, it becomes easier to speculate as to the molecular trigger that may cause a particular type of change. For example, site-directed mutagenesis of residues on the side of the $\alpha 5$ helix opposite the $\beta 6$ strand significantly increases the basal nucleotide exchange rate of G$\alpha$ (Marin *et al.*, 2001b). Perhaps the receptor disrupts these interactions in a similar manner. Assuming the general mechanism of receptor-catalyzed GDP release is conserved among receptor–G protein couplings, sequence analysis of conserved residues in these proteins may further serve to illustrate the important interactions for this process. One model suggests that there is a conserved binding site between the $\alpha 5$ helix and the adjacent $\beta 2/\beta 3$ loop for the conserved arginine in the DRY motif of receptors and that this interaction may underlie the mechanism for R*-mediated conformational changes in G$\alpha$ (Oliveira *et al.*, 1999). Much work remains to be done, however, to identify the structural basis of G protein activation, characterize the nucleotide-free conformation of the G protein, and incorporate these data into improved models of the receptor–G protein complex.

## References


Aris, L., Gilchrist, A., Rens-Domiano, S., Meyer, C., Schatz, P. J., Dratz, E. A., and Hamm, H. E. (2001). Structural requirements for the stabilization of metarhodopsin II by the C terminus of the $\alpha$ subunit of transducin. *J. Biol. Chem.* **276,** 2333–2339.

Azpiazu, I., and Gautam, N. (2001). G protein $\gamma$ subunit interaction with a receptor regulates receptor-stimulated nucleotide exchange. *J. Biol. Chem.* **276,** 41742–41747.

Azpiazu, I., Cruzblanca, H., Li, P., Linder, M., Zhuo, M., and Gautam, N. (1999). A G protein $\gamma$ subunit-specific peptide inhibits muscarinic receptor signaling. *J. Biol. Chem.* **274,** 35305–35308.

Bae, H., Anderson, K., Flood, L. A., Skiba, N. P., Hamm, H. E., and Graber, S. G. (1997). Molecular determinants of selectivity in 5-hydroxytryptamine$_{1B}$ receptor-G protein interactions. *J. Biol. Chem.* **272,** 32071–32077.

Bae, H., Cabrera-Vera, T. M., Depree, K. M., Graber, S. G., and Hamm, H. E. (1999). Two amino acids within the $\alpha 4$ helix of G$\alpha_{i1}$ mediate coupling with 5-hydroxytryptamine$_{1B}$ receptors. *J. Biol. Chem.* **274,** 14963–14971.

Baneres, J. L., and Parello, J. (2003). Structure-based analysis of GPCR function: Evidence for a novel pentameric assembly between the dimeric leukotriene $B_4$ receptor BLT1 and the G-protein. *J. Mol. Biol.* **329,** 815–829.

Blahos, J., II, Mary, S., Perroy, J., de Colle, C., Brabet, I., Bockaert, J., and Pin, J. P. (1998). Extreme C terminus of G protein $\alpha$-subunits contains a site that discriminates between G$_i$-coupled metabotropic glutamate receptors. *J. Biol. Chem.* **273,** 25765–25769.

Blahos, J., Fischer, T., Brabet, I., Stauffer, D., Rovelli, G., Bockaert, J., and Pin, J. P. (2001). A novel site on the G$\alpha$-protein that recognizes heptahelical receptors. *J. Biol. Chem.* **276,** 3262–3269.

Bourne, H. R. (1997). How receptors talk to trimeric G proteins. *Curr. Opin. Cell Biol.* **9,** 134–142.

Butkerait, P., Zheng, Y., Hallak, H., Graham, T. E., Miller, H. A., Burris, K. D., Molinoff, P. B., and Manning, D. R. (1995). Expression of the human 5-hydroxytryptamine$_{1A}$ receptor in Sf9 cells. Reconstitution of a coupled phenotype by co-expression of mammalian G protein subunits. *J. Biol. Chem.* **270,** 18691–18699.

Cabrera-Vera, T. M., Vanhauwe, J., Thomas, T. O., Medkova, M., Preininger, A., Mazzoni, M. R., and Hamm, H. E. (2003). Insights into G protein structure, function, and regulation. *Endocr. Rev.* **24,** 765–781.

Cai, K., Itoh, Y., and Khorana, H. G. (2001). Mapping of contact sites in complex formation between transducin and light-activated rhodopsin by covalent crosslinking: Use of a photoactivatable reagent. *Proc. Natl. Acad. Sci. USA* **98,** 4877–4882.

Ceruso, M. A., Periole, X., and Weinstein, H. (2004). Molecular dynamics simulations of transducin: Interdomain and front to back communication in activation and nucleotide exchange. *J. Mol. Biol.* **338,** 469–481.

Chabre, M., and le Maire, M. (2005). Monomeric G-protein-coupled receptor as a functional unit. *Biochemistry* **44,** 9395–9403.

Cherfils, J., and Chabre, M. (2003). Activation of G-protein G$\alpha$ subunits by receptors through G$\alpha$–G$\beta$ and G$\alpha$–G$\gamma$ interactions. *Trends Biochem. Sci.* **28,** 13–17.

Ciarkowski, J., Witt, M., and Slusarz, R. (2005). A hypothesis for GPCR activation. *J. Mol. Model.* **11,** 407–415.

Coleman, D. E., Berghuis, A. M., Lee, E., Linder, M. E., Gilman, A. G., and Sprang, S. R. (1994). Structures of active conformations of $G_{i\alpha 1}$ and the mechanism of GTP hydrolysis. *Science* **265,** 1405–1412.

Conklin, B. R., Farfel, Z., Lustig, K. D., Julius, D., and Bourne, H. R. (1993). Substitution of three amino acids switches receptor specificity of $G_q\alpha$ to that of $G_i\alpha$. *Nature* **363,** 274–276.

Conklin, B. R., Herzmark, P., Ishida, S., Voyno-Yasenetskaya, T. A., Sun, Y., Farfel, Z., and Bourne, H. R. (1996). Carboxyl-terminal mutations of $G_{q\alpha}$ and $G_{s\alpha}$ that alter the fidelity of receptor activation. *Mol. Pharmacol.* **50,** 885–890.

Denker, B. M., Schmidt, C. J., and Neer, E. J. (1992). Promotion of the GTP-liganded state of the $G_{o\alpha}$ protein by deletion of the C terminus. *J. Biol. Chem.* **267,** 9998–10002.

Denker, B. M., Boutin, P. M., and Neer, E. J. (1995). Interactions between the amino- and carboxyl-terminal regions of G$\alpha$ subunits: Analysis of mutated $G\alpha_o/G\alpha_{i2}$ chimeras. *Biochemistry* **34,** 5544–5553.

Downes, G. B., and Gautam, N. (1999). The G protein subunit gene families. *Genomics* **62,** 544–552.

Dratz, E. A., Furstenau, J. E., Lambert, C. G., Thireault, D. L., Rarick, H., Schepers, T., Pakhlevaniants, S., and Hamm, H. E. (1993). NMR structure of a receptor-bound G-protein peptide. *Nature* **363,** 276–281.

Ernst, O. P., Meyer, C. K., Marin, E. P., Henklein, P., Fu, W. Y., Sakmar, T. P., and Hofmann, K. P. (2000). Mutation of the fourth cytoplasmic loop of rhodopsin affects binding of transducin and peptides derived from the carboxyl-terminal sequences of transducin $\alpha$ and $\gamma$ subunits. *J. Biol. Chem.* **275,** 1937–1943.

Farrens, D. L., Altenbach, C., Yang, K., Hubbell, W. L., and Khorana, H. G. (1996). Requirement of rigid-body motion of transmembrane helices for light activation of rhodopsin. *Science* **274,** 768–770.

Figler, R. A., Graber, S. G., Lindorfer, M. A., Yasuda, H., Linden, J., and Garrison, J. C. (1996). Reconstitution of recombinant bovine $A_1$ adenosine receptors in Sf9 cell membranes with recombinant G proteins of defined composition. *Mol. Pharmacol.* **50,** 1587–1595.

Figler, R. A., Lindorfer, M. A., Graber, S. G., Garrison, J. C., and Linden, J. (1997). Reconstitution of bovine $A_1$ adenosine receptors and G proteins in phospholipid vesicles: $\beta\gamma$-subunit composition influences guanine nucleotide exchange and agonist binding. *Biochemistry* **36,** 16288–16299.

Fletcher, J. E., Lindorfer, M. A., DeFilippo, J. M., Yasuda, H., Guilmard, M., and Garrison, J. C. (1998). The G protein $\beta5$ subunit interacts selectively with the $G_q\alpha$ subunit. *J. Biol. Chem.* **273,** 636–644.

Ford, C. E., Skiba, N. P., Bae, H., Daaka, Y., Reuveny, E., Shekter, L. R., Rosal, R., Weng, G., Yang, C. S., Iyengar, R., Miller, R. J., Jan, L. Y., *et al.* (1998). Molecular basis for interactions of G protein $\beta\gamma$ subunits with effectors. *Science* **280,** 1271–1274.

Fotiadis, D., Liang, Y., Filipek, S., Saperstein, D. A., Engel, A., and Palczewski, K. (2003). Atomic-force microscopy: Rhodopsin dimers in native disc membranes. *Nature* **421,** 127–128.

Fotiadis, D., Liang, Y., Filipek, S., Saperstein, D. A., Engel, A., and Palczewski, K. (2004). The G protein-coupled receptor rhodopsin in the native membrane. *FEBS Lett.* **564,** 281–288.

Fredriksson, R., and Schiöth, H. B. (2005). The repertoire of G-protein-coupled receptors in fully sequenced genomes. *Mol. Pharmacol.* **67,** 1414–1425.

Gether, U. (2000). Uncovering molecular mechanisms involved in activation of G protein-coupled receptors. *Endocr. Rev.* **21,** 90–113.

Gether, U., Asmar, F., Meinild, A. K., and Rasmussen, S. G. (2002). Structural basis for activation of G-protein-coupled receptors. *Pharmacol. Toxicol.* **91,** 304–312.

Gilchrist, R. L., Ryu, K. S., Ji, I., and Ji, T. H. (1996). The luteinizing hormone/chorionic gonadotropin receptor has distinct transmembrane conductors for cAMP and inositol phosphate signals. *J. Biol. Chem.* **271,** 19283–19287.

Grishina, G., and Berlot, C. H. (1998). Mutations at the domain interface of $G_{s\alpha}$ impair receptor-mediated activation by altering receptor and guanine nucleotide binding. *J. Biol. Chem.* **273,** 15053–15060.

Grishina, G., and Berlot, C. H. (2000). A surface-exposed region of $G_{s\alpha}$ in which substitutions decrease receptor-mediated activation and increase receptor affinity. *Mol. Pharmacol.* **57,** 1081–1092.

Hamm, H. E. (1998). The many faces of G protein signaling. *J. Biol. Chem.* **273,** 669–672.

Hamm, H. E., Deretic, D., Arendt, A., Hargrave, P. A., Koenig, B., and Hofmann, K. P. (1988). Site of G protein binding to rhodopsin mapped with synthetic peptides from the $\alpha$ subunit. *Science* **241,** 832–835.

Hedin, K. E., Duerson, K., and Clapham, D. E. (1993). Specificity of receptor-G protein interactions: Searching for the structure behind the signal. *Cell. Signal.* **5,** 505–518.

Heydorn, A., Ward, R. J., Jorgensen, R., Rosenkilde, M. M., Frimurer, T. M., Milligan, G., and Kostenis, E. (2004). Identification of a novel site within G protein $\alpha$ subunits important for specificity of receptor-G protein interaction. *Mol. Pharmacol.* **66,** 250–259.

Higashijima, T., Ferguson, K. M., Smigel, M. D., and Gilman, A. G. (1987a). The effect of GTP and $Mg^{2+}$ on the GTPase activity and the fluorescent properties of $G_o$. *J. Biol. Chem.* **262,** 757–761.

Higashijima, T., Ferguson, K. M., Sternweis, P. C., Smigel, M. D., and Gilman, A. G. (1987b). Effects of $Mg^{2+}$ and the $\beta\gamma$-subunit complex on the interactions of guanine nucleotides with G proteins. *J. Biol. Chem.* **262,** 762–766.

Ho, M. K., and Wong, Y. H. (2000). The amino terminus of $G\alpha_z$ is required for receptor recognition, whereas its $\alpha4/\beta6$ loop is essential for inhibition of adenylyl cyclase. *Mol. Pharmacol.* **58,** 993–1000.

Hou, Y., Azpiazu, I., Smrcka, A., and Gautam, N. (2000). Selective role of G protein $\gamma$ subunits in receptor interaction. *J. Biol. Chem.* **275,** 38961–38964.

Hou, Y., Chang, V., Capper, A. B., Taussig, R., and Gautam, N. (2001). G protein $\beta$ subunit types differentially interact with a muscarinic receptor but not adenylyl cyclase type II or phospholipase C-$\beta$2/3. *J. Biol. Chem.* **276,** 19982–19988.

Hubbell, W. L., Altenbach, C., Hubbell, C. M., and Khorana, H. G. (2003). Rhodopsin structure, dynamics, and activation: A perspective from crystallography, site-directed spin labeling, sulfhydryl reactivity, and disulfide cross-linking. *Adv. Protein Chem.* **63,** 243–290.

Iiri, T., Herzmark, P., Nakamoto, J. M., van Dop, C., and Bourne, H. R. (1994). Rapid GDP release from $G_{s\alpha}$ in patients with gain and loss of endocrine function. *Nature* **371,** 164–168.

Iiri, T., Farfel, Z., and Bourne, H. R. (1998). G-protein diseases furnish a model for the turn-on switch. *Nature* **394,** 35–38.

Itoh, Y., Cai, K., and Khorana, H. G. (2001). Mapping of contact sites in complex formation between light-activated rhodopsin and transducin by covalent cross-linking: Use of a chemically preactivated reagent. *Proc. Natl. Acad. Sci. USA* **98,** 4883–4887.

Janz, J. M., and Farrens, D. L. (2004). Rhodopsin activation exposes a key hydrophobic binding site for the transducin $\alpha$-subunit C terminus. *J. Biol. Chem.* **279,** 29767–29773.

Javitch, J. A. (2004). The ants go marching two by two: Oligomeric structure of G-protein-coupled receptors. *Mol. Pharmacol.* **66,** 1077–1082.

Jian, X., Sainz, E., Clark, W. A., Jensen, R. T., Battey, J. F., and Northup, J. K. (1999). The bombesin receptor subtypes have distinct G protein specificities. *J. Biol. Chem.* **274,** 11573–11581.

Jian, X., Clark, W. A., Kowalak, J., Markey, S. P., Simonds, W. F., and Northup, J. K. (2001). G$\beta\gamma$ affinity for bovine rhodopsin is determined by the carboxyl-terminal sequences of the $\gamma$ subunit. *J. Biol. Chem.* **276,** 48518–48525.

Johnston, C. A., Willard, F. S., Jezyk, M. R., Fredericks, Z., Bodor, E. T., Jones, M. B., Blaesius, R., Watts, V. J., Harden, T. K., Sondek, J., Ramer, J. K., and Siderovski, D. P. (2005). Structure of $G\alpha_{i1}$ bound to a GDP-selective peptide provides insight into guanine nucleotide exchange. *Structure* **13,** 1069–1080.

Kenakin, T. (2003). Ligand-selective receptor conformations revisited: The promise and the problem. *Trends Pharmacol. Sci.* **24,** 346–354.

Kisselev, O., and Gautam, N. (1993). Specific interaction with rhodopsin is dependent on the $\gamma$ subunit type in a G protein. *J. Biol. Chem.* **268,** 24519–24522.

Kisselev, O., Pronin, A., Ermolaeva, M., and Gautam, N. (1995). Receptor-G protein coupling is established by a potential conformational switch in the $\beta\gamma$ complex. *Proc. Natl. Acad. Sci. USA* **92,** 9102–9106.

Kisselev, O. G., and Downs, M. A. (2003). Rhodopsin controls a conformational switch on the transducin $\gamma$ subunit. *Structure* **11,** 367–373.

Kisselev, O. G., Ermolaeva, M. V., and Gautam, N. (1994). A farnesylated domain in the G protein $\gamma$ subunit is a specific determinant of receptor coupling. *J. Biol. Chem.* **269,** 21399–21402.

Kisselev, O. G., Kao, J., Ponder, J. W., Fann, Y. C., Gautam, N., and Marshall, G. R. (1998). Light-activated rhodopsin induces structural binding motif in G protein $\alpha$ subunit. *Proc. Natl. Acad. Sci. USA* **95,** 4270–4275.

Kisselev, O. G., Meyer, C. K., Heck, M., Ernst, O. P., and Hofmann, K. P. (1999). Signal transfer from rhodopsin to the G-protein: Evidence for a two-site sequential fit mechanism. *Proc. Natl. Acad. Sci. USA* **96,** 4898–4903.

Koenig, B. W., Kontaxis, G., Mitchell, D. C., Louis, J. M., Litman, B. J., and Bax, A. (2002). Structure and orientation of a G protein fragment in the receptor bound state from residual dipolar couplings. *J. Mol. Biol.* **322,** 441–461.

Konig, B., Arendt, A., McDowell, J. H., Kahlert, M., Hargrave, P. A., and Hofmann, K. P. (1989). Three cytoplasmic loops of rhodopsin interact with transducin. *Proc. Natl. Acad. Sci. USA* **86,** 6878–6882.

Kostenis, E., Degtyarev, M. Y., Conklin, B. R., and Wess, J. (1997a). The N-terminal extension of $G\alpha_q$ is critical for constraining the selectivity of receptor coupling. *J. Biol. Chem.* **272,** 19107–19110.

Kostenis, E., Gomeza, J., Lerche, C., and Wess, J. (1997b). Genetic analysis of receptor-$G\alpha_q$ coupling selectivity. *J. Biol. Chem.* **272,** 23675–23681.

Kostenis, E., Zeng, F. Y., and Wess, J. (1998). Functional characterization of a series of mutant G protein $\alpha_q$ subunits displaying promiscuous receptor coupling properties. *J. Biol. Chem.* **273,** 17886–17892.

Kostenis, E., Martini, L., Ellis, J., Waldhoer, M., Heydorn, A., Rosenkilde, M. M., Norregaard, P. K., Jorgensen, R., Whistler, J. L., and Milligan, G. (2005). A highly conserved glycine within linker I and the extreme C terminus of G protein $\alpha$ subunits interact cooperatively in switching G protein-coupled receptor-to-effector specificity. *J. Pharmacol. Exp. Ther.* **313,** 78–87.

Kristiansen, K. (2004). Molecular mechanisms of ligand binding, signaling, and regulation within the superfamily of G-protein-coupled receptors: Molecular modeling and mutagenesis approaches to receptor structure and function. *Pharmacol. Ther.* **103,** 21–80.

Lambright, D. G., Noel, J. P., Hamm, H. E., and Sigler, P. B. (1994). Structural determinants for activation of the $\alpha$-subunit of a heterotrimeric G protein. *Nature* **369,** 621–628.

Lambright, D. G., Sondek, J., Bohm, A., Skiba, N. P., Hamm, H. E., and Sigler, P. B. (1996). The 2.0 Å crystal structure of a heterotrimeric G protein. *Nature* **379,** 311–319.

Lee, C. H., Katz, A., and Simon, M. I. (1995). Multiple regions of $G_{\alpha 16}$ contribute to the specificity of activation by the C5a receptor. *Mol. Pharmacol.* **47,** 218–223.

Liang, Y., Fotiadis, D., Filipek, S., Saperstein, D. A., Palczewski, K., and Engel, A. (2003). Organization of the G protein-coupled receptors rhodopsin and opsin in native membranes. *J. Biol. Chem.* **278,** 21655–21662.

Lichtarge, O., Bourne, H. R., and Cohen, F. E. (1996). Evolutionarily conserved $G_{\alpha\beta\gamma}$ binding surfaces support a model of the G protein-receptor complex. *Proc. Natl. Acad. Sci. USA* **93,** 7507–7511.

Lim, W. K., Myung, C. S., Garrison, J. C., and Neubig, R. R. (2001). Receptor-G protein $\gamma$ specificity: $\gamma$11 shows unique potency for $A_1$ adenosine and 5-$HT_{1A}$ receptors. *Biochemistry* **40,** 10532–10541.

Lindorfer, M. A., Myung, C. S., Savino, Y., Yasuda, H., Khazan, R., and Garrison, J. C. (1998). Differential activity of the G protein $\beta_5\gamma_2$ subunit at receptors and effectors. *J. Biol. Chem.* **273,** 34429–34436.

Liu, J., Conklin, B. R., Blin, N., Yun, J., and Wess, J. (1995). Identification of a receptor/G-protein contact site critical for signaling specificity and G-protein activation. *Proc. Natl. Acad. Sci. USA* **92,** 11642–11646.

Majumdar, S., Ramachandran, S., and Cerione, R. A. (2004). Perturbing the linker regions of the $\alpha$-subunit of transducin: A new class of constitutively active GTP-binding proteins. *J. Biol. Chem.* **279,** 40137–40145.

Marin, E. P., Krishna, A. G., Zvyaga, T. A., Isele, J., Siebert, F., and Sakmar, T. P. (2000). The amino terminus of the fourth cytoplasmic loop of rhodopsin modulates rhodopsin-transducin interaction. *J. Biol. Chem.* **275,** 1930–1936.

Marin, E. P., Krishna, A. G., Archambault, V., Simuni, E., Fu, W. Y., and Sakmar, T. P. (2001a). The function of interdomain interactions in controlling nucleotide exchange rates in transducin. *J. Biol. Chem.* **276,** 23873–23880.

Marin, E. P., Krishna, A. G., and Sakmar, T. P. (2001b). Rapid activation of transducin by mutations distant from the nucleotide-binding site. Evidence for a mechanistic model of receptor-catalyzed nucleotide exchange by G proteins. *J. Biol. Chem.* **276,** 27400–27405.

Marin, E. P., Krishna, A. G., and Sakmar, T. P. (2002). Disruption of the $\alpha 5$ helix of transducin impairs rhodopsin-catalyzed nucleotide exchange. *Biochemistry* **41,** 6988–6994.

Martin, E. L., Rens-Domiano, S., Schatz, P. J., and Hamm, H. E. (1996). Potent peptide analogues of a G protein receptor-binding region obtained with a combinatorial library. *J. Biol. Chem.* **271,** 361–366.

Mazzoni, M. R., and Hamm, H. E. (1996). Interaction of transducin with light-activated rhodopsin protects it from proteolytic digestion by trypsin. *J. Biol. Chem.* **271,** 30034–30040.

McIntire, W. E., MacCleery, G., and Garrison, J. C. (2001). The G protein $\beta$ subunit is a determinant in the coupling of $G_s$ to the $\beta_1$-adrenergic and A2a adenosine receptors. *J. Biol. Chem.* **276,** 15801–15809.

McLaughlin, J. N., Shen, L., Holinstat, M., Brooks, J. D., Dibenedetto, E., and Hamm, H. E. (2005). Functional selectivity of G protein signaling by agonist peptides and thrombin for the protease-activated receptor-1. *J. Biol. Chem.* **280,** 25048–25059.

Milligan, G. (2004). G protein-coupled receptor dimerization: Function and ligand pharmacology. *Mol. Pharmacol.* **66,** 1–7.

Mixon, M. B., Lee, E., Coleman, D. E., Berghuis, A. M., Gilman, A. G., and Sprang, S. R. (1995). Tertiary and quaternary structural changes in $G_{i\alpha 1}$ induced by GTP hydrolysis. *Science* **270,** 954–960.

Mukhopadhyay, S., and Howlett, A. C. (2005). Chemically distinct ligands promote differential $CB_1$ cannabinoid receptor-Gi protein interactions. *Mol. Pharmacol.* **67,** 2016–2024.

Myung, C. S., Lim, W. K., Defilippo, J., Yasuda, H., Neubig, R., and Garrison, J. C. (2005). Regions in the G protein $\gamma$ subunit important for interaction with receptors and effectors. *Mol. Pharmacol.* **69,** 877–887.

Nanoff, C., Koppensteiner, R., Yang, Q., Fuerst, E., Ahorn, H., and Freissmuth, M. (2006). The carboxyl terminus of the G$\alpha$-subunit is the latch for triggered activation of heterotrimeric G proteins. *Mol. Pharmacol.* **69,** 397–405.

Natochin, M., Muradov, K. G., McEntaffer, R. L., and Artemyev, N. O. (2000). Rhodopsin recognition by mutant $G_s\alpha$ containing C-terminal residues of transducin. *J. Biol. Chem.* **275,** 2669–2675.

Natochin, M., Moussaif, M., and Artemyev, N. O. (2001). Probing the mechanism of rhodopsin-catalyzed transducin activation. *J. Neurochem.* **77,** 202–210.

Natochin, M., Gasimov, K. G., Moussaif, M., and Artemyev, N. O. (2003). Rhodopsin determinants for transducin activation: A gain-of-function approach. *J. Biol. Chem.* **278,** 37574–37581.

Nikiforovich, G. V., and Marshall, G. R. (2003). Three-dimensional model for meta-II rhodopsin, an activated G-protein-coupled receptor. *Biochemistry* **42,** 9110–9120.

Noel, J. P., Hamm, H. E., and Sigler, P. B. (1993). The 2.2 Å crystal structure of transducin-$\alpha$ complexed with GTP$\gamma$S. *Nature* **366,** 654–663.

Oldham, W. M., Van Eps, N., Preininger, A. M., Hubbell, W. L., and Hamm, H. E. (2006). Mechanism of the receptor-catalyzed activation of heterotrimeric G proteins. *Nat. Struct. Mol. Biol.* **13,** 772–777.

Oliveira, L., Paiva, A. C., and Vriend, G. (1999). A low resolution model for the interaction of G proteins with G protein-coupled receptors. *Protein Eng.* **12,** 1087–1095.

Onrust, R., Herzmark, P., Chi, P., Garcia, P. D., Lichtarge, O., Kingsley, C., and Bourne, H. R. (1997). Receptor and $\beta\gamma$ binding sites in the $\alpha$ subunit of the retinal G protein transducin. *Science* **275,** 381–384.

Osawa, S., and Weiss, E. R. (1995). The effect of carboxyl-terminal mutagenesis of $G_t\alpha$ on rhodopsin and guanine nucleotide binding. *J. Biol. Chem.* **270,** 31052–31058.

Pereira, R., and Cerione, R. A. (2005). A switch 3 point mutation in the $\alpha$ subunit of transducin yields a unique dominant-negative inhibitor. *J. Biol. Chem.* **280,** 35696–35703.

Perez, D. M., and Karnik, S. S. (2005). Multiple signaling states of G-protein-coupled receptors. *Pharmacol. Rev.* **57,** 147–161.

Perez, D. M., Hwa, J., Gaivin, R., Mathur, M., Brown, F., and Graham, R. M. (1996). Constitutive activation of a single effector pathway: Evidence for multiple activation states of a G protein-coupled receptor. *Mol. Pharmacol.* **49,** 112–122.

Posner, B. A., Mixon, M. B., Wall, M. A., Sprang, S. R., and Gilman, A. G. (1998). The A326S mutant of $G_{i\alpha1}$ as an approximation of the receptor-bound state. *J. Biol. Chem.* **273,** 21752–21758.

Richardson, M., and Robishaw, J. D. (1999). The $\alpha_{2A}$-adrenergic receptor discriminates between $G_i$ heterotrimers of different $\beta\gamma$ subunit composition in Sf9 insect cell membranes. *J. Biol. Chem.* **274,** 13525–13533.

Rondard, P., Iiri, T., Srinivasan, S., Meng, E., Fujita, T., and Bourne, H. R. (2001). Mutant G protein $\alpha$ subunit activated by $G\beta\gamma$: A model for receptor activation? *Proc. Natl. Acad. Sci. USA* **98,** 6150–6155.

Schmidt, C. J., Thomas, T. C., Levine, M. A., and Neer, E. J. (1992). Specificity of G protein $\beta$ and $\gamma$ subunit interactions. *J. Biol. Chem.* **267,** 13807–13810.

Schwindinger, W. F., Miric, A., Zimmerman, D., and Levine, M. A. (1994). A novel $G_s\alpha$ mutant in a patient with albright hereditary osteodystrophy uncouples cell surface receptors from adenylyl cyclase. *J. Biol. Chem.* **269,** 25387–25391.

Sheikh, S. P., Zvyaga, T. A., Lichtarge, O., Sakmar, T. P., and Bourne, H. R. (1996). Rhodopsin activation blocked by metal-ion-binding sites linking transmembrane helices C and F. *Nature* **383,** 347–350.

Simon, M. I., Strathmann, M. P., and Gautam, N. (1991). Diversity of G proteins in signal transduction. *Science* **252,** 802–808.

Slessareva, J. E., Ma, H., Depree, K. M., Flood, L. A., Bae, H., Cabrera-Vera, T. M., Hamm, H. E., and Graber, S. G. (2003). Closely related G-protein-coupled receptors use multiple and distinct domains on G-protein $\alpha$-subunits for selective coupling. *J. Biol. Chem.* **278,** 50530–50536.

Slusarz, R., and Ciarkowski, J. (2004). Interaction of class A G protein-coupled receptors with G proteins. *Acta Biochim. Pol.* **51,** 129–136.

Sondek, J., Bohm, A., Lambright, D. G., Hamm, H. E., and Sigler, P. B. (1996). Crystal structure of a G-protein $\beta\gamma$ dimer at 2.1 Å resolution. *Nature* **379,** 369–374.

Sullivan, K. A., Miller, R. T., Masters, S. B., Beiderman, B., Heideman, W., and Bourne, H. R. (1987). Identification of receptor contact site involved in receptor-G protein coupling. *Nature* **330,** 758–760.

Taylor, J. M., Jacob-Mosier, G. G., Lawton, R. G., Remmers, A. E., and Neubig, R. R. (1994). Binding of an $\alpha_2$ adrenergic receptor third intracellular loop peptide to G$\beta$ and the amino terminus of G$\alpha$. *J. Biol. Chem.* **269,** 27618–27624.

Taylor, J. M., Jacob-Mosier, G. G., Lawton, R. G., VanDort, M., and Neubig, R. R. (1996). Receptor and membrane interaction sites on G$\beta$. A receptor-derived peptide binds to the carboxyl terminus. *J. Biol. Chem.* **271,** 3336–3339.

Teller, D. C., Okada, T., Behnke, C. A., Palczewski, K., and Stenkamp, R. E. (2001). Advances in determination of a high-resolution three-dimensional structure of rhodopsin, a model of G-protein-coupled receptors (GPCRs). *Biochemistry* **40,** 7761–7772.

Thomas, T. C., Schmidt, C. J., and Neer, E. J. (1993). G-protein $\alpha_o$ subunit: Mutation of conserved cysteines identifies a subunit contact surface and alters GDP affinity. *Proc. Natl. Acad. Sci. USA* **90,** 10295–10298.

Thomas, T. O., Bae, H., Medkova, M., and Hamm, H. E. (2001). An intramolecular contact in Ga transducin that participates in maintaining its intrinsic GDP release rate. *Mol. Cell Biol. Res. Commun.* **4,** 282–291.

Wall, M. A., Coleman, D. E., Lee, E., Iniguez-Lluhi, J. A., Posner, B. A., Gilman, A. G., and Sprang, S. R. (1995). The structure of the G protein heterotrimer $G_{i\alpha1}\beta_1\gamma_2$. *Cell* **83,** 1047–1058.

Warner, D. R., and Weinstein, L. S. (1999). A mutation in the heterotrimeric stimulatory guanine nucleotide binding protein $\alpha$-subunit with impaired receptor-mediated activation because of elevated GTPase activity. *Proc. Natl. Acad. Sci. USA* **96,** 4268–4272.

Warner, D. R., Weng, G., Yu, S., Matalon, R., and Weinstein, L. S. (1998). A novel mutation in the switch 3 region of $G_s\alpha$ in a patient with albright hereditary osteodystrophy impairs GDP binding and receptor activation. *J. Biol. Chem.* **273,** 23976–23983.

Wess, J. (1997). G-protein-coupled receptors: Molecular mechanisms involved in receptor activation and selectivity of G-protein recognition. *FASEB J.* **11,** 346–354.

Wess, J. (1998). Molecular basis of receptor/G-protein-coupling selectivity. *Pharmacol. Ther.* **80,** 231–264.

West, R. E., Jr., Moss, J., Vaughan, M., Liu, T., and Liu, T. Y. (1985). Pertussis toxin-catalyzed ADP-ribosylation of transducin. Cysteine 347 is the ADP-ribose acceptor site. *J. Biol. Chem.* **260,** 14428–14430.

Yasuda, H., Lindorfer, M. A., Woodfork, K. A., Fletcher, J. E., and Garrison, J. C. (1996). Role of the prenyl group on the G protein $\gamma$ subunit in coupling trimeric G proteins to $A_1$ adenosine receptors. *J. Biol. Chem.* **271,** 18588–18595.

# SOME MECHANISTIC INSIGHTS INTO GPCR ACTIVATION FROM DETERGENT-SOLUBILIZED TERNARY COMPLEXES ON BEADS

By TIONE BURANDA,* ANNA WALLER,* YANG WU,* PETER C. SIMONS,* SEAN BIGGS,* ERIC R. PROSSNITZ,† AND LARRY A. SKLAR*

*Department of Pathology and Cancer Center, University of New Mexico Health Science Center, Albuquerque, New Mexico 87131;
†Department of Cell Biology and Physiology, University of New Mexico Health Science Center, Albuquerque, New Mexico 87131



*ADVANCES IN PROTEIN CHEMISTRY, Vol. 74*
DOI: 10.1016/S0065-3233(07)74003-2




0065-3233/07 $35.00

## Abstract

The binding of full and partial agonist ligands (L) to G protein–coupled receptors (GPCRs) initiates the formation of ternary complexes with G proteins [ligand–receptor-G protein (LRG) complexes]. Cyclic ternary complex models are required to account for the thermodynamically plausible complexes. It has recently become possible to assemble solubilized formyl peptide receptor (FPR) and $\beta_2$-adrenergic receptor ($\beta_2$AR) ternary complexes for flow cytometric bead-based assays. In these systems, soluble ternary complex formation of the receptors with G proteins allows direct quantitative measurements which can be analyzed in terms of three-dimensional concentrations (molarity). In contrast to the difficulty of analyzing comparable measurements in two-dimensional membrane systems, the output of these flow cytometric experiments can be analyzed via ternary complex simulations in which all of the parameters can be estimated. An outcome from such analysis yielded lower affinity for soluble ternary complex assembly by partial agonists compared with full agonists for the $\beta_2$AR. In the four-sided ternary complex model, this behavior is consistent with distinct ligand-induced conformational states for full and partial agonists. Rapid mix flow cytometry is used to analyze the subsecond dynamics of guanine nucleotide-mediated ternary complex disassembly. The modular breakup of ternary complex components is highlighted by the finding that the fastest step involves the departure of the ligand-activated GPCR from the intact G protein heterotrimer. The data also show that, under these experimental conditions, G protein subunit dissociation does not occur within the time frame relevant to signaling. The data and concepts are discussed in the context of a review of current literature on signaling mechanism based on structural and spectroscopic (FRET) studies of ternary complex components.

## I. Perspectives

G protein–coupled receptors (GPCRs) represent the largest family of transmembrane-signaling molecules in the human genome with about 600 GPCR genes identified so far (Downes and Gautam, 1999; Robishaw and Berlot, 2004; Venter *et al.*, 2001). From their location at the cell surface, GPCRs are activated by interacting with specific ligands (hormones, neurotransmitters, growth factors, glycoproteins, cytokines, odorants, and photons). These stimuli elicit conformational changes in the receptor, leading to binding and activation of intracellular G protein heterotrimers. G proteins consist of 27$\alpha$, 5$\beta$, and 13$\gamma$ subunit proteins (Venter *et al.*, 2001). To date, the G protein family of the human genome consists of 16$\alpha$, 5$\beta$, and 12$\gamma$ known genes (Cabrera-Vera *et al.*, 2003;

Downes and Gautam, 1999). The remaining 11 G$\alpha$ protein subunits are composed of splice variants. G protein $\alpha$ subunits are classified into families (viz., $\alpha_i$, $\alpha_s$, $\alpha_q$, $\alpha_{12}$, and so on) and in many cases can couple to more than one receptor, though with differing affinities (Hall, 2000). The specific combination of $\alpha$, $\beta$ and $\gamma$ subunits dictates not only receptor specificity but also downstream effector specificity, providing the means of targeting specific physiological processes in response to specific external stimuli (Robishaw and Berlot, 2004).

G protein–mediated signal transduction is initiated by binding of a ligand to a GPCR (Christopoulos *et al.*, 2004). Ligand–receptor interaction causes a change in conformation of the latter which elicits a higher affinity toward the G protein heterotrimer, thus forming a ternary complex. In cells, the ternary complex of ligand, GPCR, and heterotrimeric G protein complexes, has a very short transient lifetime, which is regulated by the kinetics of nucleotide exchange between the inactive GDP-bound form and the active GTP-bound form (Hamm, 1998, 2001). The exchange is initiated by the contact between the heterotrimeric G protein entity and a ligand-activated receptor. The ternary complex thus formed survives for the short duration of the dissociative interchange between GDP and GTP. Based on the prevailing model, subsequent steps include the physical dissociation of the G$\alpha$ subunit from the G protein heterotrimer to enable specific interactions between the separated G proteins with effectors (Hamm, 1998, 2001).

The stable ternary complex is formed in the absence of guanine nucleotides (Hamm, 1998, 2001). Factors affecting the equilibrium formation and stability of the ternary complex include variable affinities of ligands for the free receptor, and subsequent coupling to the (nucleotide free) G protein (Burstein *et al.*, 1995, 1996; Lefkowitz *et al.*, 1993). The ternary complexes have been rationalized with the simple ternary complex model (De Lean *et al.*, 1980), and its more complex progeny (extended ternary model, cubic ternary model) (Christopoulos, 2002; Christopoulos *et al.*, 2004; Kenakin, 2001; Kenakin and Morgan, 1989). Much of the experimental work on ternary complexes has involved cell membranes or detergent-solubilized receptors, where the presence of guanine nucleotides is easily regulated (De Lean *et al.*, 1980; Lefkowitz *et al.*, 1993; Shi *et al.*, 2003; Simons *et al.*, 2003, 2004, 2005; Sklar *et al.*, 2000, 2002, 2003).

This current mechanistic understanding of GPCR-initiated reactivity is based on biochemical (Fung, 1983) and x-ray crystallographic data (Boniface *et al.*, 1999; Hamm, 1998, 2001; Lambright *et al.*, 1994, 1996; Sprang, 1997; Tesmer *et al.*, 2005). Much recent work has benefited from the development of fluorescent tags (Hoffmann *et al.*, 2005; Sklar *et al.*, 2002; Tsien, 1998; Zhang *et al.*, 2002) that have enabled the use of fluorescence spectroscopy, fluorescence microscopy (Hein *et al.*, 2005; Hoffmann *et al.*, 2005), fluorometry (Gether *et al.*, 1995; Lohse *et al.*, 2003; Swaminath *et al.*, 2004), and flow cytometry (Simons *et al.*, 2003, 2004, 2005; Sklar *et al.*, 2002) to probe some dynamic aspects of GPCR activation. Other spectroscopic techniques that have been employed included NMR (Abdulaev *et al.*, 2005; Ridge *et al.*, 2006) and EPR (Preininger *et al.*, 2003). In some of the more recent work involving fluorescence resonance energy transfer (FRET) measurements *in vivo*, it has emerged that not all G proteins undergo subunit dissociation (Bunemann *et al.*, 2003; Frank *et al.*, 2005; Hein *et al.*, 2005).

The intent of this chapter is to summarize how particle-based soluble receptor systems report the steps analogous to those that define how membrane GPCRs transduce ligand-initiated activation into intracellular responses. We will review the emergence of detergent-solubilized systems to directly and quantitatively evaluate receptor and nucleotide-free G protein interactions (Simons *et al.*, 2003, 2004). Soluble receptors allow access to both the extracellular ligand-binding site and the intracellular G protein-binding site of the receptor. Two different GPCRs were chosen for these studies. $\beta_2$-adrenergic receptors ($\beta_2$ARs) contribute to cardiovascular and pulmonary function, and on binding with catecholamines activate G$\alpha_s$ proteins. The *N*-formyl peptide receptor (FPR) binds to *N*-formyl methionine-containing peptides resulting from bacterial and mitochondrial protein synthesis, as well as other hydrophobic peptides (Gao *et al.*, 1994). The prototypical ligand, *N*-formyl-Met-Leu-Phe, is capable of initiating leukocyte inflammatory responses such as adhesion, chemotaxis, superoxide generation, and degranulation (Browning *et al.*, 1997). The receptor systems have been solubilized and assembled with ligands and G proteins to form a high agonist-affinity ternary complex, both in solution (Basani *et al.*, 2001) and on beads, and investigated by flow cytometry (Simons *et al.*, 2003). These detergent-solubilized complexes preserve native ligand-binding affinity and exhibit evidence of appropriate G protein assembly that discriminates full and partial agonists (Simons *et al.*, 2003).

The extent to which detergent-solubilized molecular assemblies have enabled the elucidation of some mechanistic aspects of GPCR activation, outside the complexity of cellular machinery, is highlighted in this chapter. Material is arranged in this chapter according to the following operational logic: (a) What is the nature of detergent-solubilized GPCR molecular assemblies? (b) How is the ternary complex model applicable to the analysis of detergent-solubilized molecular assemblies, with respect to the comparative effects of strong agonists and weak agonists? (c) What are the key early steps involved in nucleotide-mediated disassembly of detergent-solubilized ternary complexes?

## II. Survey of Experimental Approaches and Representative Data

### A. *Flow Cytometric Approaches to Assess GPCR Function* In Vivo

Flow cytometry is a well-established technique that enables the continuous homogeneous assaying of cell- or particle-associated fluorescence without separation of unbound fluorophore (i.e., wash steps) (Nolan *et al.*, 1999; Sklar *et al.*, 2002). Analysis of real-time events and fixed time points can be used to assess ligand–receptor interactions as well as receptor processing and receptor-mediated cell activation. Generally, the behavior of a GPCR that is expressed on the surface of a cell or immobilized on a bead can be probed by: (a) binding of a fluoresceinated ligand, (b) binding of fluorescently tagged antibodies, and (c) cell expression (or surface coverage on a bead) of a GPCR–GFP (green-fluorescence protein) fusion. Because of the ready availability of fluorescent ligands displaying affinities from micromolar to picomolar, the FPR is a model system where flow cytometry has been used to study ligand binding on cells, membranes, and detergent-solubilized molecular assemblies.

An important parameter is the affinity of the derivatized ligand as poor affinities ($K_d$ values greater than ~50 nM) can lead to excessive background (nonspecific) fluorescence signals. Nonspecific binding is determined by incubating the cells with excess nonfluorescent ligand. Alternatively, the nonspecific binding can be analyzed by incubation of the parental cells, which lack the transfected receptor, in the presence of fluoresceinated peptide alone. Further information regarding states of the receptor can be obtained through the study of permeabilized cells in which the states of the receptor can be manipulated (Gilbert *et al.*, 1999; Simons *et al.*, 2005). Absolute numbers of receptors expressed on cells can be determined by comparing the fluorescence intensity to reference beads containing known amounts of fluorescein- or antibody-binding sites, such as Simply Cellular calibration beads (Bangs Laboratories, Fiskers, IN) (Nolan *et al.*, 1999; Sklar *et al.*, 2002). The measurements of FPR ligand affinities were made via soluble, competitive binding assays with fluorescent and nonfluorescent ligands, and yielded ligand affinities for solubilized receptors that correspond with whole cell or membrane-bound receptor assays (Sklar *et al.*, 1985).

### B. *Rapid Mix Flow Cytometry*

Typical kinetic flow cytometer measurements require the removal of the sample tube from the cytometer, manual reagent mixing, and resumption of data collection, resulting in a dead time of 5–10 s or more (Buranda *et al.*, 1999; Chigaev *et al.*, 2001). To minimize the dead time, several online mixing schemes have been developed. These include approaches capable of subsecond mixing and delivery of reagents to the cytometer (Sklar *et al.*, 2002). The authors have previously assembled ternary complexes of soluble GPCRs on beads to examine individual steps associated with GTP-mediated ternary complex disassembly. These studies suggested that the slow dissociation of G$\alpha$ and G$\beta\gamma$ subunits was unlikely to be an essential component of cell activation (Simons *et al.*, 2003). These studies did not have adequate time resolution to define precisely the disassembly kinetics. A syringe-driven rapid mix system that is optimized for small sample sizes ($<$100 $\mu$l) and subsecond mixing of reagents has been developed (Wu, 2004; Wu *et al.*, 2005a,b, 2007a). This advance now makes it possible to study the early events associated with guanine nucleotide activation of GPCR assemblies on beads. Recent data obtained with the rapid mix flow cytometer has enabled the authors to analyze the subsecond dynamics of ternary complex disassembly (Wu, 2004; Wu *et al.*, 2007a). The data have shown that the fastest molecular separation event associated with guanine nucleotide activation of ternary complexes involves the GPCR-G$\alpha$ interaction (*vide infra*).

### C. *Modular Assembly of Molecular Complexes on Beads*

A series of empirical schemes targeting the assembly of ternary complexes on beads have been attempted with varying degrees of success (reviewed in Simons *et al.*, 2005). In Fig. 1, the molecular assemblies that are the subject of this chapter are based on the tethering of G protein heterotrimers to beads by an epitope tag (either hexahistidine or FLAG$^{TM}$) fused to the N-terminus of the $\gamma$ subunit, and projected away from its binding partners in three-dimensional reconstructions.

Modular construction of the ternary complex begins with the epitope tag immobilization of G$\beta\gamma$ subunits on beads. Scheme A represents the display of G$\beta_i\gamma_2$ ($i = 1, 4$) subunits tethered to the bead, by an epitope tag (hexahistidine or FLAG), with soluble $\alpha_i$ subunit completing the heterotrimer assembly (G-beads). The typical G-bead sample preparation involves the initial capture and immobilization of the $\beta\gamma$ subunits, on beads bearing chelated nickel or biotinylated M2 anti-FLAG antibodies. The bead samples are then washed and resuspended in buffer. For the anti-FLAG

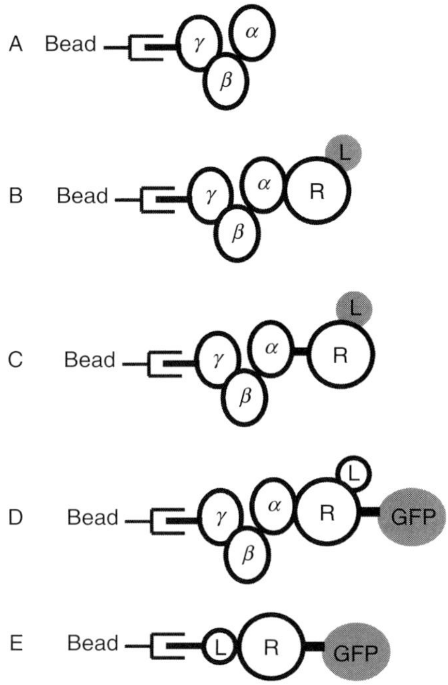


FIG. 1. Various schematics of bead display for molecular assemblies on beads. The $\beta\gamma$ subunits of the G protein (circles labeled with $\beta$ and $\gamma$) are fused with either FLAG or hexahistidine tag, which recognizes the biotinylated M2 anti-FLAG antibodies on streptavidin-coated beads or chelated nickel on the dextran-treated beads. A socket and plug connecter is utilized to depict the very high-affinity interaction of the epitope tag. This modular setup allows for either $\alpha_i$ subunit (for capturing FPR-GFP) or $\alpha_s$ subunit (for capturing $\beta_2$AR-GFP) to be coupled with the $\beta\gamma$ subunit to form the complete G protein coating the bead. Fluorescent components such as GFP or ligand are indicated in green. See text for details. (See Color Insert.)

beads, the surface density of the $\beta\gamma$ subunits can be quantitatively determined on the basis of a standard calibration protocol that relies on a fluorescently labeled FLAG peptide. Briefly, the number of M2 antibody sites on the anti-FLAG beads has been measured to be on the order of 9 million by parallel flow cytometry and centrifugation assays of a fluorescent FLAG peptide (Buranda *et al.*, 2001; Simons *et al.*, 2003). The site density of subunits is then derived from analyzing the fluorescence intensity of calibrated anti-FLAG beads (i.e., beads with known maximal surface density of fluorescent FLAG peptide sites) relative to those beads that are covered with a mixture of nonfluorescent FLAG tagged $\beta\gamma$ subunits using the fluorescent FLAG peptide.

Scheme B shows wild-type FPR and a fluorescent ligand ($L^F$) forming an $L^F$RG assembly on the bead. The superscripted *F* refers to the fluorescent component of the molecular assembly; the nomenclature, $L^F$, $R^F$ has been adopted to be consistent with previous literature (Simons *et al.*, 2003a). In a prototypical assembly, the surface coverage of $\beta\gamma$ can vary from 45% to 70% of the total available 9 million anti-FLAG-binding sites depending on the stoichiometry of the beads and G$\beta\gamma$ subunits (Wu, 2004; Wu *et al.*, 2007a,b). G$\alpha$ subunits are added to an aliquot of $\beta\gamma$-beads at 4 °C for 1 h. The beads are washed and resuspended in appropriate buffer, mixed with detergent-solubilized GPCR and ligand for 2 h at 4 °C, and analyzed at the flow cytometer (Simons *et al.*, 2003). The quantitation of the ternary complexes on beads displaying G protein is carried out using standard calibration beads. For example, the typical surface coverage of a wild-type FPR ($L^F$RG) assembly is in the 20–30,000 range with $\alpha_{i3}$ versus <10,000 sites using $\alpha_{i2}$, where the initial concentration of the solubilized GPCR is typically <100 nM. The low level of total binding may be related to the $K_d$ of the soluble ternary complex compared to the concentration of the soluble receptor, and the presence of G proteins in the soluble receptor preparation.

Scheme C shows the assembly of a fusion protein of FPR with the $\alpha_{i2}$ subunit, and a fluorescent ligand, combined to form an $L^F$R-$\alpha_{i2}\beta_1\gamma_2$ assembly. Scheme D utilizes a fusion protein of FPR with enhanced green fluorescent protein ($R^F$) bound to a nonfluorescent ligand to form an L$R^F$G assembly on the bead. All assemblies involving the $\beta_2$AR were in the Scheme D format due to the unavailability of fluorescent ligands. It is worth noting that there is a single literature report known to the authors, of a fluorescent $\beta_2$AR ligand where an Alexa-532-labeled norepinephrine (NE) was synthesized (Hegener *et al.*, 2004). Despite several attempts, the compound our group synthesized was not functional.

$\beta_2$AR-GFP, FPR, and FPR-GFP solubilized in 1% dodecyl maltoside (DOM) bind both to their ligands and G protein (Bennett *et al.*, 2001; Simons *et al.*, 2003, 2004; Sklar *et al.*, 2000). The GPCR-GFP chimerae's binding affinity to G protein is not significantly affected by the presence of the fused GFP protein and is comparable to the wild-type FPR in signaling, trafficking, and assembly (Bennett *et al.*, 2001; Key *et al.*, 2001; Shi *et al.*, 2003).

Detergent-solubilized receptors originate from cells that are transfected to overexpress the receptor of choice. The unpurified solubilized extracts contain other proteins including endogenous G proteins that are present in much higher concentrations than even the overexpressed target receptor. In normal cells, the expression of G proteins can be as much as 10–100 times higher than receptors (Ransnas and Insel, 1988). For the overexpressed receptor systems described here, it was determined by Western blots that the amount of $G\alpha_s$ for the $\beta_2$AR-GFP-expressing cells was about three times as high as $\beta_2$AR-GFP (i.e., 300 nM $G\alpha_s$ vs 100 nM $\beta_2$AR-GFP in the soluble extract), whereas the soluble $G\alpha_i$ concentration was estimated to be two times higher than the FPR-GFP for the FPR-GFP-expressing cells (i.e., 200 nM $G\alpha_i$ vs 100 nM FPR).

The interactions between GPCRs and ligands (agonists or partial agonists) were quantitatively evaluated by spectrofluorometry, for FPR (Simons *et al.*, 2003), and by flow cytometry for $\beta_2$AR (Simons *et al.*, 2004). The determination of FPR ligand affinities was accomplished through a soluble ligand competition assay (Simons *et al.*, 2003). Soluble FPR was equilibrated with fluorescently tagged ligand, *N*-formyl-methionyl-leucyl-phenylalanyl-lysine-FITC (fMLFK-FITC). The relative concentrations of bound and free ligand were determined using an anti-FITC antibody–binding assay (Sklar *et al.*, 1981). The assay can resolve the fraction of bound from free ligand on the basis of the fast rate by which the free ligand is quenched by the antibody, relative to the bound ligand, which is slower (cf. Fig. 6A *vide infra*). The affinities of nonfluorescent FPR ligands (Table I) were calculated from their competitive binding curves with fMLFK-FITC for FPR, together with a measured $K_d$ of 4.8 nM for fMLFK-FITC (Simons *et al.*, 2003). The FPR ligands were *N*-formyl-methionyl-leucine (fML), *N*-formyl-methionyl-leucyl-phenylalanine (fMLF), and *N*-formyl-methionyl-leucyl-phenylalanyl-phenylalanine (fMLFF).

The quantitative flow cytometric assay for evaluating $\beta_2$AR ligand affinities utilizes ligand-coated beads (ligand beads) to display the binary complex (Scheme E in Fig. 1). Dihydroalprenolol (DHA), an antagonist for $\beta_2$AR, was conjugated to the surface of dextran beads (Simons *et al.*, 2004). Solubilized $\beta_2$AR-GFP bound to DHA, resulting in fluorescent beads, and a $K_d$ of 3 nM was obtained. The addition of soluble ligands to the suspension of ligand beads and soluble $\beta_2$AR-GFP resulted in decreased bead fluorescence. The soluble ligand affinities were calculated from these competition curves. The $\beta_2$AR ligands used in this study were isoproterenol (ISO), epinephrine (EPI), norepinephrine(NE), salbutamol (SAL), and dobutamine (DOB).

The characteristics of the dose–response curves (Fig. 3) for receptors binding to the G protein–coated beads (G-beads) as reported by Simons *et al.* (2003, 2004) are summarized and listed in Table I. It is worth noting that the maximum site coverage of complexes varied with full and partial agonists, and that no receptor–G protein complexes were detected in the absence of ligand, under these experimental conditions. Ternary complex assembly followed schemes shown in Fig. 1D. Our characterization of full and partial agonist was based on the ability to form a ternary complex on beads and compared well to other studies that examined the ability of these ligands to elicit cellular responses such as cAMP production (Kent *et al.*, 1980). The FPR ligand *N*-formyl-methionyl-phenylalanine (fMF) has

TABLE I
Molecular Assembly Data from Soluble Ligand, Ligand-Bead, and G-Bead Data

| Receptor | Ligand | Receptor affinity[a] ($\mu$M) | EC50 on G-bead[b] ($\mu$M) | Max receptor on G-bead[c] (No. per bead) | Classification[d] |
|---|---|---|---|---|---|
| $\beta_2$AR | No ligand | | | No precoupling | |
| | ISO | 0.22 | 0.10 | 32,700 | Full |
| | EPI | 0.68 | 0.18 | 27,500 | Full |
| | NE | 19 | 21 | 30,900 | Full |
| | SAL | 2.3 | 1.1 | 16,500 | Partial |
| | DOB | 2.4 | 3.2 | 3700 | Partial |
| FPR | No ligand | | | No precoupling | |
| | fMLFF | 0.056 | 0.013 | 18,400 | Full |
| | fMLF | 2.6 | 0.26 | 18,900 | Full |
| | fML | 37 | 4.9 | 13,000 | ~Partial |

[a]Evaluated from $K_i$ values based on data from DHA-beads for $\beta_2$AR with $K_d$ of DHA set at 3 nM (for ISO, EPI, NE, SAL, and DOB) or from soluble competitive binding assays for FPR with $K_d$ of fMLFF-FITC set at 4.8 nM (for fMLFF, fMLF, and fML).

[b]Evaluated from sigmoidal dose–response curve fitted to G-bead assembly data, $G\alpha_s$-bead for $\beta_2$AR-GFP and $G\alpha_i$-bead for FPR-GFP.

[c]Based on maximum mean channel number for a bead evaluated for highest ligand concentration and calibrated to molecular equivalent soluble fluorescence via Bang Labs calibration beads.

[d]General classification of agonists as full or partial depending on their ability to elicit cellular responses.

Summary of ligand, G protein, and receptor-binding data from previous studies with receptor proteins extracted from U937 cell membranes and solubilized in 1% DOM (Simons *et al.*, 2003, 2004).

previously been proposed to be a partial agonist (Sklar *et al.*, 1985) and the fML homologue elicited somewhat lower FPR-GFP coverage of the $G\alpha_i$-beads compared to the other ligands (Simons *et al.*, 2003).

## III. Analysis of Soluble Receptor Ternary Complex Assemblies

### A. *General Considerations*

Historically, the development of the model of GPCR activation began with the postulate of a "collisional coupling" model, which was based on the premise that the encounters between receptors and G proteins occur under conditions of free lateral diffusion within the cell membrane, and, therefore, only activated receptors could interact productively with

G proteins (Orly and Schramm, 1976). The collision coupling model was later modified to account for how the signal from the receptor to its cognate effector could be rapidly and specifically relayed (Neubig, 1994). As a result, the "precoupled model" which proposes that receptor and G protein may be coupled even in the absence of agonist (Neubig *et al.*, 1988) was put forth in part to account for the signaling specificity. Later models sought to account for the rapid signaling kinetics by postulating that receptors, their G proteins, and effectors are compartmentalized in microdomains (Head *et al.*, 2005; Insel *et al.*, 2005a,b; Neubig, 1994; Ostrom and Insel, 2004) where they diffuse freely and couple on agonist stimulation; however, only signaling compounds within one such domain should be able to interact with each other, thereby providing the necessary specificity. The concentrations of signaling components should be locally very high and thus permit subsecond kinetics observed with GPCR signaling (Gross and Lohse, 1991; Hein *et al.*, 2005).

Ternary complex assembly on beads allows direct quantitative measurements, which can be analyzed in terms of measurable concentrations of soluble components. This approach allows one to circumvent the near-intractable problem of attempting to define receptor–protein concentrations in the cell membranes, where spatial interactions are lateral and, therefore, two-dimensional. In addition, the lateral distribution of receptor systems is heterogeneous due to microdomains (Head *et al.*, 2005; Insel *et al.*, 2005a,b; Neubig, 1994; Oh and Schnitzer, 2001; Ostrom and Insel, 2004; Simons and Ikonen, 1997; Simons and Toomre, 2000), variable stoichiometries in the expression levels of GPCRs and G proteins (Hakak *et al.*, 2003; Ransnas and Insel, 1988), as well as the growing belief that GPCRs form oligomers (Nobles *et al.*, 2005). All of this complexity makes it difficult to truly test the applicability of a simple model in cell membranes because the essential equilibrium-binding parameters cannot be extracted from experiments. The bead systems facilitate the independent measurement of affinities associated with agonists (L), GPCRs (R), and G proteins (G) because the reagent stoichiometry and concentrations are well defined. We are thus able to use the experimentally derived data to test the applicability of the simple ternary complex model.

### *B. Simple Ternary Complex Model: General Considerations*

The simple ternary complex model (Fig. 2A) describes the binding of ligands to GPCRs, leading to the activation of G protein (De Lean *et al.*, 1980; Lefkowitz *et al.*, 1993). This model is based on the four equilibrium reactions that account for all of the thermodynamically possible interactions between the three species: ligand, GPCR, and G protein.

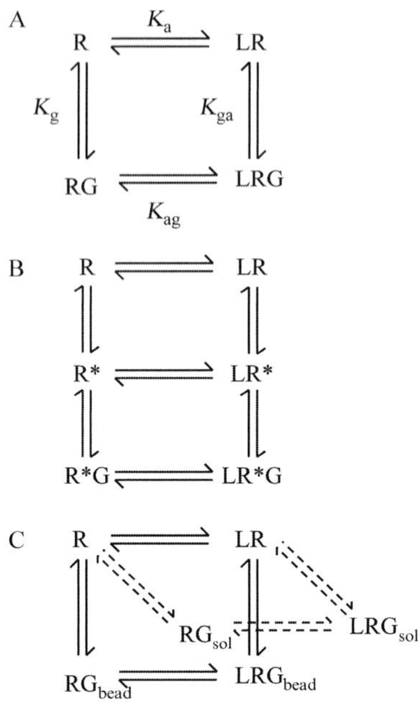


FIG. 2. Diagram of ternary complex models. (A) Simple ternary complex model or four-sided model includes the plausible interactions between receptor (R), ligand (L), and G protein (G). The equilibrium dissociation constants governing the different interactions are: $K_a$ for interacting R and L, $K_g$ for interacting R and G resulting in precoupled receptors, $K_{ga}$ for interacting LR and G, and $K_{ag}$ for interacting RG and A. (B) Extended ternary complex model or seven-sided model with isomerization of receptor from inactive (R) to active (R*) states, both of which can bind with agonist (A). However, only the active state R* can interact with G protein (G). (C) Schematic of the ternary complex model with the addition of soluble G protein ($G_{sol}$) competing for the soluble receptor with the bead-bound G protein ($G_{bead}$). The labeling scheme for the equilibrium dissociation constants of this model is the same as in panel A. Note flow cytometric experimental data only measures fluorescently labeled receptors bound on bead (i.e., $RG_{bead}$ + $LRG_{bead}$).

The formation of the ternary complex of ligand-receptor-G protein is accomplished via two pathways. One starts with the binding of receptor (R) to agonist (L) and the other with the binding of receptor to G protein (G). The affinities describing these binary interactions are independent of each other, while the affinities describing the subsequent interactions with G protein or ligand for the different pathways, respectively, are governed by the constraints imposed by microscopic reversibility. The four different binding states of the receptor depicted in this ternary complex model need not formally represent fixed conformations, but could reflect a distribution of conformations such that a given equilibrium dissociation constant represents the average binding energy of that distribution. The equilibrium dissociation constants describing these affinities are listed in Eqs. (1)–(4). $K_a$ and $K_g$ are unconditional and independent parameters involved in bimolecular processes. $K_{ga}$ and $K_{ag}$ are conditional in that they describe the binding of a third component to a complex formed by the other two and are interdependent.

$$K_a = \frac{[L][R]}{[LR]} \tag{1}$$

$$K_g = \frac{[R][G]}{[RG]} \tag{2}$$

$$K_{ga} = \frac{K_g}{\alpha} = \frac{[LR][G]}{[LRG]} \tag{3}$$

$$K_{ag} = \frac{K_a}{\alpha} = \frac{[RG][L]}{[LRG]} \tag{4}$$

In Eqs. (3) and (4), $\alpha$ is defined as a cooperativity factor that represents the efficiency with which the ligand elicits a cellular response. For positive cooperativity, as displayed by agonists, $\alpha > 1$; for inverse agonists, which exhibit negative cooperativity by hindering receptor activation of G protein, $\alpha < 1$ (Christopoulos, 2002; Christopoulos and Kenakin, 2002; Weiss *et al.*, 1996a,b).

The ternary complex model accounts for the fact that GPCRs exhibit distinct ligand affinities associated with the binary LR complex and the ternary LRG complexes. While the LRG assembly propagates the cellular response, the model also allows for intrinsic or constitutive activation of G proteins independent of ligand binding due to the formation of RG complexes (Lefkowitz *et al.*, 1993).

An extended (seven-sided) ternary complex model (Fig. 2B) was proposed by Samama *et al.* (1993) to accommodate mutant receptors that exhibited constitutive activity and to link receptor affinity with efficacy. This model includes the isomerization of the receptor between two conformational states, inactive (R) and active (R*), and only allows for the active R* conformational state to interact with the G protein. Conceptually, the model allows for the receptor to toggle between on and off states where ligand or G protein manipulates the population size of these two conformational states, rather than affecting the activation strength of a particular conformational state. The different types of ligands influence this equilibrium between the two receptor states. Activating ligands or agonists bind strongly to the R* state, while nonactivating ligands such as inverse agonists preferably bind with the R state. In this model, the distinction between agonist eliciting full responses versus partial responses lies in their different binding affinities for the R* state, with a full agonist having a higher affinity relative to its affinity for R than a partial agonist. Although both the four-sided and the seven-sided models provide mechanistic insight for ligand receptor–G protein interactions, application of either model to an experimental system has been hindered by the inability to quantitatively measure receptor–G protein interactions in membranes. In this soluble receptor system, there is the potential to visualize the alternate pathway directly, where $K_g$ is the receptor affinity for G protein in the absence of ligand, and $K_{ag}$ is the affinity of precoupled receptor (RG) for ligand.

A measurement system that is able to quantitatively determine the interactions of receptor and G protein has the potential for more detailed testing of ternary complex models. The soluble receptor systems, ($\beta_2$AR and FPR) described in Section II, allow for the direct and quantitative evaluation of receptor and G protein interactions (Simons *et al.*, 2003, 2004). Soluble receptors allow access to both the extracellular ligand-binding site and the intracellular G protein-binding site of the receptor. As the site densities on the particles are typically lower than those that support rebinding (Goldstein *et al.*, 1989), simple three-dimensional concentrations are appropriate for the components. Thus, by applying molar units for all the reaction components in the definitions listed in Fig. 2A, the units for the equilibrium dissociation constants are molar, not moles per square meter as for membrane-bound receptor interactions. These assemblies are also suitable for kinetic analysis of ternary complex disassembly.

### C. *Application of Ternary Complex Model*

GPCR interactions with ligand and G protein are represented by the ternary complex formalism (Fig. 2A; Christopoulos and Kenakin, 2002; De Lean *et al.*, 1980; Samama *et al.*, 1993). The quantitative analysis of the soluble assembly system formally requires inclusion of soluble G protein due to the use of a crude receptor preparation. These soluble G proteins compete with the G protein attached to the G-beads for the solubilized receptor as shown in Fig. 2C (Simons *et al.*, 2003, 2004). Experimental values from G-beads (Fig. 3) were fitted with the calculations of bead-bound receptors ($RG_{bead} + ARG_{bead}$) based on this model, which includes soluble G proteins. Simulations were made by Mathematica®, numerically solving the series of equations that describe the model in Fig. 1C for each ligand concentration of the dose–response curve with fixed values for $K_a$ (based on ligand identity) and $K_g$, and a series of varied values of $K_{ga}$. These simulations were compared to experimental data to evaluate $K_g$ and $K_{ga}$, and subsequently the values of $K_{ag}$ were determined based on the thermodynamic constraint ($K_{ag} = K_a \times K_{ga}/K_g$).

It is worth noting that in one pathway toward ternary complex formation, the equilibrium constant $K_a$ represents the receptor affinity for ligand and $K_{ga}$ represents the affinity of ligand-bound receptor for G protein. In this soluble receptor system, there is an opportunity to visualize the alternate pathway directly, where $K_g$ is the receptor affinity for G protein in the absence of ligand, and $K_{ag}$ is the affinity of precoupled receptor (RG) for ligand.

### D. *Landscapes of G-Bead Assemblies Based on Application of Experimentally Derived Binding Constants to the Ternary Complex Model*

Ternary complex simulations including the soluble G protein made using Mathematica® result in three-dimensional landscapes. The landscapes show calculated values of bead-bound receptors ($RG_{bead} + ARG_{bead}$) versus ligand concentration and affinity of the ligand-bound receptor for G protein (log $K_{ga}$). The landscapes were based on fixed values of ligand affinity ($K_a$) and G protein affinity ($K_g$). Examples of two such landscapes for $\beta_2$AR-GFP binding with ISO ($K_a = 220$ nM) for $K_g = 10^{-6}$ M (A) and $10^{-7}$ M (B) are depicted in Fig. 4 in order to illustrate the interplay between the different affinities defining the ternary complex model. The receptor-dependent variable is $K_g$, which describes the affinity of free receptor for G protein and is formally independent of the particular ligand binding to the receptor. The three other equilibrium dissociation variables ($K_a$, $K_{ag}$, and $K_{ga}$) are dependent on the ligand. The specific affinity of receptor for ligand ($K_a$) was experimentally measured and this value was held constant for the specific ligand during fitting to experimental data.

The calculated precoupling of receptor to G protein, formation of RG, was directly affected by the input value of $K_g$. Thus, as the value of $K_g$ decreases (the affinity increases), the number of precoupled receptors increases. This is evident when comparing Fig. 4A with B and there is a significant amount of bead-bound receptors at very low ligand concentration for Fig. 4B ($K_g = 10^{-7}$ M). In Fig. 4A, at a minimal concentration of agonists, log ligand concentration $= -11$, for $K_g = 10^{-6}$ M there is $\sim 2 \times 10^{-11}$ M of RG. In contrast in Fig. 4B when $K_g = 10^{-7}$ M, there is an order of magnitude more RG, $2 \times 10^{-10}$ M. However for $K_g = 10^{-7}$ M (Fig. 4B), at higher $K_{ga}$ values

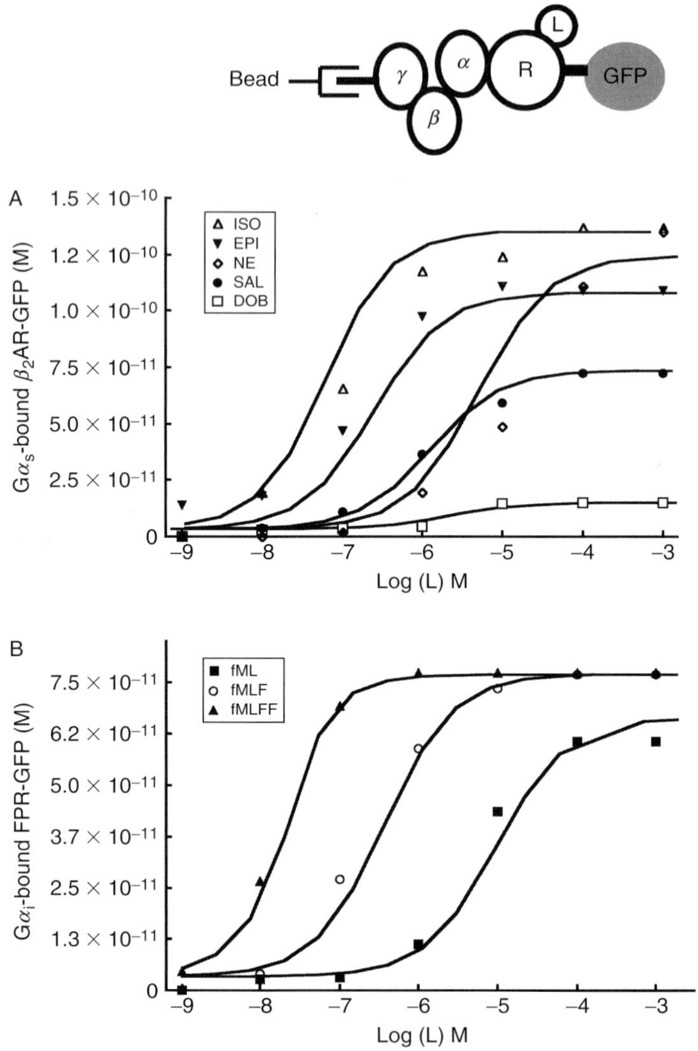


FIG. 3. Experimental dose–response data on G-beads from previous work (Simons *et al.*, 2003, 2004) fitted to simulations of the ternary complex model including soluble G protein (Fig. 1C). The inclusion of soluble G protein in the model (Fig. 1C) is required due to the presence of extra G protein from the solubilized receptors and without which resulted in simulations that overestimated bead-bound receptors. Note that the same equilibrium dissociation constant values were used for the interactions with G protein on bead as with soluble G protein ($Gtot_{bead}$ and $Gtot_{sol}$). Although the individual kinetic reaction rate constants for the interactions with soluble G protein might be faster than those for the bead-bound G protein, their ratios (the equilibrium dissociation constants) are expected to remain the same. The calibrated GFP per bead as

(e.g., log $K_{ga} \cong -5$), there is a *decrease* in the LRG plus RG complexes at higher ligand concentration in the landscape of bead-bound receptors. At this particular combination of dissociation constant and ligand concentration, the majority of receptors are ligand-bound (LR) and very few LRG and RG complexes (the bead-bound receptors) are formed due to the low affinity of LR to G proteins. In other words, when the value of $K_{ga}$ is greater than $K_g$, the ternary complex model simulates "inverse agonism" because mechanistically the binding of the inverse agonist actually decreases the coupling of the receptor to G proteins. Hence, as long as the affinity for the ligand is higher than the affinity for G protein (i.e., $K_g > K_a$), the receptor remains in a ligand-bound state (LR) rather than forming a ternary complex (LRG). This inverse agonism dip in the landscape at high $K_{ga}$ values and high ligand concentrations is not seen in the simulations with higher $K_g$ value (i.e., lower receptor affinity for G protein, Fig. 3A) because there is no significant basal level of precoupled receptors, RG. Note that the maximal levels of bead-bound receptors are the same for both $K_g$ scenarios because the G protein on the bead is the limiting element for complex formation. Only when $K_g$ values are above $\sim 5 \times 10^{-6}$ M do the landscapes correspond to both the experimental dose–response curves and the vanishingly low levels of RG precoupling (see below).

### E. *Ternary Complex Analysis of Soluble Receptor Assemblies*

The ternary complex model with both bead-bound as well as soluble G proteins (Fig. 2C) reproduces the experimental data from the G-beads (Fig. 3). Because the amount of precoupled receptors on beads at even the highest concentrations of receptors (up to 100 nM) is insignificant, a low

evaluated by flow cytometry was converted to molar concentration of $LRG_{bead}$ and $RG_{bead}$. (A) Data for $\beta_2$AR-GFP bound to $G\alpha_s$ protein beads (data points) fitted to simulations (lines) with an appropriate $K_a$ value for the different ligands and the $K_g$ value fixed at $5 \times 10^{-6}$ M for a suitable precoupling (RG) level. The $\beta_2$AR ligands are isoproterenol (ISO), epinephrine (EPI), norepinephrine (NE), salbutamol (SAL), and dobutamine (DOB). (B) Data of bound FPR-GFP to $G\alpha_i$ protein beads (data points) fitted to ternary complex model via simulations (solid lines) with $K_g$ set at $5 \times 10^{-6}$ M and $K_a$ value fixed to specific ligand receptor affinity. The FPR ligands were *N*-formyl-methionyl-leucine (fML), *N*-formyl-methionyl-leucyl-phenylalanine (fMLF), and *N*-formyl-methionyl-leucyl-phenylalanyl-phenylalanine (fMLFF). The qualitative behavior of full agonists and partial agonists was reproduced in at least three separate experiments. The data are fit to a single representative experiment in which all ligands were tested at the same time. (See Color Insert.)

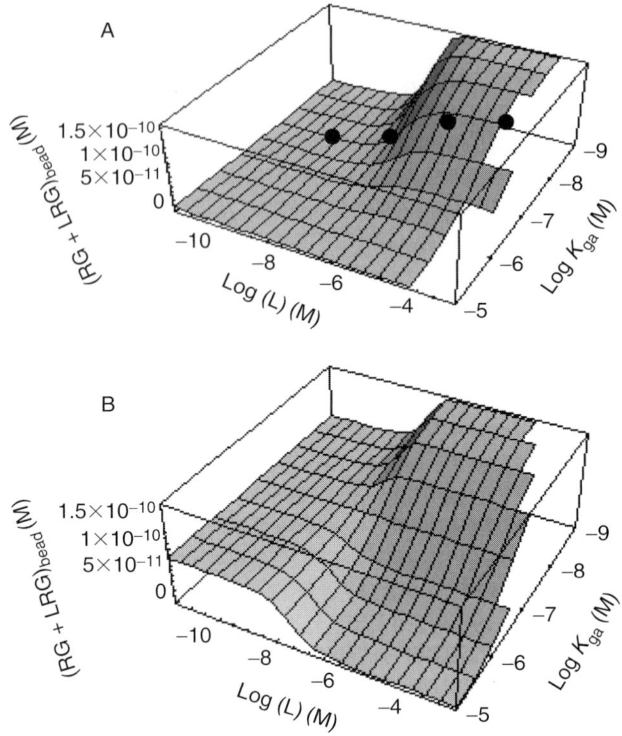


FIG. 4. Landscape of bead-bound receptors based on ternary complex model. Graphs result from simulations of ternary complex model including soluble G protein (shown in Fig. 2D) using Mathematica®. Total concentration of bead-bound receptor ($LRG_{bead}$ + $RG_{bead}$) is graphed versus ligand concentration for various $K_{ga}$ values. (A) Landscape simulation for ISO, using the following parameters; $K_g = 5 \times 10^{-6}$ M, $K_a$ = 220 nM, total receptor concentration of 30 nM, and G protein concentrations on the bead of 0.6 nM and in solution, 100 nM. These concentrations correspond with experimental conditions for the bead suspensions in 10 $\mu$l volumes in 96-well plates. The isotherm corresponding to the experimental data is indicated with dots. The correlation between experiment in Fig. 3 and simulation is clearly shown. (B) $K_g$ set at $10^{-7}$ M. Note the appearance of an inverse agonism dip at higher log $K_{ga}$ values which becomes visible due to the higher concentration of precoupled receptors (RG) when the RG affinity is increased (i.e., lower $K_g$ value compared to panel A). (See Color Insert.)

affinity (large value for $K_g$, greater than 1 $\mu$M) for the interaction between receptors and G protein is required. In addition, this value is independent of the ligands. Thus, the basal level of precoupled receptors for these detergent-solubilized receptors has to be small, so as not to obscure the low level of binding shown by some partial agonists (e.g., DOB). In addition, the value for $K_g$ is a lower limit value that represents the highest affinity of receptor to G protein that could describe the G-bead data.

It is worth noting that to date, $K_g$ has not been directly measured in membranes, but is instead implicitly dependent on R*/R in numerical analyses. However, a recent study, in living cells, has attempted to directly measure the extent of precoupling of a cell transfection titration of YFP-tagged $\alpha_{2A}$ adrenergic receptors ($\alpha_{2A}$-YFP) and CFP-tagged G proteins in the absence of an agonist (Hein *et al.*, 2005). In this study, FRET was used as a measure of the association of the receptor and the cognate G protein heterotrimeric complex. An increased FRET signal was measured in the presence of NE. The dependence of the kinetics as a function of G protein transfection was interpreted in terms of collisional coupling without significant precoupling.

Once the $K_g$ value is estimated for both $\beta_2$AR-GFP and FPR-GFP, each individual ligand dose–response curve could be fitted to calculate $K_{ga}$ with the $K_a$ value fixed to the affinity for that particular ligand (Table I). Figure 3 shows the resulting fits to dose–response curves of (A) $\beta_2$AR-GFP binding to G$\alpha_s$ protein-coated beads with $K_g = 5 \times 10^{-6}$ M and (B) FPR-GFP binding to G$\alpha_i$ protein-coated beads with $K_g = 5 \times 10^{-6}$ M. At these $K_g$ values and experimental conditions, with no ligand present, less than 200 receptors are bound to each G-bead at the detection limit of a flow cytometer.

The calculated values for $K_{ga}$ and $K_{ag}$ corresponding to these fits are listed in Table II. For $\beta_2$AR-GFP, full agonists (ISO, EPI, NE) have similar $K_{ga}$ values (log $K_{ga}$ is approximately $-7.2$), suggesting that once the receptor is bound by a full agonist, the receptor exhibits a conformational state which has a similar, high G protein affinity. On the other hand, when $\beta_2$AR-GFP is bound with a partial agonist (SAL or DOB), the receptor has a lower affinity (i.e., log $K_{ga} > -7.2$) for the G protein. Similarly, for FPR-GFP, full agonists (fMLF and fMLFF) have equivalent $K_{ga}$ values (log $K_{ga} = -7.54$). The potential partial agonist (fML) results in a slightly lower affinity (log $K_{ga} = -7.2$). Additionally, for the $\beta_2$AR ligand ISO, the value for $K_{ga}$ (log $K_{ga} = -7.6$) is similar to the value found experimentally based entirely on competition between soluble and bead-bound G protein (log $K_{ga} \cong -7$) (Simons *et al.*, 2004).

The ternary complex model allows for cooperative interactions among ligand, receptor, and G protein as defined by the cooperativity factor $\alpha$. With a lower limit of $K_g$ established from the levels of RG precoupling, the comparisons of $K_{ag}$ with $K_a$ indicate for both $\beta_2$AR and FPR that precoupled receptors (RG) would exhibit higher affinity for an agonist

TABLE II
Equilibrium Dissociation Constants of Ternary Complex Model from G-Bead Data

| Ligand | Log $K_a$[a] | Log $K_{ga}$[b] | Log $K_{ag}$[c] | Log $\alpha$[d] |
|---|---|---|---|---|
| ISO | −6.66 | −7.6 | −9.0 | 2.3 |
| EPI | −6.16 | −7.27 | −8.13 | 2.0 |
| NE | −4.72 | −7.46 | −6.87 | 2.1 |
| SAL | −5.54 | −6.9 | −7.14 | 1.6 |
| DOB | −5.62 | −5.97 | −6.30 | 0.7 |
| fMLFF | −7.25 | −7.54 | −9.47 | 2.2 |
| fMLF | −5.58 | −7.53 | −7.81 | 2.2 |
| fML | −4.43 | −7.20 | −6.33 | 1.9 |

[a]For ligands ISO through DOB experimentally measured via ligand beads. For fMLFF, fMLF, and fML experimentally measured via competitive binding in a soluble assay.

[b]Evaluated from fitting data to ternary complex model.

[c]Determined from microscopic reversibility.

[d]Calculated using Eqs. 3 and 4, respectively, using derived data in the Table.

Simulations of ternary complex model were performed with total receptor concentration of 30 nM, soluble G protein concentration of 100 nM for $\beta_2$AR or 200 nM for FPR, and bead-bound G protein concentration of 0.6 nM per experimental conditions. Values for $K_a$ were fixed from previously experimental values; see receptor affinity in Table I (Simons *et al.*, 2003, 2004). Best fits for all partial and full agonists were achieved with the value of $K_g$ fixed at $5 \times 10^{-6}$ M (log $K_g = -5.3$), which resulted in insignificant amount of precoupled receptor (RG) in the absence of ligand.

than uncoupled receptors (R). For full agonists, the presence of G protein increases the agonist affinity by at least two orders of magnitude (i.e., $\log \alpha = \log K_a - \log K_{ag} > 2$). This $\alpha$-value would increase as RG affinity (precoupling) decreases. Additionally, this difference corresponds to a change in binding energy $\Delta\Delta G$ of at least $-2.7$ kCal/mol for $\beta_2$AR and $-2.8$ kCal/mol for FPR. For partial agonists, the conformational state of the G protein-bound receptor (RG) is less favorable (i.e., approximately one order of magnitude increased agonist affinity or $\log \alpha \cong 1$ or $\Delta\Delta G < 2$ kCal/mol). The relative magnitudes of the energy differences for full and partial agonists are independent of $K_g$. These differences in ligand affinities ($K_a$ and $K_{ag}$) correspond with findings from Seifert *et al.* (2001) of low- and high-affinity states of $\beta_2$AR. They reported biphasic ligand binding to membrane-bound $\beta_2$AR (e.g., combination of L + R and L + RG) and monophasic, low-affinity ligand binding in the presence of GTP$\gamma$S (e.g., only L + R) (Seifert *et al.*, 2001). In this ternary complex analysis, the conformational state induced by G protein binding to GPCR enhances the ability of a receptor to bind a full agonist as compared with a partial agonist. Due to the thermodynamic constraint of the ternary complex model ($K_a \times K_{ga} = K_g \times K_{ag}$), there is a similar increase in the affinity of ligand-bound receptor (LR) with G protein as compared to free receptor (R) with G protein (i.e., $\log \alpha = \log K_g - \log K_{ga}$). This difference is reduced for partial agonists.

### *F. Are Unique Conformational Changes in Receptors Elicited by Interactions with Ligands Resulting in Varied G Protein Interactions by the Ligand-Bound Receptor?*

It has been suggested that conformational states of GPCRs are selected by the binding of ligands and G proteins. There is growing evidence that GPCR exhibit multiple conformations (i.e., conformational pleiotropy) and that these may initiate different signal transduction pathways (Gether *et al.*, 1995; Krumins and Barber, 1997; Perez *et al.*, 1996; Schwartz *et al.*, 1995; Seifert *et al.*, 2001; Wiens *et al.*, 1998). For example, studies by Schwartz *et al.* have shown that the tachykinin NK1 receptor exhibits conformational states that activate either $G\alpha_q$ or $G\alpha_s$ in response to endogenous ligands such as substance P, neurokinin A and B (Holst *et al.*, 2001; Schwartz and Rosenkilde, 1996). All of these data suggest that ligands induce receptor conformational changes that can result in more than one active state.

Conformational variation in response to different ligands (Gether *et al.*, 1995) also results in variation in the G protein coupling sites for $\beta_2$AR (Ghanouni *et al.*, 2001; Kobilka, 2002). Moreover, an agonist-dependent motion of the sixth transmembrane domain of $\beta_2$AR has been measured by the quenching of labeled Cys-265. Partial agonists (DOB and SAL) only moderately quenched the fluorescence compared with the full agonist (ISO). This result was interpreted in terms of distinct receptor conformational states being stabilized by partial and full agonists. Finally, a comparison of wild-type $\beta_2$AR receptors, constitutively active mutants, and receptors fused with $\alpha$ subunit of the G protein (Seifert *et al.*, 2001) suggested that conformations stabilized by full agonists were indistinguishable from one another other and distinct from those stabilized by the partial agonists.

Analysis of the soluble G-bead assembly provides a complementary classification of full and partial agonists, based on their distinct abilities to assemble ternary complexes (LRG). It appears that the behavior of receptors and entire ligand families can be described by the simple ternary complex model alone (Fig. 2A). The analysis provides estimates for the ligand-dependent equilibrium constants that govern the simple ternary complex model. Unique, potentially intermediate, conformational states of the receptor defined by interactions with a particular ligand are characterized by individual binding constants. While these data do not directly show these different conformational states, the bead system appears to act as a sensor of thermodynamic changes that support the notion that different conformational states are induced by different ligands.

Figure 5 compares the distribution of the conformational states of a receptor elicited by full or partial agonists in the four-sided and seven-sided (extended) ternary complex models. It intends to account for the varied abilities of the states to interact with G proteins. In the four-sided model, there is a gradation of receptor conformational states. Uncoupled receptor, R, binds very weakly with G protein as compared to full agonist-bound receptor ($L_fR$), which exhibits a higher affinity for G protein. Based on the simple ternary complex analysis, a partial agonist bound to a receptor ($L_pR$) induces an intermediate conformation (solid light green line) compared with a full agonist. In contrast, a dual distribution (dashed light green line), with both low-affinity and high-affinity G protein binders, would result from the extended ternary complex model (Fig. 2B). Extending the conformations into additional dimensions in this representation would provide for selectivity for different G proteins.

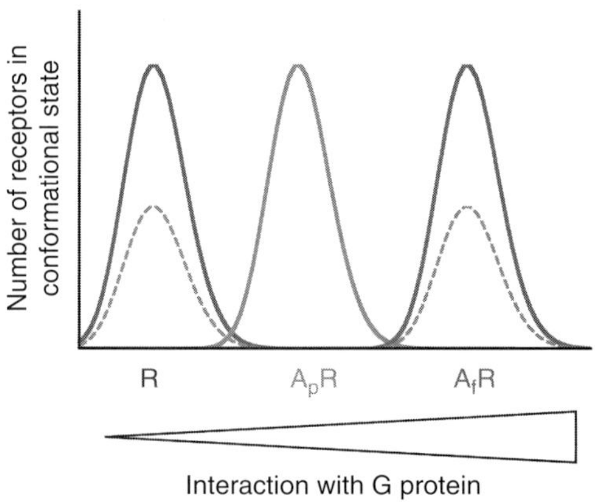


FIG. 5. Conceptual schematic of the receptor conformational states elicited by binding to partial ($L_p$) or full ($L_f$) agonists, and a depiction of the correlation between the various conformational states and their ability to bind with G proteins. Solid lines show the conformational distributions hypothesized from soluble ternary complex data analyzed by the simple ternary complex model. When a partial agonist binds with a receptor ($L_pR$) in this model, the receptor forms a conformational state which has an intermediate affinity for G protein, consequently leading to formation of intermediate amounts of $L_pRG$. On the other hand, the dotted line represents the potential receptor conformations induced by a partial agonist consistent with the extended ternary complex model, which includes the isomerization of receptor between R and R*, the only receptor conformation allowed to bind with G protein. For this model, the interactions of a partial agonist with a receptor would result in two populations of ligand-bound receptors with only one (LR*) able to bind with G protein. The *x*-axis is analogous to the cooperativity factor $\alpha$. (See Color Insert.)

These classifications of agonists, however, do not explicitly address how ligand-induced receptor conformations relate to the *activation* of G protein. Thus, ternary complexes formed by partial agonists could be less efficient in the initiation of the G protein signal transduction pathway. Tota and Schimerlik (1990) evaluated GTPase activity as a function of ternary complex formation for full and partial agonist binding to the muscarinic receptor. They found that once ternary complex was formed, the complexes had the same capability to activate GTPase. They concluded that full and partial agonists differ in the amount of ternary complex formed. The soluble ternary complex analysis presented here would account only for the reduced number of ternary complexes formed by partial agonists in terms of reduced affinity of receptor for G protein. To this end, we observed that ternary complexes formed by both full and partial agonists were disassembled by GTP$\gamma$S at the same rate, suggesting that the rate-limiting step of guanine nucleotide activation is therefore independent of the affinity of the ligand. This issue is developed in the following section.

## IV. Guanine Nucleotide Activation of Ternary Complex: Some Dynamic Aspects of Structure and Reactivity

### A. *General Considerations*

Over the past several decades, a canonical view of G protein activation has included the dissociation of the G protein heterotrimer, thus allowing each of the subunits to interact independently with effectors. Evidence from biochemistry (Fung, 1983), and structural studies of activated and resting G protein subunits by x-ray crystallography has supported this view (Davis *et al.*, 2005; Hamm, 1998, 2001; Lambright *et al.*, 1994, 1996; Sprang, 1997; Tesmer *et al.*, 2005). However, more recent experimental evidence has led to questions about the absolute necessity of including subunit dissociation as an element of activation. For example, Levitzki *et al.* (Klein *et al.*, 2000; Levitzki and Klein, 2002) used a G$\alpha$ subunit fused to its cognate G$\beta\gamma$ subunit, to show that both the fusion of subunits and therefore absence of dissociation of subunits did not impair signaling function. Other evidence has come from FRET assays using fluorescently tagged G$\alpha$ and G$\beta\gamma$ subunits (*vide infra*) (Bunemann *et al.*, 2001, 2003; Frank *et al.*, 2005; Hein *et al.*, 2005; Hoffmann *et al.*, 2005).

In this section, we will present some aspects of recent progress in the study of GPCR-mediated G protein activation, where much discussion continues. Some poorly defined elements that continue to be debated include, in temporal order: (a) How ligand-activated GPCRs affect the guanine nucleotide exchange at G$\alpha$ (Cherfils and Chabre, 2003; Iiri *et al.*, 1998; Nanoff *et al.*, 2006); (b) Whether G$\alpha$, G$\beta\gamma$, and GPCRs remain associated after activation (Azpiazu and Gautam, 2004; Frank *et al.*, 2005; Rebois and Hebert, 2003; Rebois *et al.*, 1997). (c) How the G protein subunits and their effectors are spatially arranged during signaling (Ridge *et al.*, 2006; Vetter and Wittinghofer, 2001).

### *B. Structural Studies*

Crystal structures of separated $\alpha$ (Coleman *et al.*, 1994; Tesmer *et al.*, 1997) and $\beta\gamma$ (Sondek *et al.*, 1996) and complexed G protein subunits (Hamm, 1998; Lambright *et al.*, 1996; Rensdomiano and Hamm, 1995) have provided important insights into the structural rearrangements accompanying GPCR-mediated guanine nucleotide exchange and the GTPase cycle. The G$\alpha$ subunit is composed of two domains: a guanine nucleotide-binding domain with high structural homology to the Ras family of GTPases and an all $\alpha$ helical domain that, in combination with the GTPase domain, helps to form a deep pocket for binding the guanine nucleotide (Ridge *et al.*, 2003). The subunit contains three flexible regions designated Switch I, Switch II, and Switch III that have been shown to adopt different conformations in the presence of GDP, GTPS, and other nucleotide adducts and analogues (Lambright *et al.*, 1996; Sondek *et al.*, 1996). In the presence of GDP, Switches I and II make extensive contacts with the G$\beta\gamma$-propeller base and the N-terminal G$\alpha$ subunit sequence lies over the barrel side. The resulting complex is very stable, does not readily allow GDP dissociation, and is inactive as far as signaling is concerned. With GTP inside the G$\alpha$-binding pocket, two hydrogen bonds are likely formed between $\gamma$ phosphate oxygens and the highly conserved main chain Thr and Gly residues located in Switch I and II, respectively (Vetter and Wittinghofer, 2001; Wittinghofer and Pai, 1991). The onset of hydrogen bonding connecting the flexible switch regions triggers the dissociation of Switch II from G$\beta\gamma$ (Vetter and Wittinghofer, 2001; Wittinghofer and Pai, 1991). The global allosteric effect on the Switch domains I, II, and III is the adoption of a conformation appropriate for binding effectors and regulators of G protein signaling (RGS) proteins (Tesmer *et al.*, 2005).

The signaling process appears to involve a series of molecular *ON* and *OFF* switches that are driven by binding interactions between components of the ternary complex, guanine nucleotides, and effectors (Vetter and Wittinghofer, 2001). In cells, the transient interaction of a ligand-activated receptor with a heterotrimeric G protein causes the rapid ejection of an otherwise stably bound GDP nucleotide from the nucleotide-binding pocket on the G$\alpha$ subunit, to be replaced with GTP which is present in the cytoplasm at higher concentration. It is worth noting that although both nucleotides have comparable affinities for the nucleotide-binding pocket, $10^{-11}$ M (measured at 4°C) (Goody *et al.*, 1991), the GDP/GTP exchange turns *ON* the processes that enable the effectors to bind to G proteins. A recent paper by Tesmer *et al.* (2005) has reported on an x-ray crystallographic study that has provided structurally correlated "snapshots" of inactive and activated heterotrimeric $G\alpha_q\beta\gamma$ protein. A salient feature of the activated snapshot is that the G$\alpha$ subunit was shown to be fully dissociated from the G$\beta\gamma$, reoriented from its position in the inactive G$\alpha\beta\gamma$ heterotrimer and now bound to a G protein–coupled receptor kinase 2 (GRK2). The authors suggest that these snapshots represent the initial and final frame of a "signaling movie," where the intervening details are at present ill defined or unknown. Some recent attempts to fill in the intervening gaps in the "signaling movie" are presented below.

### C. *Some Dynamic Aspects of GPCR Activation*

The mechanism by which ligand-activated GPCRs catalyze the exchange of GDP for GTP bound to the G$\alpha$-component of the G$\alpha\beta\gamma$-complex is not clear (Hamm, 2001). Because the switch regions are at least 30 Å removed from the cytoplasmic peptide loops of the receptor, it has been proposed that the receptor must achieve the catalytic feat by means of "action at a distance" (Iiri *et al.*, 1998; Rondard *et al.*, 2001). It has been suggested that the GPCR manages the nucleotide exchange through "lever-arm" and "gearshift" mechanisms. These models are derived from a structural analogy that exists between the G protein heterotrimer and the complexes of monomeric GTP-binding proteins and their cognate guanine nucleotide exchange factors (Cherfils and Chabre, 2003; Rondard *et al.*, 2001). The suggested mechanism predicts that the ligand-activated receptor affects the release of GDP by using the G$\beta\gamma$ subunit to directly (the lever-arm) (Rondard *et al.*, 2001) or indirectly (the gearshift) (Cherfils and Chabre, 2003) distort the G$\alpha$ subunit switch regions (I and II) and allow nucleotide release and recognition (Davis *et al.*, 2005). Another hypothesis suggests that the activated receptor may use a "latch" on the surface of the G$\alpha$ subunit and thus prompt GDP release (Sprang, 1997). The "latch" hypothesis has been examined (Nanoff *et al.*, 2006).

A FRET study (Hein *et al.*, 2005) has examined the early events in the dynamics of the NE-induced interaction between YFP-labeled $\alpha_{2A}$ adrenergic receptors ($\alpha_{2A}$-YFP) with CFP-labeled G proteins ($G\alpha_i$-CFP) in live cells. Using the onset of FRET as a real-time indicator of G protein activation, the authors were able to measure a limiting activation switch

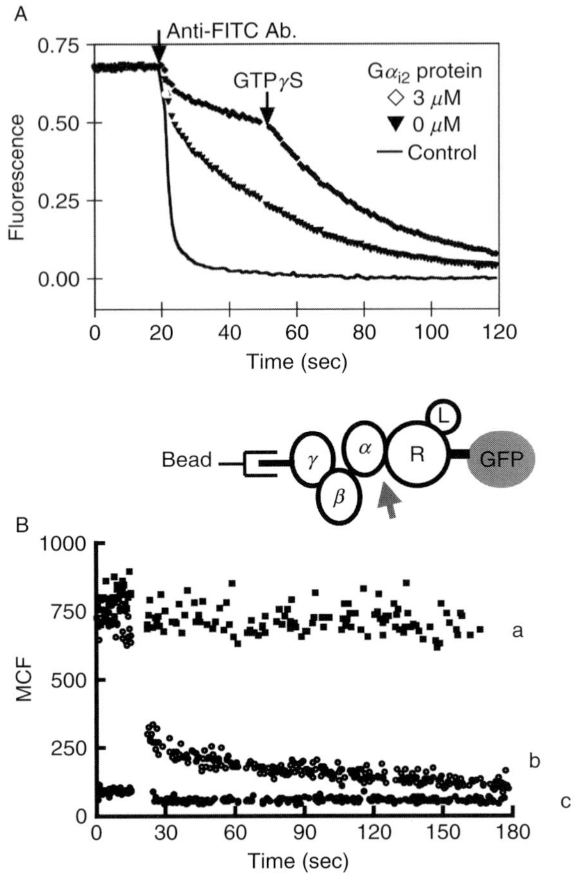


FIG. 6. (A) Analysis of the dissociation of fMLFK-FITC ($L^F$) from a detergent-solubilized G protein–coupled and –uncoupled receptor, a fluorometric anti-FITC antibody assay (see text Section II.C. for details). The assay is based on a fluorescein-conjugated ligand used along with an antifluorescein antibody that rapidly quenches the fluorescein of the free ligand on binding (e.g., blocked sample). The antibody is able to interact only with the fluorescein on the free ligand, as FPR-ligand complexes sterically inhibit the antibody from binding the fluorescein. Data are similar to results shown in Figure 4 of Bennett *et al.* (2001). Reconstitution of the FPR with exogenous G protein titrations: Purified G$\alpha_{i3}$ subunits were combined with $\beta\gamma$ in an equimolar ratio (10 $\mu$M) for 15 min on ice. Solubilized receptor that had been cleared of endogenous G proteins was mixed with the G proteins and with 10-nM fMLFK-FITC in a volume of 12 $\mu$l and incubated for 2 h at 4 °C. Prior to analysis, samples were diluted to 200 $\mu$l and equilibrated to room temperature for 2 min. Anti-FITC-Ab was added at 20 s and GTP$\gamma$S added at 50 s. An increase in the concentration of G protein added to the GPCR reconstitution is accompanied by an increase in the fraction of tightly bound slowly dissociating ligand prior to the addition of GTP$\gamma$S. Analysis of ligand dissociation before

time constant of 44 ms. The switch time for G protein activation was found to be dependent on agonist concentration. The authors also showed that contact between the receptor and the G$\alpha$ subunit was dependent on the concentration of G$\alpha$ in the cell. This dependence on the concentration of G$\alpha$ subunits was taken to be consistent with the absence of agonist independent precoupled receptor–G protein complexes, and that the rate of encounter was solely governed by the lateral diffusion of $\alpha_{2A}$-YFP and G$\alpha_i$-CFP. The study also examined the dissociation kinetics of $\alpha_{2A}$-YFP and G$\alpha_i$-CFP from FRET diminution data, caused by the washing away of ligand. A FRET decay constant of $13 \pm 2.1$ s was determined. We have examined the kinetics of GTP$\gamma$S-induced disassembly of stable ''nucleotide-empty pocket'' ternary complexes (Wu, 2004; Wu *et al.*, 2007a). The data show that the GPCR dissociates from the G protein with a subsecond time constant. These issues are discussed in the following section.

### D. *Modular Disassembly of the Ternary Complexes: Guanine Nucleotide Activation Causes the Rapid Separation of the GPCR and Ligand from the G Protein*

In earlier studies, FPR ternary complexes reconstituted in detergent were analyzed in terms of the sensitivity of ligand dissociation kinetics, to the presence of guanine nucleotides (Bennett *et al.*, 2001; Simons *et al.*, 2003; Wu *et al.*, 2007a). The results of the original study (Fig. 6A; Bennett *et al.*, 2001) showed differential dissociation kinetics of a fluorescent ligand from a G protein–coupled ($t_{1/2} \leq 100$ s) and uncoupled FPR ($t_{1/2} \approx 20$ s). This work was followed by an effort where ternary complexes of detergent-solubilized GPCRs were assembled on beads (cf. schemes B–D in Fig. 1) to examine individual steps associated with GTP-mediated ternary complex disassembly (Simons *et al.*, 2003). As shown in Fig. 6B, these studies did not have adequate time resolution to define precisely the disassembly kinetics. The bead data suggested that the dissociation of the ligand from the G protein–uncoupled receptor was preceded by a much faster process, which involved either the breakup of LR from G or G$\alpha$ from G$\beta\gamma$ (Simons *et al.*, 2003).

and after addition of GTP$\gamma$S gave $t_{1/2}$ of $\approx$20 s and $\approx$100 s, respectively. (B) Disassembly of LR$^F$G after manual addition of GTP$\gamma$S to beads on a flow cytometer. Data show (a) the fluorescence time course of the complex after dilution with buffer, (b) dissociation from beads of LR$^F$ after rapid mixing with 0.2-mM GTP$\gamma$S, and (c) background fluorescence of LR$^F$G prepared in the presence of GTP$\gamma$S. The data corresponding to the rapid kinetics of disassembly are lost during the 10-s dead time, the arrow indicates the break point of the complex. (See Color Insert.)

Rapid mix flow cytometry and modular molecular assemblies of ternary complexes on beads provide a unique platform to examine the kinetics of disassembly at specific junction points of the complex (Wu, 2004; Table III). A rapid mix device enables one to capture the fast subsecond kinetics of the disassembly of the LRG complex in the presence of 0.1-mM GTP$\gamma$S or 0.1-mM GDP. The disassembly of the LRG complex following activation by guanine nucleotides can in principle occur at three junction points: through the separation of L from RG, through the separation of LR from G, or through the separation of LR$\alpha$ from $\beta\gamma$. These junction points essentially correspond to the three classes of receptor assemblies involving the wild-type and fusion proteins (Fig. 1B–D). In cells, the nucleotide concentrations are maintained at relatively high levels (i.e., in the millimolar range). The GPCR ternary complex exists transiently, during the time period when the nucleotide-binding pocket is empty and GDP is exchanged for GTP under mass transfer considerations. Under our experimental conditions, *stable* LRG assemblies on beads are created in circumstances where the GDP is ejected without nucleotide rebinding,

TABLE III
Summary of Experimental Results of LRG Dissociation Kinetics Measured by Small Volume Rapid Mix Flow Cytometry

| | Ligand | Receptor | G$\alpha$ | $\beta\gamma$ | $\tau_{,\text{ fast}}$ (s) | Junction[c] |
|---|---|---|---|---|---|---|
| 1[a] | $L^F$ | R-G$\alpha_{i2}$ | | $\beta_1\gamma_2$ | 18.3 (GTP$\gamma$S)<br>33.8 (GDP) | L-RG |
| 2[a] | $L^F$ | R-G$\alpha_{i2}$ | | $\beta_4\gamma_2$ | 12.4 (GTP$\gamma$S) | L-RG |
| 3[b] | $L^F$ | R | G$\alpha_{i2}$ | $\beta_1\gamma_2$ | 0.8 ± 0.2 (GTP$\gamma$S)<br>1.5 ± 0.4 (GDP) | LR-G |
| 4[b] | $L^F$ | R | G$\alpha_{i2}$ | $\beta_4\gamma_2$ | 0.7 ± 0.1 (GTP$\gamma$S) | LR-G |
| 5[b] | L | $R^F$ | G$\alpha_{i2}$ | $\beta_1\gamma_2$ | 0.8 ± 0.1 (GTP$\gamma$S)<br>1.1 (GDP) | LR-G |
| 6[b] | L | $R^F$ | G$\alpha_{i2}$ | $\beta_4\gamma_2$ | 0.6 ± 0.05 (GTP$\gamma$S) | LR-G |
| 7[b] | $L^F$ | R | G$\alpha_{i3}$ | $\beta_1\gamma_2$ | 3.1 ± 0.2 (GTP$\gamma$S)<br>5.8 (GDP) | LR-G |
| 8[b] | L | $R^F$ | G$\alpha_{i3}$ | $\beta_1\gamma_2$ | 2.4 ± 0.2 (GTP$\gamma$S)<br>5.2 ± 0.6 (GDP) | LR-G |

[a]Kinetic data analyzed using a single-phase exponential model.

[b]Kinetic data analyzed using a two-phase exponential model. The data only show the results of the analysis of the fast component. The slower component was typically two orders of magnitude or more slower than the fast component (cf. Fig. 3). We have attributed the slow component to receptor misfolding (see text for details). Dissociation produced by GTP$\gamma$S was always faster and to a greater extent than GDP.

[c]Experimentally measured point of dissociation in the ternary complex.

due to the low nanomolar concentrations of soluble nucleotides, thus leaving an empty pocket. It is worth noting that in these experiments, formation of ternary complexes was completely inhibited to a similar extent, if attempted in the presence of millimolar concentrations of GDP or GTP$\gamma$S (Fig. 6B).

Figure 7 shows a representative data set collected by rapid mix flow cytometry. The dissociation kinetics of the ligand, fMLFK-FITC ($L^F$) from an intact complex $L^FR\text{-}\alpha_{i2}\beta_1\gamma_2$ was measured after activation with either GTP$\gamma$S or GDP where the junctions between $L^F$ and R or R-$\alpha$ and $\beta\gamma$ can be broken. In Fig. 7A, the traces show (a) the fluorescence time course of the complex after dilution with buffer, (b) dissociation from beads of $L^F$ after rapid mixing with 0.2 mM GTP$\gamma$S, and (c) background fluorescence of $L^FR\text{-}\alpha_{i2}\beta_1\gamma_2$ ternary complex prepared in the presence of GTP$\gamma$S a negative control. Figure 7B shows the dissociation of $L^F$ corrected for dilution effects and background (Wu, 2004). The traces (a) and (b) correspond to dissociation caused by GDP ($t_{1/2} = 18$ s) and GTP$\gamma$S ($t_{1/2} = 34$ s), respectively. The dissociation rate is similar to the ligand dissociation rate measured from an isolated LR complex (Bennett *et al.*, 2001; Simons *et al.*, 2003). Corroborating evidence in favor of ligand dissociation, rather than the alternative pathway of the dissociation of the $\alpha$ subunit from the $\beta\gamma$ subunit, was that when the nucleotides were added in the presence of excess fMLFK-FITC, the loss of fluorescence was inhibited for the entire time frame of the experiment (Wu *et al.*, 2007a). In any event the separation of R-$\alpha$ and $\beta\gamma$ can be no faster than the ligand dissociation rate.

Since the measured separation of R-$\alpha$ and $\beta\gamma$ can be no faster than the dissociation of $L^F$ from $L^FR$, the subsecond dissociation of the ternary complex of Fig. 1D must reflect the dissociation of FPR-GFP from G (Table III, rows 5 and 6; Wu *et al.*, 2007a). Under conditions of the rapid mix flow cytometer, addition of a large excess (0.1 mM) of either GTP$\gamma$S or GDP to the ternary complexes on beads produces essentially the same rapid dissociation of LR from the heterotrimer (Table, III rows 5–8), with notable differences between $\alpha_{i2}$ and $\alpha_{i3}$ (*vide infra*).

Finally, the behavior of wild-type FPR with the fluorescent ligand generally matched that of FPR-GFP, indicating that the receptor dissociates from the G protein prior to the departure of the fluorescent ligand from the receptor (Wu *et al.*, 2007a). The kinetics we report for FPR-GFP dissociation from G$\alpha_i$ in the detergent system is much faster than that reported in cells by FRET for $\beta_2$AR and $G_s$ by Hein *et al.* (2005). We do not know whether this reflects differences between the intact cells and detergent, the difference between specific receptors and G proteins, the possibility that their measurement has contributions from a higher order topographical redistribution, or the possibility that ligand dissociation and

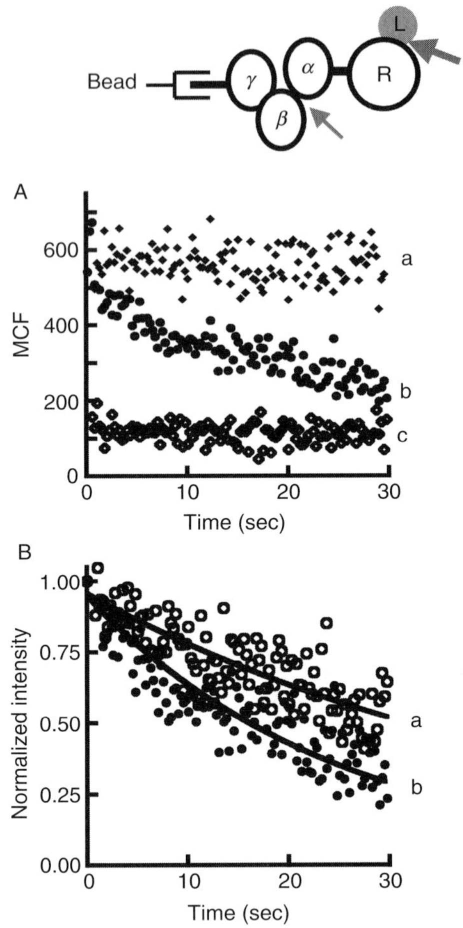


FIG. 7. Disassembly of FPR-G$\alpha_{i2}$ ternary complex. Red arrows in the scheme show junction points of disassembly, with the larger arrow indicating the weaker link that actually is shown to break in the experiment. (A) Data show (a) the fluorescence time course of the complex after dilution with buffer, (b) dissociation from beads of $L^F$ after rapid mixing with 0.2-mM GTP$\gamma$S, and (c) background fluorescence of $L^F$R-$\alpha_{i2}\beta_1\gamma_2$ prepared in the presence of GTP$\gamma$S. (B) Dilution and background-corrected data showing measurements of fMLFK-FITC dissociation kinetics from G$\alpha_{i2}$-FPR complexed to $\beta_1\gamma_2$ subunits on beads after the addition of 0.1-mM GDP (a) or GTP$\gamma$S (b). The data were fit to a single exponential model; the halftime $t_{1/2}$ of GDP is $\approx$34 s, compared to that of GTP$\gamma$S ($\approx$18 s). (See Color Insert.)

rebinding (Abbott and Nelsestuen, 1988; Christopoulos and Kenakin, 2002; Shea *et al.*, 1997) is rate-limiting in their measurement of RG dissociation.

Hein *et al.* (2005) also observed that exchanging the wild-type $G\alpha_i$ subunit with a mutant $G\alpha_i$ND (where the mutant differs from the wild type at a single amino acid located in the GDP-binding pocket) leads to a slower dissociation ($t_{1/2} = 30.3 \pm 2.9$ s vs 13 s for the wild type). In Section III.F. of this chapter, we discuss how GPCRs can access multiple conformational states by binding to different ligands and G proteins. In the case of G proteins, a variety of subunit isotypes can induce the receptor to achieve conformational states which display a higher affinity for the ligand (Vauquelin and Liefde, 2005). In Table III, our data show that the substitution of $\beta_1$ for $\beta_4$ the $G\beta\gamma$ dimer has a measurable effect ($t_{1/2} = 18.3$ s vs 12.4 s) on the dissociation rate of the ligand, after the receptor uncouples from the G protein ($t_{1/2} < 1$ s). This curious finding might imply that the memory of the conformational state induced by the G protein lingers after the G protein interaction is disconnected. Interestingly, our equilibrium-binding data (Simons *et al.*, 2003) and G protein activation (Table III, rows 1 and 2) show that the stability of the complexes and dissociation rates of receptor or ligand depend on the isotype of the G proteins. Other data have consistently shown that the stability of the LRG complexes improved according to the rank $\alpha_{i1} < \alpha_{i2} < \alpha_{i3}$ (Bennett *et al.*, 2001; Simons *et al.*, 2003; Wu *et al.*, 2007a). In Table III, the rate of nucleotide mediated separation of receptor and G protein appears to be similar with either a $\beta_1$ or $\beta_4$ component. However, the same process appears to be sensitive to the isotype of the $\alpha$ subunit, where $\alpha_{i2}$-containing ternary complexes separate at rates that are several times faster than $\alpha_{i3}$-containing complexes.

### E. *GDP Activity?*

GDP is generally believed to ''turnoff'' receptor-mediated activation of the ternary complex (Gilman, 1987). However, there have been some scattered literature reports where GDP was shown to induce activation in a process attributed to the conversion of GDP to GTP *in vivo* (Kikkawa *et al.*, 1990; Kimura and Shimada, 1983; Ugur *et al.*, 2005). Recent data have suggested that GDP is capable of activating $G\alpha_s$ to the degree that subsequent interaction with adenylyl cyclase was measured (Ugur *et al.*, 2005). The authors of that report also suggest that such a result might imply that nucleotide exchange may not be an essential component of GPCR activation (Ugur *et al.*, 2005). While our kinetic results demonstrate that both GDP and GTP have comparable effects on the ternary complex, we do not know if downstream effector activity would result. Based on our data so far, it is our view that while GDP appears to have close to the same potency of GTP in terms of the ability to flip the switch that causes the dissociation of the receptor from the $\alpha$ subunit, it is likely not to have the same "signaling efficacy" of GTP. In addition, the structural data (Vetter and Wittinghofer, 2001) that has elucidated the role played by GTP's $\gamma$ phosphate (or by $AlF^{4-}$ when GDP is bound) in the manipulation of Switch regions I and II makes the suggestion that GDP and GTP could be interchangeable in their signaling roles seem untenable. Furthermore, from the viewpoint that a regulated signaling cascade is essential, it would be difficult to imagine the need of GPCR activity and the need to maintain a higher intracellular concentration of GTP if nucleotide exchange was not a necessary part of activation.

### *F. G$\alpha$ Subunit Dissociation?*

The experimental evidence presented here is consistent with a model of G protein activation that limits the contribution of subunit dissociation in early events associated with cell activation. That we do not observe subunit dissociation could be due to the absence of effectors that could potentially be responsible for turning *ON* a switch mechanism that may cause the dissociation of the $\alpha$ subunit. Its worth being reminded that, under basal conditions, GDP which has a reported affinity of $10^{-11}$ M (Goody *et al.*, 1991) for the $\alpha$ subunit is readily ejected from the binding pocket on contact between the ligand-activated receptor and the heterotrimeric G protein. Our model system is much simpler than the cellular environment, where membranes and effector proteins that regulate G protein activity could potentially influence the dynamic reactivity of the ternary complexes (Bunemann *et al.*, 2003; Ford *et al.*, 1998; Kukkonen *et al.*, 2001; Lindorfer *et al.*, 1998; McCormick and Wittinghofer, 1996; Myung *et al.*, 1999; Neubig and Siderovski, 2002). For example, the ability of RGS proteins in membranes to interact with a GTP-loaded $\alpha$ subunit could also affect the stability of the heterotrimeric complex on a time frame associated with termination of the response. Smrcka and coworkers (Davis *et al.*, 2005; Goubaeva *et al.*, 2003) have reported on a consensus G protein regulatory peptide from AGS3 that was shown to induce the dissociation of G$\alpha$ from G$\beta\gamma$ by simply binding to a "hot spot," thus actively promoting subunit dissociation independently of nucleotide exchange.

More recent data have supported the view that the activated G protein heterotrimers can function as intact units, where conformational changes alone provide a mechanism by which effector-interacting domains at the interface of $\alpha$ and $\beta\gamma$ are thus exposed (Bunemann *et al.*, 2003; Klein *et al.*, 2000; Vilardaga *et al.*, 2003). The binding of the nucleotide is allosteric to the switch regions of the $\alpha$ subunit, and it is the changes in conformation that have been measured by FRET (Bunemann *et al.*, 2003). It has also emerged that subunit dissociation might be governed by a region that is found in G$\alpha_i$ subunits and not G$\alpha_s$ subunits (Frank *et al.*, 2005).

### G. *Outlook*

The modular assembly of G proteins, detergent-solubilized receptors, and ligands on beads has the potential to offer unprecedented insight into the thermodynamics and kinetics of GPCR complexes. Much of the advantage of this platform is the enabling of the practitioner to define reagent stoichiometry and identity, for example, isotypes of promiscuous G protein subunits can be clearly defined. The assembly of ternary complexes on beads has the potential of meeting the challenge of a systematic study that could lead to the elucidation of differences in dynamic and equilibrium roles played by different isotypes of $\alpha$ and $\beta\gamma$ subunits (Kukkonen *et al.*, 2001; Lindorfer *et al.*, 1998; Mayeenuddin *et al.*, 2006; McIntire *et al.*, 2001, 2002; Simons *et al.*, 2003). The studies described here have been limited by availability of purified protein targets and have so far used the limited isotypes available to the authors. This chapter has documented the problems associated with attempts to define reagent concentrations in cell membranes. Included in that discussion is the existence of lipid microdomains where GPCRs can cluster to produce high local concentrations of ternary complexes (Head *et al.*, 2005; Insel *et al.*, 2005a,b; Neubig, 1994; Ostrom and Insel, 2004). One can imagine that the clustering of GPCRs in rafts could potentially influence the conclusions drawn from the *in vitro* FRET the measurements. The FRET data have suggested that G$\alpha$ subunits do not dissociate from G$\beta\gamma$ but rather undergo conformational changes (Bunemann *et al.*, 2003; Vilardaga *et al.*, 2003). Because lipid microdomains are submicroscopic (perhaps as small as tens of nanometers wide) (Edidin, 2003; Kusumi *et al.*, 2004) if the G$\alpha$ subunit were to indeed dissociate, it might not necessarily laterally diffuse away, such that the distance of closest approach between a donor and acceptor pair associated with a G$\beta\gamma$-bound or free subunit cannot be readily resolved. On beads (cf. Fig. 6), such an experiment produces unequivocal results as the G$\alpha$ subunit is able to fall off the bead. At this time, the availability of purified reagents is the limiting factor in the use of the beads to further elucidate GPCR-signaling pathways.

## Acknowledgments

This work was supported by NIH Grants K25AI060036 (TB), AI 36357(ERP) EB00264, GM3956, U54MH074425 (LAS). The New Mexico Tate Cigarette Tax to the UNM Cancer Center.

## References

Abbott, A. J., and Nelsestuen, G. L. (1988). The collisional limit—an important consideration for membrane-associated enzymes and receptors. *FASEB J.* **2,** 2858–2866.

Abdulaev, N. G., Ngo, T., Zhang, C., Dinh, A., Brabazon, D. M., Ridge, K. D., and Marino, J. P. (2005). Heterotrimeric G-protein alpha-subunit adopts a "preactivated" conformation when associated with beta gamma-subunits. *J. Biol. Chem.* **280,** 38071–38080.

Azpiazu, I., and Gautam, N. (2004). A fluorescence resonance energy transfer-based sensor indicates that receptor access to a G protein is unrestricted in a living mammalian cell. *J. Biol. Chem.* **279,** 27709–27718.

Basani, R. B., D'Andrea, G., Mitra, N., Vilaire, G., Richberg, M., Kowalska, M. A., Bennett, J. S., and Poncz, M. (2001). RGD-containing peptides inhibit fibrinogen binding to platelet alpha(IIb)beta 3 by inducing an allosteric change in the amino-terminal portion of alpha(IIb). *J. Biol. Chem.* **276,** 13975–13981.

Bennett, T. A., Key, T. A., Gurevich, V. V., Neubig, R., Prossnitz, E. R., and Sklar, L. A. (2001). Real-time analysis of G protein-coupled receptor reconstitution in a solubilized system. *J. Biol. Chem.* **276,** 22453–22460.

Boniface, J. J., Reich, Z., Lyons, D. S., and Davis, M. M. (1999). Thermodynamics of T cell receptor binding to peptide-MHC: Evidence for a general mechanism of molecular scanning. *Proc. Natl. Acad. Sci. USA* **96,** 11446–11451.

Browning, D. D., Pan, Z. K., Prossnitz, E. R., and Ye, R. D. (1997). Cell type and developmental stage-specific activation of NF-kappa B by fMet-Leu-Phe in myeloid cells. *J. Biol. Chem.* **272,** 7995–8001.

Bunemann, M., Bucheler, M. M., Philipp, M., Lohse, M. J., and Hein, L. (2001). Activation and deactivation kinetics of alpha(2A)- and alpha(2C)-adrenergic receptor-activated G protein-activated inwardly rectifying $K^+$ channel currents. *J. Biol. Chem.* **276,** 47512–47517.

Bunemann, M., Frank, M., and Lohse, M. J. (2003). $G_i$ protein activation in intact cells involves subunit rearrangement rather than dissociation. *Proc. Natl. Acad. Sci. USA* **100,** 16077–16082.

Buranda, T., Jones, G., Nolan, J., Keij, J., Lopez, G. P., and Sklar, L. A. (1999). Ligand receptor dynamics at streptavidin coated particle surfaces: A flow cytometric and spectrofluorimetric study. *J. Phys. Chem. B.* **103,** 3399–3410.

Buranda, T., Lopez, G. P., Simons, P., Pastuszyn, A., and Sklar, L. A. (2001). Detection of epitope-tagged proteins in flow cytometry: Fluorescence resonance energy transfer-based assays on beads with femtomole resolution. *Anal. Biochem.* **298,** 151–162.

Burstein, E. S., Spalding, T. A., Braunerosborne, H., and Brann, M. R. (1995). Constitutive activation of muscarinic receptors by the G-protein G(Q). *FEBS Lett.* **363,** 261–263.

Burstein, E. S., Spalding, T. A., and Brann, M. R. (1996). Constitutive activation of chimeric m2/m5 muscarinic receptors and delineation of G-protein coupling selectivity domains. *Biochem. Pharmacol.* **51,** 539–544.

Cabrera-Vera, T. M., Vanhauwe, J., Thomas, T. O., Medkova, M., Preininger, A., Mazzoni, M. R., and Hamm, H. E. (2003). Insights into G protein structure, function, and regulation. *Endocr. Rev.* **24,** 765–781.

Cherfils, J., and Chabre, M. (2003). Activation of G-protein G alpha subunits by receptors through G alpha-G beta and G alpha-G gamma interactions. *Trends Biochem. Sci.* **28,** 13–17.

Chigaev, A., Blenc, A. M., Braaten, J. V., Kumaraswamy, N., Kepley, C. L., Andrews, R. P., Oliver, J. M., Edwards, B. S., Prossnitz, E. R., Larson, R. S., and Sklar, L. A. (2001). Real time analysis of the affinity regulation of alpha 4-integrin. The physiologically activated receptor is intermediate in affinity between resting and Mn($^{2+}$) or antibody activation. *J. Biol. Chem.* **276,** 48670–48678.

Christopoulos, A. (2002). Allosteric binding sites on cell-surface receptors: Novel targets for drug discovery. *Nat. Rev. Drug Discov.* **1,** 198–210.

Christopoulos, A., and Kenakin, T. (2002). G protein-coupled receptor allosterism and complexing. *Pharmacol. Rev.* **54,** 323–374.

Christopoulos, A., May, L. T., Avlani, V. A., and Sexton, P. M. (2004). G-protein-coupled receptor allosterism: The promise and the problem(s). *Biochem. Soc. Trans.* **32,** 873–877.

Coleman, D. E., Berghuis, A. M., Lee, E., Linder, M. E., Gilman, A. G., and Sprang, S. R. (1994). Structures of active conformations of G(I-Alpha-1) and the mechanism of Gtp hydrolysis. *Science* **265,** 1405–1412.

Davis, T. L., Bonacci, T. M., Sprang, S. R., and Smrcka, A. V. (2005). Structural and molecular characterization of a preferred protein interaction surface on G protein beta gamma subunits. *Biochemistry* **44,** 10593–10604.

De Lean, A., Stadel, J. M., and Lefkowitz, R. J. (1980). A ternary complex model explains the agonist-specific binding properties of the adenylate cyclase-coupled beta-adrenergic receptor. *J. Biol. Chem.* **255,** 7108–7117.

Downes, G. B., and Gautam, N. (1999). The G protein subunit gene families. *Genomics* **62,** 544–552.

Edidin, M. (2003). The state of lipid rafts: From model membranes to cells. *Annu. Rev. Biophys. Biomol. Struct.* **32,** 257–283.

Ford, C. E., Skiba, N. P., Bae, H. S., Daaka, Y. H., Reuveny, E., Shekter, L. R., Rosal, R., Weng, G. Z., Yang, C. S., Iyengar, R., Miller, R. J., Jan, L. Y., *et al.* (1998). Molecular basis for interactions of G protein beta gamma subunits with effectors. *Science* **280,** 1271–1274.

Frank, M., Thumer, L., Lohse, M. J., and Bunemann, M. (2005). G protein activation without subunit dissociation depends on a G alpha$_{(i)}$-specific region. *J. Biol. Chem.* **280,** 24584–24590.

Fung, B. K. K. (1983). Characterization of transducin from bovine retinal rod outer segments.1. Separation and reconstitution of the subunits. *J. Biol. Chem.* **258,** 10495–10502.

Gao, J. L., Becker, E. L., Freer, R. J., Muthukumaraswamy, N., and Murphy, P. M. (1994). A high potency nonformylated peptide agonist for the phagocyte N-formylpeptide chemotactic receptor. *J. Exp. Med.* **180,** 2191–2197.

Gether, U., Lin, S., and Kobilka, B. K. (1995). Fluorescent labeling of purified beta 2 adrenergic receptor. Evidence for ligand-specific conformational changes. *J. Biol. Chem.* **270,** 28268–28275.

Ghanouni, P., Steenhuis, J. J., Farrens, D. L., and Kobilka, B. K. (2001). Agonist-induced conformational changes in the G-protein-coupling domain of the beta 2 adrenergic receptor. *Proc. Natl. Acad. Sci. USA* **98,** 5997–6002.

Gilbert, T. L., Prossnitz, E. R., and Sklar, L. A. (1999). The uncoupled state of the human formyl peptide receptor. *J. Recept. Signal Transduct. Res.* **19,** 327–340.

Gilman, A. G. (1987). G-Proteins—transducers of receptor-generated signals. *Annu. Rev. Biochem.* **56,** 615–649.

Goldstein, B., Posner, R. G., Torney, D. C., Erickson, J., Holowka, D., and Baird, B. (1989). Competition between solution and cell surface receptors for ligand. Dissociation of hapten bound to surface antibody in the presence of solution antibody. *Biophys. J.* **56,** 955–966.

Goody, R. S., Frech, M., and Wittinghofer, A. (1991). Affinity of guanine nucleotide binding-proteins for their ligands-facts and artefacts. *Trends Biochem. Sci.* **16,** 327–328.

Goubaeva, F., Ghosh, M., Malik, S., Yang, J., Hinkle, P. M., Griendling, K. K., Neubig, R. R., and Smrcka, A. V. (2003). Stimulation of cellular signaling and G protein subunit dissociation by G protein betagamma subunit-binding peptides. *J. Biol. Chem.* **278,** 19634–19641.

Gross, W., and Lohse, M. J. (1991). Mechanism of activation of A2 adenosine receptors.2. A restricted collision-coupling model of receptor-effector interaction. *Mol. Pharmacol.* **39,** 524–530.

Hakak, Y., Shrestha, D., Goegel, M. C., Behan, D. P., and Chalmers, D. T. (2003). Global analysis of G-protein-coupled receptor signaling in human tissues. *FEBS Lett.* **550,** 1–3.

Hall, A. (Ed.) (2000). ''GTPases.''Oxford University Press, Oxford.

Hamm, H. E. (1998). The many faces of G protein signaling. *J. Biol. Chem.* **273,** 669–672.

Hamm, H. E. (2001). How activated receptors couple to G proteins. *Proc. Natl. Acad. Sci. USA* **98,** 4819–4821.

Head, B. P., Patel, H. H., Roth, D. M., Lai, N. C., Niesman, I. R., Farquhar, M. G., and Insel, P. A. (2005). G-protein-coupled receptor signaling components localize in both sarcolemmal and intracellular caveolin-3-associated microdomains in adult cardiac myocytes. *J. Biol. Chem.* **280,** 31036–31044.

Hegener, O., Prenner, L., Runkel, F., Baader, S. L., Kappler, J., and Haberlein, H. (2004). Dynamics of b2-adrenegic receptor-ligand complexes on living cells. *Biochemistry* **43,** 6190–6199.

Hein, P., Frank, M., Hoffmann, C., Lohse, M. J., and Bunemann, M. (2005). Dynamics of receptor/G protein coupling in living cells. *EMBO J.* **24,** 4106–4114.

Hoffmann, C., Gaietta, G., Bunemann, M., Adams, S. R., Oberdorff-Maass, S., Behr, B., Vilardaga, J. P., Tsien, R. Y., Eisman, M. H., and Lohse, M. J. (2005). A FlAsH-based FRET approach to determine G protein-coupled receptor activation in living cells. *Nat. Methods* **2,** 171–176.

Holst, B., Hastrup, H., Raffetseder, U., Martini, L., and Schwartz, T. W. (2001). Two active molecular phenotypes of the tachykinin NK1 receptor revealed by G-protein fusions and mutagenesis. *J. Biol. Chem.* **276,** 19793–19799.

Iiri, T., Farfel, Z., and Bourne, H. R. (1998). G-protein diseases furnish a model for the turn-on switch. *Nature* **394,** 35–38.

Insel, P. A., Head, B. P., Ostrom, R. S., Patel, H. H., Swaney, J. S., Tang, C. M., and Roth, D. M. (2005a). Caveolae and lipid rafts—G protein-coupled receptor signaling microdomains in cardiac myocytes. *Ann. N.Y. Acad. Sci.* **1047,** 166–172.

Insel, P. A., Head, B. P., Patel, H. H., Roth, D. M., Bundey, R. A., and Swaney, J. S. (2005b). Compartmentation of G-protein-coupled receptors and their signalling components in lipid rafts and caveolae. *Biochem. Soc. Trans.* **33,** 1131–1134.

Kenakin, T. P. (2001). Quantitation in receptor pharmacology. *Recept. Channels* **7,** 371–385.

Kenakin, T. P., and Morgan, P. H. (1989). Theoretical effects of single and multiple transducer receptor coupling proteins on estimates of the relative potency of agonists. *Mol. Pharmacol.* **35,** 214–222.

Kent, R. S., De Lean, A., and Lefkowitz, R. J. (1980). A quantitative analysis of beta-adrenergic receptor interactions: Resolution of high and low affinity states of the receptor by computer modeling of ligand binding data. *Mol. Pharmacol.* **17,** 14–23.

Key, T. A., Bennett, T. A., Foutz, T. D., Gurevich, V. V., Sklar, L. A., and Prossnitz, E. R. (2001). Regulation of formyl peptide receptor agonist affinity by reconstitution with arrestins and heterotrimeric G proteins. *J. Biol. Chem.* **276,** 49204–49212.

Kikkawa, S., Takahashi, K., Takahashi, K., Shimada, N., UI, M., Kimura, N., and Katada, T. (1990). Kinase on the GTP-binding proteins. *J. Biol. Chem.* **265,** 21536–21540.

Kimura, N., and Shimada, N. (1983). GDP does not mediate but rather inhibits hormonal signal to adenylate-cyclase. *J. Biol. Chem.* **258,** 2278–2283.

Klein, S., Reuveni, H., and Levitzki, A. (2000). Signal transduction by a nondissociable heterotrimeric yeast G protein. *Proc. Natl. Acad. Sci. USA* **97,** 3219–3223.

Kobilka, B. K. (2002). Agonist-induced conformational changes in the beta2 adrenergic receptor. *J. Pept. Res.* **60,** 317–321.

Krumins, A. M., and Barber, R. (1997). The stability of the agonist beta2-adrenergic receptor-$G_s$ complex: Evidence for agonist-specific states. *Mol. Pharmacol.* **52,** 144–154.

Kukkonen, J. P., Nasman, J., and Akerman, K. E. O. (2001). Modelling of promiscuous receptor-$G_{(i)}/G_{(s)}$—protein coupling and effector response. *Trends Pharmacol. Sci.* **22,** 616–622.

Kusumi, A., Koyama-Honda, I., and Suzuki, K. (2004). Molecular dynamics and interactions for creation of stimulation-induced stabilized rafts from small unstable steady-state rafts. *Traffic* **5,** 213–230.

Lambright, D. G., Noel, J. P., Hamm, H. E., and Sigler, P. B. (1994). Structural determinants for activation of the alpha-subunit of a heterotrimeric G-protein. *Nature* **369,** 621–628.

Lambright, D. G., Sondek, J., Bohm, A., Skiba, N. P., Hamm, H. E., and Sigler, P. B. (1996). The 2.0 angstrom crystal structure of a heterotrimeric G protein. *Nature* **379,** 311–319.

Lefkowitz, R. J., Cotecchia, S., Samama, P., and Costa, T. (1993). Constitutive activity of receptors coupled to guanine-nucleotide regulatory proteins. *Trends Pharmacol. Sci.* **14,** 303–307.

Levitzki, A., and Klein, S. (2002). G-protein subunit dissociation is not an integral part of G-protein action. *Chembiochem.* **3,** 815–818.

Lindorfer, M. A., Myung, C. S., Savino, Y., Yasuda, H., Khazan, R., and Garrison, J. C. (1998). Differential activity of the G protein beta(5)gamma(2) subunit at receptors and effectors. *J. Biol. Chem.* **273,** 34429–34436.

Lohse, M. J., Vilardaga, J. P., and Bunemann, M. (2003). Direct optical recording of intrinsic efficacy at a G protein-coupled receptor. *Life Sci.* **74,** 397–404.

Mayeenuddin, L. H., McIntire, W. E., and Garrison, J. C. (2006). Differential sensitivity of P-Rex1 to isoforms of G protein beta gamma dimers. *J. Biol. Chem.* **281,** 1913–1920.

McCormick, F., and Wittinghofer, A. (1996). Interactions between Ras proteins and their effectors. *Curr. Opin. Biotechnol.* **7,** 449–456.

McIntire, W. E., MacCleery, G., and Garrison, J. C. (2001). The G protein beta subunit is a determinant in the coupling of G(s) to the beta(1)-adrenergic and A2a adenosine receptors. *J. Biol. Chem.* **276,** 15801–15809.

McIntire, W. E., Myung, C. S., MacCleery, G., Wang, Q., and Garrison, J. C. (2002). Reconstitution of G protein-coupled receptors with recombinant G protein alpha and beta gamma subunits. *G Protein Pathways, Pt A, Receptors* **343,** 372–393.

Myung, C. S., Yasuda, H., Liu, W. W., Harden, T. K., and Garrison, J. C. (1999). Role of isoprenoid lipids on the heterotrimeric G protein gamma subunit in determining effector activation. *J. Biol. Chem.* **274,** 16595–16603.

Nanoff, C., Koppensteiner, R., Yang, Q., Fuerst, E., Ahorn, H., and Freissmuth, M. (2006). The carboxyl terminus of the G alpha-subunit is the latch for triggered activation of heterotrimeric G proteins. *Mol. Pharmacol.* **69,** 397–405.

Neubig, R. R. (1994). Membrane organization in G-protein mechanisms. *FASEB J.* **8,** 939–946.

Neubig, R. R., and Siderovski, D. P. (2002). Regulators of G-protein signalling as new central nervous system drug targets. *Nat. Rev. Drug Discov.* **1,** 187–197.

Neubig, R. R., Gantzos, R. D., and Thomsen, W. J. (1988). Mechanism of agonist and antagonist binding to alpha-2 adrenergic-receptors—evidence for a precoupled receptor guanine-nucleotide protein complex. *Biochemistry* **27,** 2374–2384.

Nobles, M., Benians, A., and Tinker, A. (2005). Heterotrimeric G proteins precouple with G protein-coupled receptors in living cells. *Proc. Natl. Acad. Sci. USA* **102,** 18706–18711.

Nolan, J. P., Lauer, S., Prossnitz, E. R., and Sklar, L. A. (1999). Flow cytometry: A versatile tool for all phases of drug discovery. *Drug Discov. Today* **4,** 173–180.

Oh, P., and Schnitzer, J. E. (2001). Segregation of heterotrimeric G proteins in cell surface microdomains: $G_{(q)}$ binds caveolin to concentrate in caveolae; whereas $G_{(i)}$ and $G_{(s)}$ target lipid rafts by default. *Mol. Biol. Cell* **12,** 685–698.

Orly, J., and Schramm, M. (1976). Coupling of catecholamine receptor from one cell with adenylate-cyclase from another cell by cell-fusion. *Proc. Natl. Acad. Sci. USA* **73,** 4410–4414.

Ostrom, R. S., and Insel, P. A. (2004). The evolving role of lipid rafts and caveolae in G protein-coupled receptor signaling: Implications for molecular pharmacology. *Br. J. Pharmacol.* **143,** 235–245.

Perez, D. M., Hwa, J., Gaivin, R., Mathur, M., Brown, F., and Graham, R. M. (1996). Constitutive activation of a single effector pathway: Evidence for multiple activation states of a G protein-coupled receptor. *Mol. Pharmacol.* **49,** 112–122.

Preininger, A. M., Van Eps, N., Yu, N. J., Medkova, M., Hubbell, W. L., and Hamm, H. E. (2003). The myristoylated amino terminus of G $alpha_{(i1)}$ plays a critical role in the structure and function of G $alpha_{(i1)}$ subunits in solution. *Biochemistry* **42,** 7931–7941.

Ransnas, L. A., and Insel, P. A. (1988). Quantitation of the guanine nucleotide binding regulatory protein Gs in S49 cell membranes using antipeptide antibodies to alphas. *J. Biol. Chem.* **263,** 9482–9485.

Rebois, R. V., and Hebert, T. E. (2003). Protein complexes involved in heptahelical receptor-mediated signal transduction. *Recept. Channels* **9,** 169–194.

Rebois, R. V., Warner, D. R., and Basi, N. S. (1997). Does subunit dissociation necessarily accompany the activation of all heterotrimeric G proteins!. *Cell. Signal.* **9,** 141–151.

Rensdomiano, S., and Hamm, H. E. (1995). Structural and functional-relationships of heterotrimeric G-proteins. *FASEB J.* **9,** 1059–1066.

Ridge, K. D., Abdulaev, N. G., Sousa, M., and Palczewski, K. (2003). Phototransduction: Crystal clear. *Trends Biochem. Sci.* **28,** 479–487.

Ridge, K. D., Abdulaev, N. G., Zhang, C., Ngo, T., Brabazon, D. M., and Marino, J. P. (2006). Conformational changes associated with receptor-stimulated guanine nucleotide exchange in a heterotrimeric G-protein alpha-subunit—NMR analysis of GTP gamma S-bound states. *J. Biol. Chem.* **281,** 7635–7648.

Robishaw, J. D., and Berlot, C. H. (2004). Translating G protein subunit diversity into functional specificity. *Curr. Opin. Cell Biol.* **16,** 206–209.

Rondard, P., Iiri, T., Srinivasan, S., Meng, E., Fujita, T., and Bourne, H. R. (2001). Mutant G protein alpha subunit activated by G beta gamma: A model for receptor activation? *Proc. Natl. Acad. Sci. USA* **98,** 6150–6155.

Samama, P., Cotecchia, S., Costa, T., and Lefkowitz, R. J. (1993). A mutation-induced activated state of the beta 2-adrenergic receptor. Extending the ternary complex model. *J. Biol. Chem.* **268,** 4625–4636.

Schwartz, T., Gether, U., Schambye, H., and Hjorth, S. (1995). Molecular mechanism of action. *Curr. Pharm. Des.* **1,** 325–342.

Schwartz, T. W., and Rosenkilde, M. M. (1996). Is there a 'lock' for all agonist 'keys' in 7TM receptors? *Trends Pharmacol. Sci.* **17,** 213–216.

Seifert, R., Wenzel-Seifert, K., Gether, U., and Kobilka, B. K. (2001). Functional differences between full and partial agonists: Evidence for ligand-specific receptor conformations. *J. Pharmacol. Exp. Ther.* **297,** 1218–1226.

Shea, L. D., Omann, G. M., and Linderman, J. J. (1997). Calculation of diffusion-limited kinetics for the reactions in collision coupling and receptor cross-linking. *Biophys. J.* **73,** 2949–2959.

Shi, M., Bennett, T. A., Cimino, D. F., Maestas, D. C., Foutz, T. D., Gurevich, V. V., Sklar, L. A., and Prossnitz, E. R. (2003). Functional capabilities of an N-formyl peptide receptor-G(alpha i2) fusion protein: Assemblies with G proteins and arrestins. *Biochemistry* **42,** 7283–7293.

Simons, K., and Ikonen, E. (1997). Functional rafts in cell membranes. *Nature* **387,** 569–572.

Simons, K., and Toomre, D. (2000). Lipid rafts and signal transduction. *Nat. Rev. Mol. Cell Biol.* **1,** 31–39.

Simons, P., Vines, C. M., Key, T. A., Potter, R. M., Shi, M., Sklar, L. A., and Prossnitz, E. R. (Eds.) (2005). *In* ''Analysis of GTP-Binding Protein-Coupled Receptor Assemblies by Flow Cytometry.''Oxford University Press, New York City.

Simons, P. C., Shi, M., Foutz, T., Cimino, D. F., Lewis, J., Buranda, T., Lim, W. K., Neubig, R. R., McIntire, W. E., Garrison, J., Prossnitz, E., and Sklar, L. A. (2003). Ligand-receptor-G-protein molecular assemblies on beads for mechanistic studies and screening by flow cytometry. *Mol. Pharmacol.* **64,** 1227–1238.

Simons, P. C., Biggs, S. M., Waller, A., Foutz, T., Cimino, D. F., Guo, Q., Neubig, R. R., Tang, W. J., Prossnitz, E., and Sklar, L. A. (2004). Real-time analysis of ternary complex on particles: Direct evidence for partial agonism at the agonist-receptor-G-protein complex assembly step of signal transduction. *J. Biol. Chem.* **279,** 13514–13521.

Sklar, L. A., Oades, Z. G., Jesaitis, J., Painter, R. G., and Cochrane, C. G. (1981). Fluoresceinated chemotactic peptide and high-affinity antifluorescein antibody as a probe of the temporal characteristics of neutrophil stimulation. *Proc. Natl. Acad. Sci. USA* **78,** 7540–7544.

Sklar, L. A., Sayre, J., McNeil, V. M., and Finney, D. A. (1985). Competitive binding kinetics in ligand-receptor-competitor systems. Rate parameters for unlabeled ligands for the formyl peptide receptor. *Mol. Pharmacol.* **28,** 323–330.

Sklar, L. A., Vilven, J., Lynam, E., Neldon, D., Bennett, T. A., and Prossnitz, E. (2000). Solubilization and display of G protein-coupled receptors on beads for real-time fluorescence and flow cytometric analysis. *Biotechniques* **28,** 976–980, 982–985.

Sklar, L. A., Edwards, B. S., Graves, S. W., Nolan, J. P., and Prossnitz, E. R. (2002). Flow cytometric analysis of ligand-receptor interactions and molecular assemblies. *Annu. Rev. Biophys. Biomol. Struct.* **31,** 97–119.

Sondek, J., Bohm, A., Lambright, D. G., Hamm, H. E., and Sigler, P. B. (1996). Crystal structure of a G(A) protein beta gamma dimer at 2.1 angstrom resolution. *Nature* **379,** 369–374.

Sprang, S. R. (1997). G protein mechanisms: Insights from structural analysis. *Annu. Rev. Biochem.* **66,** 639–678.

Swaminath, G., Xiang, Y., Lee, T. W., Steenhuis, J., Parnot, C., and Kobilka, B. K. (2004). Sequential binding of agonists to the beta2 adrenoceptor. Kinetic evidence for intermediate conformational states. *J. Biol. Chem.* **279,** 686–691.

Tesmer, J. J. G., Sunahara, R. K., Gilman, A. G., and Sprang, S. R. (1997). Crystal structure of the catalytic domains of adenylyl cyclase in a complex with G(s alpha). GTP gamma S. *Science* **278,** 1907–1916.

Tesmer, V. M., Kawano, T., Shankaranarayanan, A., Kozasa, T., and Tesmer, J. J. G. (2005). Snapshot of activated G proteins at the membrane: The G alpha(q)-GRK2-G beta gamma complex. *Science* **310,** 1686–1690.

Tota, M. R., and Schimerlik, M. I. (1990). Partial agonist effects on the interaction between the atrial muscarinic receptor and the inhibitory guanine nucleotide-binding protein in a reconstituted system. *Mol. Pharmacol.* **37,** 996–1004.

Tsien, R. Y. (1998). The green fluorescent protein. *Annu. Rev. Biochem.* **67,** 509–544.

Ugur, O., Oner, S., Molinari, P., Ambrosio, C., Sayar, K., and Onaran, H. O. (2005). Guanine nucleotide exchange-independent activation of $G_{(s)}$ protein by beta(2)-adrenoceptor. *Mol. Pharm.* **68,** 720–728.

Vauquelin, G., and Liefde, I. V. (2005). G protein-coupled receptors: A count of 1001 conformations. *Fundam. Clin. Pharmacol.* **19,** 45–56.

Venter, J. C., Adams, M. D., Myers, E. W., Li, P. W., Mural, R. J., Sutton, C. G., Smith, H. O., Yandell, M., Evans, C. A., Holt, R. A., Gocayne, J. D., Amanatides, P., *et al.* (2001). The sequence of the human genome. *Science* **291,** 1304–1351.

Vetter, I. R., and Wittinghofer, A. (2001). Signal transduction—the guanine nucleotide-binding switch in three dimensions. *Science* **9,** 1299–1304.

Vilardaga, J. P., Bunemann, M., Krasel, C., Castro, M., and Lohse, M. J. (2003). Measurement of the millisecond activation switch of G protein-coupled receptors in living cells. *Nat. Biotechnol.* **21,** 807–812.

Weiss, J. M., Morgan, P. H., Lutz, M. W., and Kenakin, T. P. (1996a). The cubic ternary complex receptor-occupancy model.1. Model description. *J. Theor. Biol.* **178,** 151–167.

Weiss, J. M., Morgan, P. H., Lutz, M. W., and Kenakin, T. P. (1996b). The cubic ternary complex receptor-occupancy model. III. resurrecting efficacy. *J. Theor. Biol.* **181,** 381–397.

Wiens, B. L., Nelson, C. S., and Neve, K. A. (1998). Contribution of serine residues to constitutive and agonist-induced signaling via the D2S dopamine receptor: Evidence for multiple, agonist-specific active conformations. *Mol. Pharmacol.* **54,** 435–444.

Wittinghofer, A., and Pai, E. F. (1991). The structure of ras protein—a model for a universal molecular switch. *Trends Biochem. Sci.* **16,** 382–387.

Wu, Y. (2004). Biomolecular assembly and design of high-throughput bioanalytical systems. *In* "Biomolecular Assembly and Design of High-Throughput Bioanalytical Systems Ph.D. Thesis Chemical Engineering." University of New Mexico, Albuquerque.

Wu, Y., Simons, P. C., Lopez, G. P., Sklar, L. A., and Buranda, T. (2005a). Dynamics of fluorescence dequenching of ostrich-quenched fluorescein biotin: A multifunctional quantitative assay for biotin. *Anal. Biochem.* **342,** 221–228.

Wu, Y., Zwartz, G., Lopez, G. P., Sklar, L. A., and Buranda, T. (2005b). Small-volume rapid-mix device for subsecond kinetic analysis in flow cytometry. *Cytometry Part A.* **67**(1), 37–44.

Wu, Y., Buranda, T., Simons, P. C., Lopez, G. P., McIntire, W. E., Garrison, J. C., Prossnitz, E. R., and Sklar, L. A. (2007a). Rapid mix flow cytometry measurements of subsecond regulation of G protein-coupled receptor ternary complex dynamics by guanine nucleotides. *Anal. Biochem.* In press.

Wu, Y., Campos, S. K., Lopez, G. P., Ozbun, M. A., Sklar, L. A., and Buranda, T. (2007b). The devolpment of quantum dot calibration beads and quantitative multicolor bioassays in flow cytometry and microscopy. *Anal. Biochem.* **364,** 180–192.

Zhang, J., Campbell, R. E., Ting, A. Y., and Tsien, R. Y. (2002). Creating new fluorescent probes for cell biology. *Nat. Rev. Mol. Cell. Biol.* **3,** 906–918.

# ACTIVATION OF G PROTEIN–COUPLED RECEPTORS

**By XAVIER DEUPI AND BRIAN KOBILKA**

**Department of Molecular and Cellular Physiology, Stanford University School of Medicine, Stanford, California 94305**



## Abbreviations

| | |
|---|---|
| $\beta_2$AR | $\beta_2$-adrenoceptor |
| GPCR | G protein–coupled receptor |
| TM | transmembrane segment |

## Abstract

G protein–coupled receptors (GPCRs) mediate responses to hormones and neurotransmitters, as well as the senses of sight, smell, and taste. These remarkably versatile signaling molecules respond to structurally diverse ligands. Many GPCRs couple to multiple G protein subtypes, and several have been shown to activate G protein–independent signaling pathways. Drugs acting on GPCRs exhibit efficacy profiles that may differ for different signaling cascades. The functional plasticity exhibited by GPCRs can be attributed to structural flexibility and the existence of multiple ligand-specific conformational states. This chapter will review our current



DOI: 10.1016/S0065-3233(07)74004-4

understanding of the mechanism by which agonists bind and activate GPCRs.

## I. Introduction

G protein–coupled receptors (GPCRs) represent the single largest class of membrane proteins in the human genome. Eukaryotic GPCRs have been classified by sequence similarity into five classes (A–F or 1–5) (Attwood and Findlay, 1994; Kolakowski, 1994); however, not all of these classes are represented in humans. A detailed analysis of the human genome reveals at least 800 unique GPCRs, of which ~460 are predicted to be olfactory receptors (Fredriksson *et al.*, 2003). Based on sequence similarity within the seven transmembrane segments (TMs) (Fredriksson *et al.*, 2003), these receptors can be clustered into five families: the rhodopsin family (701 members), the adhesion family (24 members), the frizzled/taste family (24 members), the glutamate family (15 members), and the secretin family (15 members). The physiological function of a large fraction of these 800 GPCRs remains still unknown; these receptors are referred to as orphan GPCRs. However, deorphanization of nonolfactory GPCRs is an ongoing process (Howard *et al.*, 2001), as they are a promising group of targets for the pharmaceutical industry. Therefore, the actual number of orphan GPCRs continues to decline.

GPCRs share a common structural signature of seven hydrophobic segments predicted to be membrane-spanning domains, with an extracellular N-terminus and an intracellular C-terminus (Fig. 1). While the vast majority of GPCRs have been shown to activate one or more cytoplasmic heterotrimeric GTP-binding proteins (G proteins), there is now considerable evidence that some GPCRs can activate signaling pathways that do not involve G proteins (Azzi *et al.*, 2003; Luttrell and Lefkowitz, 2002). For this reason, the terms seven-TM receptor or heptahelical receptors are also been used in place of GPCRs.

This structural and functional similarity stands in contrast to the structural diversity of the natural GPCR ligands (Ji *et al.*, 1998). These range from subatomic particles (a photon) to ions ($H^+$ and $Ca^{2+}$), small organic molecules, peptides, or proteins. The location of the ligand-binding domains for many GPCRs has been determined (Ji *et al.*, 1998). While many small organic agonists bind within the TM segments, peptide hormones and proteins often bind to the N-terminus and extracellular sequences joining the TM domains. However, size of the ligand alone cannot be used to predict the location of the binding site: for instance, glycoprotein hormones, glutamate, and $Ca^{2+}$ all activate their respective

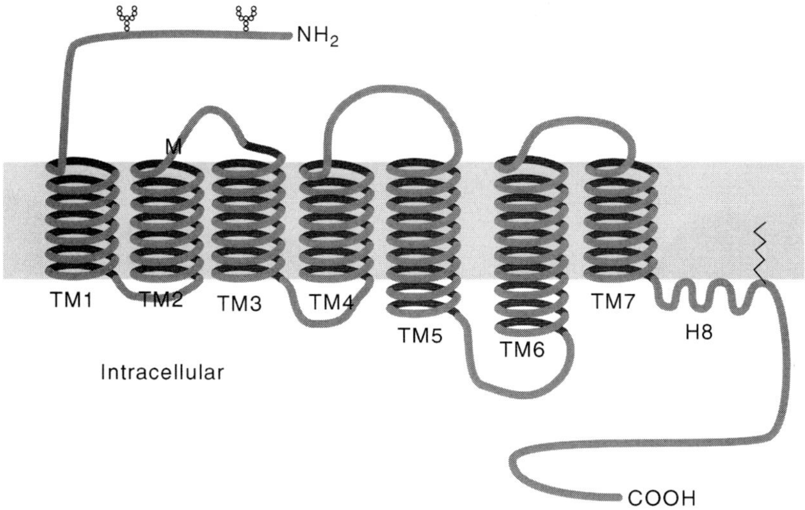


FIG. 1. Secondary structure common to GPCRs. (See Color Insert.)

receptors by binding to relatively large N-terminal domains (Ji *et al.*, 1998; Pin *et al.*, 2003). It is interesting to note that for many GPCRs that bind their native agonists on the extracellular loops or the N-terminus, it has been possible to identify small-molecular-weight allosteric modulators that bind within the TM domains (Knoflach *et al.*, 2001; Ray and Northup, 2002).

In contrast to the diversity in the size of native agonists and the location of the ligand-binding sites, the vast majority of known GPCRs have been shown to activate one or more of the 16 known G protein $\alpha$ subunits (Sprang, 1997). G proteins are structurally homologous, and the mechanism by which GPCRs activate different G proteins is expected to be similar. Therefore, it is likely that the diverse modes of agonist binding to extracellular structures and transmembrane domains result in similar structural changes in cytoplasmic domains that interface with G proteins.

This chapter will review what is known about the mechanism of transmembrane signaling by GPCRs, specifically the process by which agonist binding leads to conformational changes necessary for G protein activation. There is a paucity of experimental data that directly address this process; however, several recent studies are beginning to provide mechanistic insight. These studies suggest that a lock-and-key model of agonist binding does not apply to GPCRs, that is, many of the amino acids that interact with agonists are not optimally positioned for agonist binding in the nonliganded receptor. For agonists to bind, intramolecular interactions that keep the receptor in an inactive state must be broken. Evidence suggests that agonists bind in stages involving one or more conformational intermediates. If correct, the mechanism will have implications for understanding the physiology of GPCRs and for the development of better drug design strategies.

## II. Structural and Mechanistic Homology Among GPCRs

Before focusing on the details about the mechanism of GPCR activation, we will review some of the evidence suggesting that GPCRs are structurally homologous and probably undergo similar structural changes when activating G proteins.

### A. *Rhodopsin as a Structural Model for GPCRs*

There is a wealth of information about the structure and mechanism of activation of rhodopsin. Rhodopsin is a highly specialized GPCR in which the ligand, 11-*cis*-retinal, behaves as a covalently bound inverse agonist that is converted to a full agonist on its photoisomerization to the all-*trans* conformation. This mechanism of agonist activation is highly specific, in contrast to the vast majority of GPCRs that are activated by diffusible agonists. Nevertheless, there is evidence that at least some of the structural changes that occur in rhodopsin are similar to those observed in other GPCRs. Rhodopsin structure and what is known about its light-induced conformational changes have been the subject of several excellent reviews (Hubbell *et al.*, 2003; Ridge *et al.*, 2003; Sakmar, 2002; Sakmar *et al.*, 1991; Schertler, 2005), and some of the main points will be briefly discussed here.

The most detailed information about structural changes associated with activation of a GPCR comes from studies of rhodopsin. This is in part due to its natural abundance and biochemical stability relative to other GPCRs. Electron paramagnetic resonance spectroscopy (EPR) studies provide evidence that photoactivation of rhodopsin involves a rotation and tilting of TM6 relative to TM3 (Farrens *et al.*, 1996). Further support for motion of TM6 during rhodopsin activation was provided by chemical reactivity measurements and fluorescence spectroscopy (Dunham and Farrens, 1999), as well as by ultraviolet absorbance spectroscopy (Lin and Sakmar, 1996) and zinc cross-linking of histidines (Sheikh *et al.*, 1996). Light-induced conformational changes have also been observed in the cytoplasmic domain spanning TM1 and TM2, and the cytoplasmic end of TM7 (Altenbach *et al.*, 1999a,b, 2001).

In addition, rhodopsin is the only GPCR for which high-resolution crystal structures are available (Li *et al.*, 2004; Okada *et al.*, 2000, 2002, 2004;

Palczewski *et al.*, 2000; Teller *et al.*, 2001), with a maximum resolution of 2.2 Å (Okada *et al.*, 2004). While the structures determined by cryoelectron microscopy of 2-D crystals (Krebs *et al.*, 2003; Ruprecht *et al.*, 2004; Schertler *et al.*, 1993) are of lower resolution (up to 5.5 Å), they provide additional information about the orientation of TM segments relative to the lipid bilayer that cannot be obtained from 3-D crystals.

The currently available 3-D structures of rhodopsin correspond to an inactive form of the receptor. While these structures have been a true cornerstone for the study of GPCR structure and function, understanding the activation mechanism demands knowledge of the structure of the active form(s) of the receptor. The publication (Ruprecht *et al.*, 2004) and analysis (Schertler, 2005) of a low-resolution map of metarhodopsin I, an intermediate in the process of rhodopsin activation, reveals that its formation is not accompanied by the large rigid-body movements in TM segments shown to be involved in rhodopsin activation (Farrens *et al.*, 1996). However, a more subtle change, consisting in the rearrangement in the conformation of the Trp residue of the highly conserved CWxxP motif in TM6, has been detected in this intermediate. Thus, it seems that there is no gradual transformation of the inactive protein into the active form, but the activation is initiated through small-scale changes in the conformation of some key residues, which will presumably trigger the larger conformational changes related to the subsequent stages of the activation process. Thus, these key residues can be envisioned as molecular switches that, once turned on, lead to receptor activation.

### *B. GPCRs Activated by Diffusible Agonists*

Rhodopsin is routinely used as a model system for the study of GPCR structure and activation. While high-resolution structures for other GPCRs have not yet been obtained, there is indirect evidence that some rhodopsin family members are structurally very similar to rhodopsin. Ballesteros and Javitch found that structural insights obtained from mutagenesis data and substituted cysteine accessibility studies on monoamine receptors were consistent with the high-resolution structure of rhodopsin, suggesting that rhodopsin serves as a good template for homology modeling (Ballesteros *et al.*, 2001b).

In spite of the remarkable diversity of ligands and ligand-binding domains in the family of GPCRs, there is also considerable evidence for a common mechanism of activation. When comparing sequences, GPCRs are most similar at the cytoplasmic ends of the TMs adjacent to the second and third cytoplasmic domains, the regions known to interact with cytoplasmic G proteins (Mirzadegan *et al.*, 2003). Members of the large family of GPCRs transduce signals by activating one or more members of the relatively small family of highly homologous heterotrimeric G proteins. For example, the thyroid-stimulating hormone (TSH) receptor is activated by a large glycoprotein hormone that binds to the N-terminus while the $\beta_2$-adrenoceptor ($\beta_2$AR) is activated by adrenaline (approximately the size of a single amino acid) that binds to the TM segments; yet both of these receptors activate the same G protein (Gs), indicating that the structural changes in the cytoplasmic domains of these two receptors must be very similar. Moreover, many GPCRs exhibit promiscuous coupling to more than one G protein. For example, rhodopsin preferentially couples to transducin while the $\beta_2$AR preferentially couples to Gs; however, both are capable of activating Gi (Cerione *et al.*, 1985).

Additional evidence that GPCRs undergo similar conformational changes within TM segments and cytoplasmic domains comes from biophysical and biochemical studies. Fluorescence spectroscopic studies of $\beta_2$AR labeled with florescent probes demonstrate movement in both TM3 and TM6 on activation (Gether *et al.*, 1997b). Studies of $\beta_2$AR labeled with fluorescent probes at the cytoplasmic end of TM6 provide evidence that agonists induce a rotation or tilting movement of the cytoplasmic end of TM6 similar to that observed in rhodopsin (Ghanouni *et al.*, 2001b; Jensen *et al.*, 2000). Additional support for movement of TM3 and TM6 in the $\beta_2$AR comes from zinc cross-linking studies (Sheikh *et al.*, 1999) and chemical reactivity measurements in constitutively active $\beta_2$AR mutants (Javitch *et al.*, 1997; Rasmussen *et al.*, 1999). Cysteine cross-linking studies on the M3 muscarinic receptor provide evidence for the movement of the cytoplasmic ends of TM5 and TM6 toward each other on agonist activation (Ward *et al.*, 2002).

Despite the evident similarities between rhodopsin and the rest of Class A GPCRs, it has been proposed that this protein might not be a good template for models of more distantly related rhodopsin family members such as the cholecystokinin CCK1 receptor (Archer *et al.*, 2003). In addition, rhodopsin is unique among GPCRs because of the presence of a covalent linkage between the receptor and its ligand, retinal. Thus, the dynamic processes of agonist association and dissociation common to most GPCRs are not part of the activation mechanism of rhodopsin. Therefore, some caution is needed when extrapolating the information about rhodopsin structure and function to other GPCRs, and a more detailed knowledge of the peculiarities of each system is needed.

### C. *GPCR Oligomers*

There is a growing body of evidence that GPCRs exist as dimers (or oligomers) and that these dimers may be important for G protein activation for at least some GPCR families. This topic has been addressed in several excellent reviews (Angers *et al.*, 2002; Bulenger *et al.*, 2005; Devi, 2001; Javitch, 2004) and will only be briefly addressed here. Dimerization is clearly an important mechanism of receptor activation for the glutamate family of GPCRs (Pin *et al.*, 2003, 2004), where ligand-induced changes in the dimer interface of the N-terminal ligand-binding domain have been demonstrated by crystallography (Kunishima *et al.*, 2000; Tsuchiya *et al.*, 2002). However, the role of dimerization in the activation of rhodopsin family members is less clear. For instance, cryoelectron microscopy images suggest that rhodopsin may exist as homodimers in rod outer segment membranes (Liang *et al.*, 2003). In addition, neutron scattering studies provide evidence that a pentameric complex forms when purified leukotriene B(4) is reconstituted with purified Gi, suggesting that a receptor homodimer is needed to complex with a heterotrimeric G protein (Baneres and Parello, 2003). Nevertheless, it remains to be seen if a receptor dimer is required for G protein activation. The effect of agonist binding on the formation or disruption of dimers is not consistent among the rhodopsin family members that have been examined (Angers *et al.*, 2002). Moreover, ligands interact with individual receptor monomers, and there is currently no evidence that ligands span the interface between receptor dimers. If changes in dimerization occur, it is likely a secondary consequence of ligand-induced changes in the arrangement of the TM segments. Evidence in support of this comes from biophysical studies on leukotriene B(4) homodimers demonstrating that ligand binding to one protomer leads to conformational changes in its partner (Mesnier and Baneres, 2004). While dimers may be important for G protein activation, it is essential to understand the agonist-induced structural changes that occur in the context of individual GPCR monomers.

## III. Conformational States

Proteins are often thought of as rigid structures. The classic model of receptor function is the lock-and-key analogy, where the agonist fits precisely into a complementary pocket in the receptor protein. However, it is known that proteins are dynamic molecules that exhibit rapid, small-scale structural fluctuations. One of the best ways to discuss protein conformations is in terms of an energy diagram (Fig. 2). The basal conformational state is a low-energy state of the protein in a particular environment. The width of the energy well reflects the conformational flexibility. The probability that a protein will undergo transitions to other conformational states is a function of the energy difference between the two states, while the kinetics of the process is function of the height of

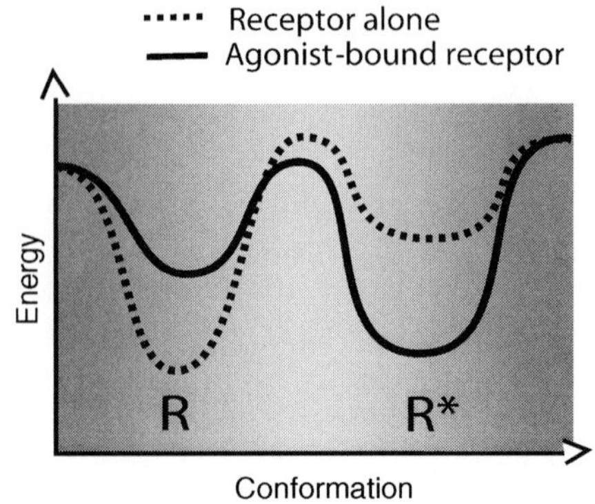


FIG. 2. Energy landscape diagram describing a possible mechanism of GPCR activation by an agonist. (See Color Insert.)

energy barrier between them. In the case of a receptor, the energy provided by ligand binding may change both the depth of an energy well and the height of an energy barrier. The schematic diagram shown in Fig. 2 depicts only two conformational states; however, as discussed below, there is evidence that GPCRs may exist in multiple states.

### A. *Basal Activity and Ligand Efficacy*

Rhodopsin has virtually no detectable basal activity in the absence of light, but can be fully activated by a single photon. But with the exception of rhodopsin, most GPCRs do not behave as simple bimodal (i.e., on–off) switches. In fact, many GPCRs have a considerable amount of basal, agonist-independent activity. The activity of receptors can be either increased or decreased by different classes of ligands (Fig. 3). The term ''efficacy'' is used to describe the effect of a ligand on the functional properties of the receptor [for a more complete discussion of efficacy, refer to Kenakin (2002)]. Agonists are defined as ligands that fully activate the receptor. Partial agonists induce submaximal activation of the G protein even at saturating concentrations. Inverse agonists inhibit basal activity. Antagonists have no effect on basal activity, but competitively block access of other ligands. Therefore, based on functional behavior, GPCRs behave more like rheostats than simple bimodal switches. Different ligands can ''dial in'' virtually any level of activity from fully active to fully inactive.

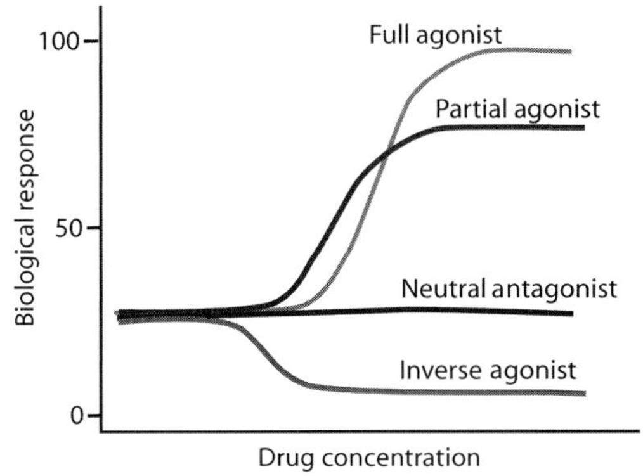


FIG. 3. Ligand efficacy. The effect of different classes of drugs on a GPCR that has some detectable basal activity. (See Color Insert.)

### B. *Multiple Agonist-Specific States*

A number of kinetic models have been developed to explain ligand efficacy, using information derived from indirect measures of receptor conformation such as ligand-binding affinity and the activation of G proteins or effector enzymes (Kenakin, 2001; Leff, 1995; Lefkowitz *et al.*, 1993; Weiss *et al.*, 1996). The simplest of them, the two-state model, proposes that a receptor exists primarily in two states, the inactive state (R) and the active state (R*). In the absence of ligands, the level of basal receptor activity is determined by the equilibrium between R and R*. The efficacy of ligands reflects their ability to alter the equilibrium between these two states. Full agonists bind to and stabilize R*, while inverse agonists bind to and stabilize R. Partial agonists have some affinity for both R and R* and are therefore less effective in shifting the equilibrium toward R*.

The two-state model can describe much of the functional behavior of GPCRs and explain the spectrum of responses to ligands of different efficacy in simple experimental systems consisting of one receptor and one G protein. However, there is a growing body of experimental evidence for the existence of multiple conformational states [summarized in Kenakin (2003)]. Within this framework, each ligand may induce or stabilize a unique conformational state that can be distinguished by the activity of that state toward different signaling molecules (G proteins, kinases, arrestins). In the case of the $\beta_2$AR, it has been possible to monitor directly some of these ligand-specific states using fluorescence spectroscopy (Ghanouni *et al.*, 2001a; Swaminath *et al.*, 2004, 2005). These studies show that $\beta_2$ARs labeled at Cys265$^{6.27}$ [1] on the cytoplasmic end of TM6, adjacent to the G protein–coupling domain, is able to report conformational changes in the G protein–coupling domain. These modifications alter the molecular environment around the fluorophore, which is translated to changes in fluorescence intensity and fluorescence lifetime. In these experiments, fluorescence lifetime analysis can detect discrete conformational states in a population of molecules, while fluorescence intensity measurements reflect their weighted average. Our findings show a single broad distribution of fluorescent lifetimes in the absence of ligands, suggesting that this domain oscillates around a single detectable conformation (Fig. 4A). We have also observed that antagonist binding reduces the width of the distribution, but does not change the mean lifetime, suggesting that, while the conformation does not change, the domain flexibility is reduced (Fig. 4A). However, binding of a full agonist to the $\beta_2$AR leads to the formation of two lifetime distributions (Fig. 4B and C), suggesting two distinct conformations. Moreover, the conformations induced by a full agonist can be distinguished from those induced by partial agonists (Fig. 4C) (Ghanouni *et al.*, 2001a; Swaminath *et al.*, 2005).

### C. *Defining the "Active State"*

As we learn more about the complexity of GPCR signaling, it is becoming more difficult to define exactly what is meant by activation and "active state." Works from several laboratories have shown that GPCRs can activate signaling pathways by G protein–independent mechanisms such as through arrestin and possibly other signaling molecules (Azzi *et al.*, 2003; Baker *et al.*, 2003; Lefkowitz and Shenoy, 2005), and that the "active state" for receptor activation of arrestin or other G protein–independent pathways may differ from that for receptor activation of a G protein (Azzi *et al.*, 2003). Thus, a drug classified as an inverse agonist when monitoring receptor activation of a G protein–dependent signaling pathway may behave as a partial agonist for a G protein–independent signaling pathway (Azzi *et al.*, 2003; Baker *et al.*, 2003). For example, a modified angiotensin

[1] Note: The residues of $\beta_2$AR are numbered according to their position in the sequence followed by the Ballesteros general number (Ballesteros *et al.*, Methods Neurosci. 1995; 25: 366–428) in superscript. In this numbering scheme, each residue is identified by two numbers: the first (1–7) corresponds to the helix where it is located; the second indicates its position relative to the most conserved residue of the helix, arbitrarily assigned to 50. For instance, Trp286$^{6.48}$ is the tryptophan in TM6 located two residues before the highly conserved proline Pro288$^{6.50}$. This general method can be applied to all rhodopsin-like GPCRs and allows easy comparison among residues in the 7-TM segments of different receptor families.

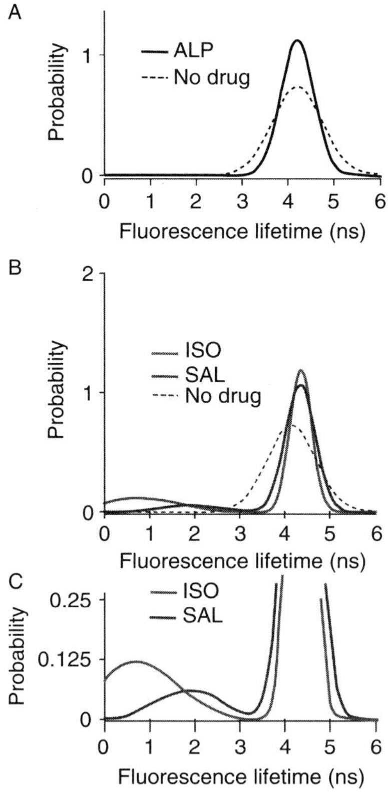


FIG. 4. Fluorescence lifetime distributions of $\beta_2$AR labeled at Cys265$^{6.27}$ with fluorescein maleimide (Ghanouni *et al.*, 2001a). (A) Single lifetime distributions are observed for unliganded receptor and receptor bound to the neutral antagonist alprenolol (ALP). (B and C) Two lifetime distributions are observed for $\beta_2$AR bound to the full agonist isoproterenol (ISO) and the partial agonist salbutamol (SAL). The short lifetime distribution for ISO is different from that for SAL, consistent with a different active conformation. (See Color Insert.)

peptide has been shown to promote arrestin-dependent activation of ERK1/2, but not activation of Gq (Wei *et al.*, 2003). Moreover, the fully active state may differ for different G proteins (Kenakin, 2003). For a GPCR capable of activating more than one G protein, a drug may act as a full agonist toward one G protein and as a partial agonist toward another. To simplify the following discussion, the activity of a particular conformational state of a GPCR will be defined here by the effect that conformational state has on the activity of the receptor's cognate (or preferred)

G protein. Thus, a full agonist maximally activates the cognate G protein while an inverse agonists maximally inhibits any basal activation of the G protein by the receptor.

## IV. Activation by Agonists

### A. *Insights from Constitutively Active Mutants*

To understand the process of receptor activation, we must first understand the properties of the basal or nonliganded state of the receptor. As outlined above, some GPCRs, such as rhodopsin and the FSH receptor (Kudo *et al.*, 1996), have little or no detectable basal activity. Conversely, other GPCRs, such as cannabinoid receptors, exhibit a high degree of basal activity in the absence of ligands (Nie and Lewis, 2001; Sharma and Sharma, 1997). This basal activity could reflect an inherent flexibility of the receptor and, thus, a tendency to exist in more than one conformational state in the absence of ligands. It could also reflect a highly constrained state that has a relatively high affinity for a G protein. The concept of basal activity and receptor activation can be considered in terms of an energy landscape (Fig. 5). In the case of a receptor with low basal activity, in the absence of agonist, the receptor may be relatively constrained into one inactive conformational state having a deep energy well (Fig. 5A). High basal activity might be explained by a smaller energy difference between the inactive and active states, with a lower energy barrier (Fig. 5B). This might also be thought of as a receptor with greater conformational flexibility (i.e., fewer conformational constraints). Alternatively, it is possible that a receptor may exist in predominantly one constrained state that has intermediate activity toward its G protein (Fig. 5C). While both of these mechanisms may apply to different receptors, there is experimental evidence linking conformational flexibility and structural instability to elevated basal activity (Gether *et al.*, 1997a).

How are these concepts translated into receptor structure? TM domains are held in the basal state by intervening loops and noncovalent interactions between side chains. However, proteolysis and split receptor studies suggest that the noncovalent interactions appear to play a greater role in determining the specific basal arrangement of the TM segments relative to each than do some of the intervening loop structures. For example, the cotransfection of a plasmid encoding the N-terminus through TM5 with a plasmid encoding TM6 through the C-terminus [i.e., excluding the third intracellular loop (IC3)] generates a functional $\beta_2$AR (Kobilka *et al.*, 1988), where the fragments assemble and are held together by noncovalent interactions. Similar

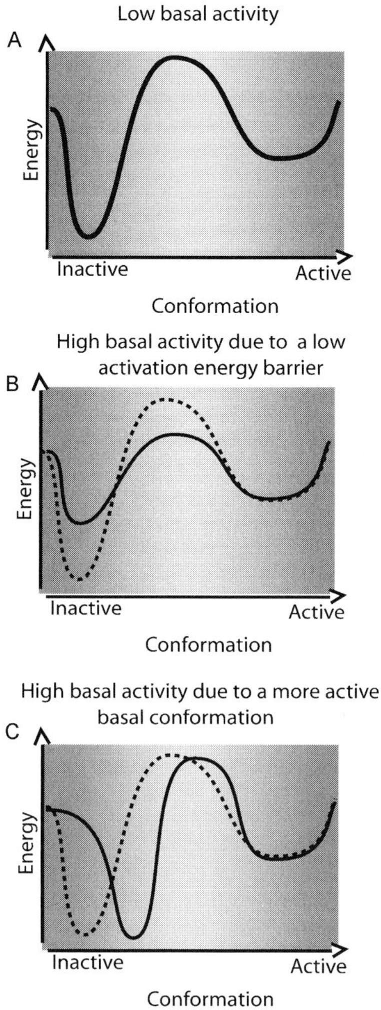


FIG. 5. Energy landscapes used to explain possible mechanisms of elevated basal activity. (A) Receptor with low basal activity. (B and C) Receptors with high basal activity, relative to panel A (shown as a dotted line). (B) The basal activity is elevated because of a reduced energy barrier separating the basal and active conformational states. (C) The basal activity is higher because the basal conformational state has a higher activity toward the G protein. (See Color Insert.)

observations have been made for the muscarinic receptor. Schoneberg and Wess were able to generate functional M3 muscarinic receptors with discontinuity within the loop connecting TM3 and TM4, the loop connecting TM4 and TM5, and the loop connecting TM5 and TM6 (Schoneberg *et al.*, 1995).

Similarly, both $\alpha_2$-adrenergic receptor (Wilson *et al.*, 1990) and the $\beta_2$AR (Swaminath and Kobilka, unpublished data) are capable of binding ligands after proteolytic cleavage of loop structures.

It has also been observed that the degree of basal activity can be dramatically enhanced by single point mutations in a variety of structural domains (Parnot *et al.*, 2002). These constitutive active mutations (CAMs) provide insight into the structural basis of basal activity and of receptor activation. For instance, the fact that CAMs can be generated in virtually any structural domain (Parnot *et al.*, 2002) suggests that in the inactive state the receptor structure is constrained by multiple intramolecular interactions that link TM segments or link TM segments with inter-TM segment loops. Thus, mutations that disrupt these interactions would increase the "flexibility" of the protein (movement of TM domains relative to each other) and the probability that the receptor can assume a more active conformation.

One might predict that mutations that lead to enhanced basal activity by disrupting intramolecular interactions could also lead to decreased structural stability. Mutation of Leu272$^{6.34}$ at the cytoplasmic end of TM6 in the $\beta_2$AR to alanine results in elevated basal activity (Samama *et al.*, 1993) as well as biochemical instability (Gether *et al.*, 1997a). Purified L272A $\beta_2$AR denatures two to three times faster than wild-type receptor (Gether *et al.*, 1997a). The increased basal activity observed in the native $\beta_2$AR at reduced pH is also associated with an increased rate of denaturation (Ghanouni *et al.*, 2000). These denaturation processes can be attenuated by both agonists and antagonists (Gether *et al.*, 1997a). Instability has also been reported in constitutive active mutants of the $\beta_1$-adrenoceptor (McLean *et al.*, 2002) and the $H_2$ histamine receptor (Alewijnse *et al.*, 2000). It is worth mentioning that ligands, both agonists and antagonists, can stabilize the receptor against denaturation and act as biochemical chaperones (McLean *et al.*, 2002; Petaja-Repo *et al.*, 2002), suggesting that they form stabilizing bridges between TM segments.

### B. *Molecular Switches*

So what do CAMs tell us about GPCR activation? We have seen how active states can be achieved by destabilizing the normal arrangement of TM domains by mutations at several different sites. As discussed above, TM domains are held in the basal state primarily by a network of noncovalent interactions between side chains. Thus, any compound that disrupts one of the many intramolecular interactions that stabilize the basal state could have, in principle, agonist activity. The process of disrupting a stabilizing intramolecular interaction can be thought of as activating a *molecular switch.* Importantly, these molecular switches can be activated by mutations or ligands. It is expected that for any given GPCR, there will be numerous molecular switches. Evidence for this is apparent in constitutively active mutants that can be further activated by agonists, suggesting that the mutation activated only one of several switches and that other switches still must be triggered by agonist binding. While some of these molecular switches will be specific for a given GPCR, common molecular switches have been proposed for members of the rhodopsin family of GPCRs. Two of these switches, a rotamer toggle switch in TM6 and an ionic lock between TM3 and TM6, are briefly described below in Section IV.B.1 and 2.

### 1. *Rotamer Toggle Switch*

Using site-directed mutagenesis studies and computer simulations, it has been suggested that rotameric positions of Cys285$^{6.47}$, Trp286$^{6.48}$, and Phe290$^{6.52}$ of the $\beta_2$AR are coupled, and modulate the bend angle of TM6 around the highly conserved proline kink at Pro288$^{6.50}$, leading to the movement of the cytoplasmic end of TM6 (Shi *et al.*, 2002) (Fig. 6). Authors

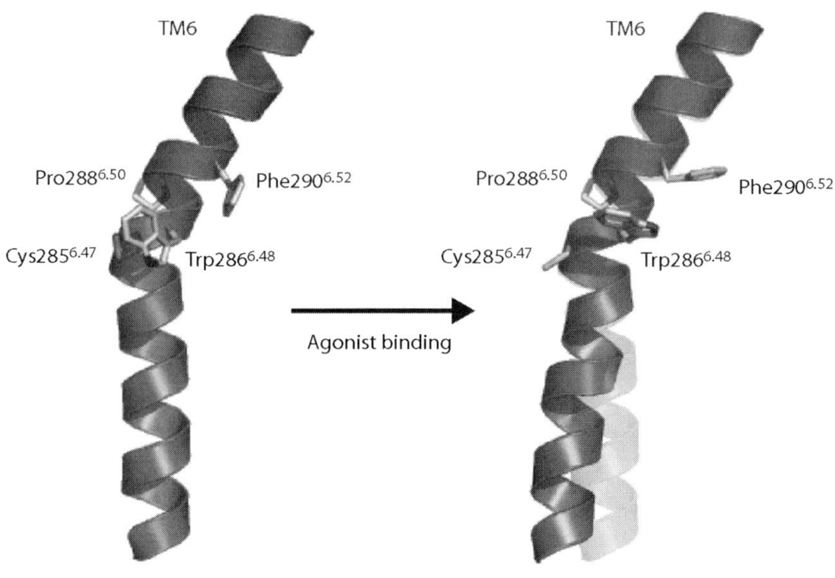


FIG. 6. The rotamer toggle switch. Agonist binding leads to changes in the rotameric states of aromatic amino acids in TM6 resulting in a change in the angle of the helical kink formed by the highly conserved Pro288$^{6.50}$. (See Color Insert.)

also proposed that interactions between the aromatic ring of catecholamine agonists and Phe290$^{6.52}$ in TM6 play a role in the stabilization of the active form of this switch. While this mechanism was initially defined for catecholamine receptors, this sequence motif is highly conserved in amine and opsin receptors, so it is expected that this step in the activation mechanism will be conserved within these families.

### 2. *Ionic Lock*

Another molecular switch, the ionic lock, involves the interaction between Glu$^{6.30}$, highly conserved in amine and opsin receptors (>93%), and the Asp$^{3.49}$/Arg$^{3.50}$ pair, in the highly conserved (D/E)RY motif found in virtually all Class A GPCRs (Ballesteros *et al.*, 2001a) (Fig. 7). This ionic interaction is proposed to hold together the cytoplasmic ends of TM3 and TM6 in the resting state of different amine receptors (Ballesteros *et al.*, 2001a; Greasley *et al.*, 2002; Shapiro *et al.*, 2002). This interaction is also observed in the crystal structures of inactive rhodopsin (Li *et al.*, 2004; Okada, 2004; Okada *et al.*, 2002; Palczewski *et al.*, 2000; Teller *et al.*, 2001), and disruption of this interaction during activation is suggested by various biophysical (Farrens *et al.*, 1996; Gether *et al.*, 1997b), biochemical (Arnis *et al.*, 1994; Ghanouni *et al.*, 2000; Sheikh *et al.*, 1996, 1999), and mutagenesis (Alewijnse *et al.*, 2000;

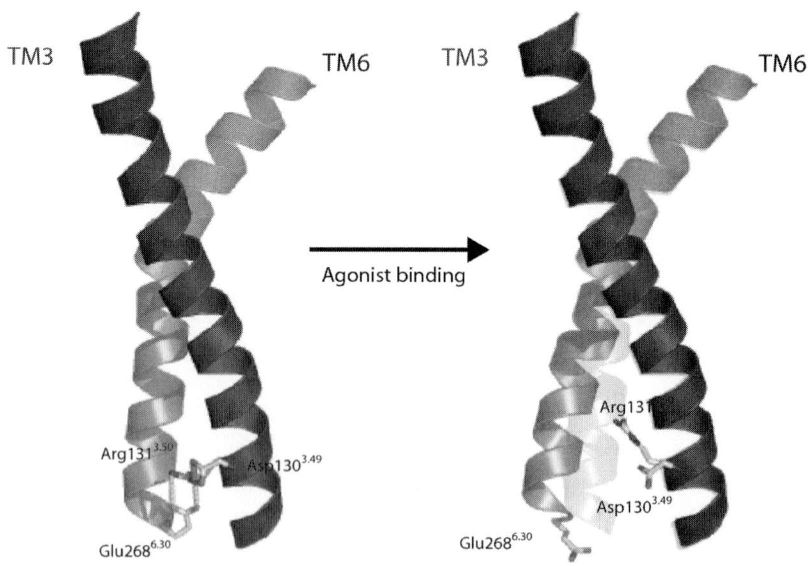


FIG. 7. The ionic lock stabilizes interactions between the cytoplasmic ends of TM3 and TM6 in the inactive state. Agonist binding disrupts these interactions. (See Color Insert.)

Gaborik *et al.*, 2003; Kim *et al.*, 1997; Rasmussen *et al.*, 1999; Scheer *et al.*, 1996) studies. This ionic lock is key in the structure of the receptor, probably in combination with other inter- and/or intrahelical interactions, as it keeps TM6 in a relatively distorted conformation with a marked decrease of the helical twist at the level of the proline kink at Pro288$^{6.50}$. As a result, the cytoplasmic end of TM6 is much closer to TM3 than would be expected from the distortion induced solely by the proline.

### C. *Activation of Molecular Switches by Ligands*

Figure 8 shows two possible ways that ligands may influence the arrangement of TM domains. Ligands may serve as bridges that stabilize new interactions between TM domains (Fig. 8A). In doing so, ligands may move specific TM domains closer to each other, push them further apart, or rotate one relative to the other. At the other end of the spectrum, ligands may act by simply disrupting existing intramolecular interactions (Fig. 8B). An example of an agonist binding to and displacing stabilizing interactions can be found in the AT1 receptor. Experimental evidence suggests that Asn111$^{3.35}$ interacts with Asn295$^{7.46}$ in TM7 to stabilize the inactive state of the receptor (Balmforth *et al.*, 1997). Other evidences suggest that Asn111$^{3.35}$ interacts with Tyr4 of angiotensin (Noda *et al.*, 1996). Thus, during activation, angiotensin would replace Asn111$^{3.35}$ as the interacting partner with Asn295$^{7.46}$. It has also been shown that

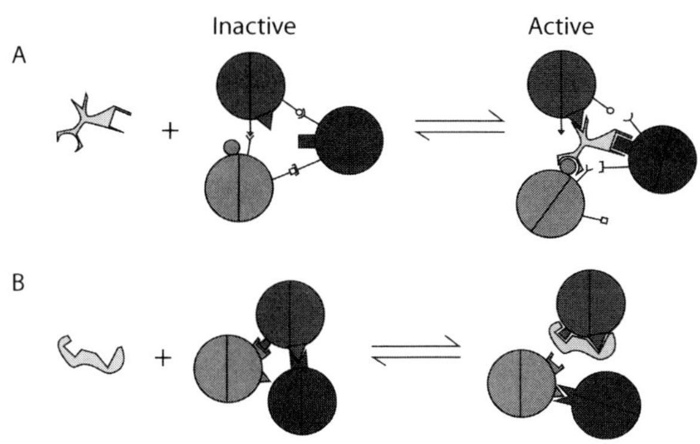


FIG. 8. Mechanisms by which agonist binding may change the relative arrangement of TM segments. (A) The agonist binding requires disruption of intramolecular interactions and the formation of new interactions with the ligand. (B) The agonist binds directly to amino acids involved in forming stabilizing intramolecular interactions. (See Color Insert.)

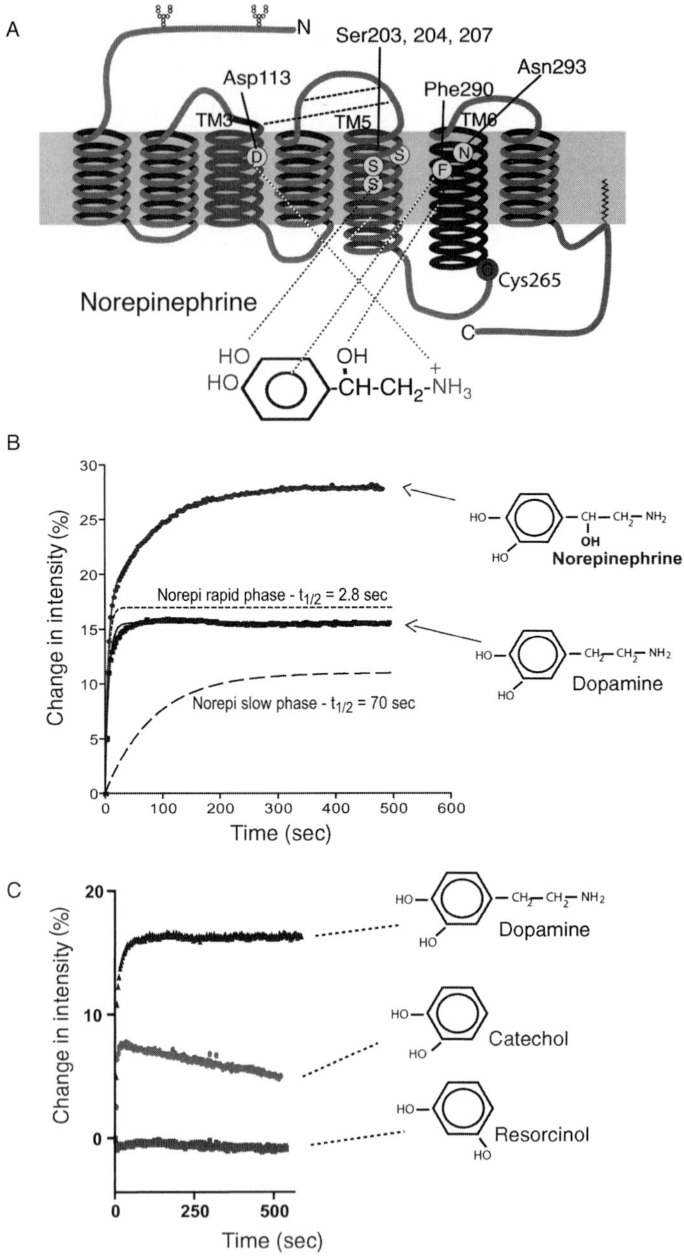


FIG. 9. Agonist-induced conformational changes in the $\beta_2$AR. (A) Sites of interaction between norepinephrine and the $\beta_2$AR identified by site-directed mutagenesis. The catecholamine nitrogen interacts with Asp113$^{3.32}$ in TM3 (Strader *et al.*, 1989a).

histamine binding to the $H_1$ receptor induces $Ser^{3.36}$ to interact with $Asn^{7.45}$ on activation (Jongejan *et al.*, 2005). It is likely that most ligands use a combination of mechanisms shown in Fig. 8.

### D. *Agonist Binding and Activation Is a Multistep Process*

In both of the models shown in Fig. 8, the ability of the ligand to bind the receptor depends on the dynamic nature of the noncovalent interactions between TM segments. That is, agonist activation cannot be explained by a simple lock-and-key model, and there is no preformed binding site for the agonist (Del Carmine *et al.*, 2004; Liapakis *et al.*, 2004). In the basal state, sites of contact for the ligand either are not optimally aligned to bind all structural components of the agonist (Fig. 8A) or are involved in intramolecular interactions (Fig. 8B). Thus, interactions must break and reform on a timescale compatible with rapid binding and activation of receptors. In the case of the model shown in Fig. 8A, ligands may first bind to one interacting site and be poised to bind to a subsequent site on disruption of intramolecular interactions between TM segments. This would involve the formation of one or more intermediate conformational states.

### E. *The $\beta_2$AR as a Model System for Ligand Binding and Activation: Biophysical Analysis of Agonist-Induced Conformational Changes*

The $\beta_2$AR is a good model system for studying agonist binding because the sites of interaction between catecholamine ligands and the $\beta_2$AR have been extensively characterized (Liapakis *et al.*, 2000; Strader *et al.*, 1989c; Wieland *et al.*, 1996) (Fig. 9A). In summary, the amine nitrogen interacts with

---

Hydroxyls on the catechol ring interact with serines $203^{5.42}$ (Liapakis *et al.*, 2000), $204^{5.43}$, and $207^{5.46}$ (Strader *et al.*, 1989b) in TM6. The chiral $\beta$-hydroxyl interacts with $Asn293^{6.55}$ in TM6 (Wieland *et al.*, 1996) and the aromatic ring interacts with $Phe290^{6.52}$ in TM6 (Strader *et al.*, 1989c). Also shown is the relative position of $Cys265^{6.27}$, the labeling site for tetramethylrhodamine. (B) Agonist-induced conformational changes in purified $\beta_2$AR labeled with tetramethylrhodamine at $Cys265^{6.27}$. Conformational response to norepinephrine and dopamine was examined by monitoring changes in fluorescence intensity as a function of time. The response to norepinephrine was best fit with a two-site exponential association function, while there was no significant difference between a one-site and a two-site fit for the response to dopamine. The rapid and slow components of the biphasic response to norepinephrine are shown as dotted lines. (See Color Insert.)

Asp113$^{3.32}$ in TM3 (Strader *et al.*, 1989a), the catechol hydroxyls interact with serines in TM5 (Liapakis *et al.*, 2000; Strader *et al.*, 1989b; Wieland *et al.*, 1996). Interactions with the aromatic ring and the chiral $\beta$-hydroxyl both have been mapped to TM6 (Wieland *et al.*, 1996).

Several studies investigating the process of catecholamine binding to the $\beta_2$AR provide evidence supporting a multistep process for agonist binding (Ghanouni *et al.*, 2001a; Swaminath *et al.*, 2004, 2005). For instance, fluorescence lifetime studies reveal the existence of at least one intermediate conformational state in the presence of the full agonist isoproterenol or partial agonists dobutamine and salbutamol (Ghanouni *et al.*, 2001a) (Fig. 4C). The existence of an intermediate conformational state can also be demonstrated kinetically in the $\beta_2$AR. In these studies, $\beta_2$AR was labeled at Cys265$^{6.27}$ at the cytoplasmic end of TM6 with tetramethylrhodamine (TMR-$\beta_2$AR). Based on homology with rhodopsin, Cys265$^{6.27}$ is located in the IC3 at the cytoplasmic end of the TM6. Mutagenesis studies have shown this region of IC3 to be important for G protein coupling (Liggett *et al.*, 1991; O' Dowd *et al.*, 1988). Moreover, TM6, along with TM3 and TM5, contains amino acids that form the agonist-binding site. Thus, an environmentally sensitive fluorophore covalently bound to Cys265$^{6.27}$ is well positioned to detect agonist-induced conformational changes relevant to G protein activation. Using this experimental system, agonist-induced conformational changes can be observed by monitoring fluorescence intensity of TMR-$\beta_2$AR over time. Our results show how binding of the catecholamine norepinephrine results in curve that is best fit by a two-site exponential association function (Fig. 9B) (Swaminath *et al.*, 2004). This suggests that on catecholamine binding, $\beta_2$ARs undergo transitions to two kinetically distinguishable conformational states through a combination of a fast and a slow conformational change. Using a panel of chemically related catechol derivatives, we identified the specific chemical groups on the agonist responsible for the rapid and slow conformational changes in the receptor. In the presence of catechol and dopamine, only rapid conformational changes were observed (Fig. 9B and C). In contrast, both rapid and slow conformational changes were observed on binding to norepinephrine, epinephrine, and isoproterenol. These results suggest that formation of interactions between the catechol ring and the amine group of the ligand with the Ser residues in TM5 and Asp113$^{3.32}$ occurs rapidly, while interactions between the $\beta$-hydroxyl of the ligand and Asn293$^{6.55}$ occur more slowly, possibly due to the need to overcome a strong stabilizing intramolecular interaction.

Interestingly, the conformational changes observed in these biophysical assays were correlated with biological responses in functional assays. Dopamine, which induces only a rapid conformational change, is efficient at activating Gs but not receptor internalization. In contrast, norepinephrine and epinephrine, which induce both rapid and slow conformational changes, are efficient at activating Gs and receptor internalization.

The process of activation by dopamine can be further dissected using catechol alone as a ligand. As discussed above, the catechol ring of catecholamines is predicted to interact with serines in TM5 and the aromatic residues of the rotamer toggle switch in TM6 (Fig. 9A). We found that catechol alone was able to induce a rapid conformational change in TMR-$\beta_2$AR similar to the response observed with dopamine (Fig. 9C). Moreover, we found that catechol is a weak partial agonist (Swaminath *et al.*, 2005). Both ligand binding and conformational studies demonstrated that catechol occupied the same binding space as the catechol component of catecholamines (Swaminath *et al.*, 2005). Based on these observations, we speculate that catechol binding is sufficient to activate the rotamer toggle switch, inducing the fast change in fluorescence, but not other molecular switches required for full activation. Catechol has a remarkably high affinity ($K_D = 160\ \mu M$, based on a conformational assay) considering its size (formula weight 110), which is consistent with an agonist fragment where a high proportion of the catechol atoms are involved in binding interactions with the receptor. Moreover, the relatively high binding affinity suggests that energetic cost of the conformational changes required for optimal interactions between the $\beta_2$AR and catechol is small.

The binding of the catechol ring of dopamine results in the same structural change that occurs on binding of catechol alone, but the interaction between the amine group and Asp113$^{3.32}$ also stabilizes a specific arrangement of TM3 relative to TM5 and TM6. This additional conformational change imparts a much greater activity toward Gs. Note that binding affinity for dopamine ($K_i = 350\ \mu M$) is similar to that for catechol. This is surprising considering that the interaction between the primary amine and Asp113$^{3.32}$ makes the strongest contribution to the binding energy. Part of the binding energy associated with the interaction between dopamine and Asp113$^{3.32}$ might be offset by the energetic cost of the conformational change needed for the binding interaction to occur. Thus, in the inactive state, TM5 and TM6 are positioned such that little energy is needed to accommodate the binding of the catechol ring. In contrast, the movement of TM3 relative to TM5 and TM6 required for binding of dopamine may involve breaking of intramolecular interactions, thereby consuming part of the energy provided by the ionic interaction between the ligand and the receptor. Evidence from unpublished studies suggests that this added energy is required to disrupt the ionic lock.

Using the information obtained through these biophysical studies on the $\beta_2$AR (Ghanouni *et al.*, 2001a; Swaminath *et al.*, 2004, 2005), we proposed a model whereby agonist binding and activation occur through a series of conformational intermediates (Fig. 10). Within this model, catechol and

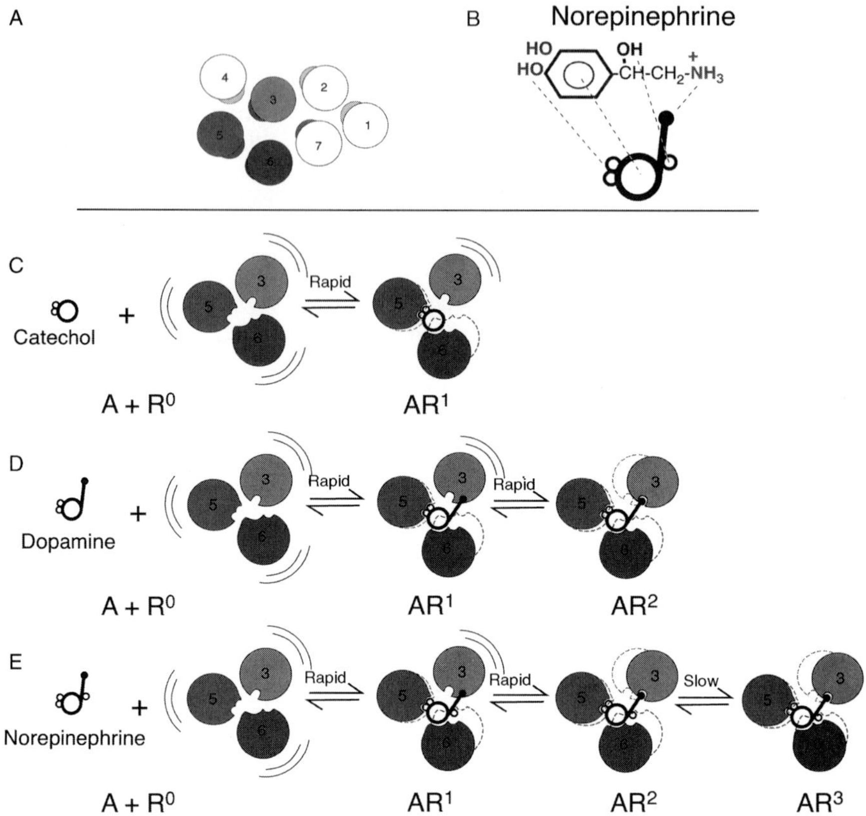


FIG. 10. Sequential binding model. (A) Arrangement of the TM domains of the $\beta_2$AR as viewed from the extracellular surface. The agonist-binding domains are shown in red (TM3), green (TM5), and blue (TM6). (B) Diagram representing structural components of norepinephrine. (C–E) In the absence of ligand, the receptor (R) is conformationally flexible. Conformational state $R^1$ is stabilized by interactions between TM5 and TM6 and the catechol ring. The transition to state $R^2$ occurs when Asp113$^{3.32}$ in TM3 binds the amine nitrogen. The transitions from R to $R^2$ are rapid. The slow transition from $R^2$ to $R^3$ involves interactions between the chiral $\beta$-hydroxyl and Asn293$^{6.55}$ on TM6. (See Color Insert.)

dopamine can be considered as fragments of full catecholamine agonists (for instance, norepinephrine), which are capable of stabilizing intermediate states that have only partial activity.

Evidences for intermediate conformational states have also been observed in other receptors. Time-resolved peptide-binding studies on the neurokinin receptor revealed that an agonist peptide binds with biphasic kinetics. The rapid binding component was associated with a cellular calcium response while the slow component was required for cAMP signaling (Palanche *et al.*, 2001). These results support a mechanistic model for GPCR activation where contacts between the receptor and structural determinants of the agonist stabilize a succession of conformational states with distinct cellular functions.

## V. Concluding Remarks

This chapter has addressed activation of GPCRs using data from only a small subset of rhodopsin family members in which the agonist-binding site is formed by the TM segments. At the other end of the spectrum are receptors for glycoprotein hormones and the glutamate family of receptors in which the ligand-binding site is found within a large N-terminal domain. Nevertheless, glycoprotein hormone receptors can be activated by mutations within TM segments, and glutamate receptor activity can also be modulated by small organic compounds that bind within the TM segments (Pin *et al.*, 2003), suggesting that agonist binding ultimately leads to the disruption of interactions that stabilize the arrangement of the TM segments. In the case of the glycoprotein hormones, evidence suggests that part of the N-terminus interacts with sequence between TM4 and TM5 to stabilize the inactive state and that this interaction is disrupted by agonist binding (Nishi *et al.*, 2002; Yi *et al.*, 2002). In the glutamate family, receptors are homodimers held together by the N-terminal Venus flytrap domain. Agonist binding leads to large structural changes in the Venus flytrap motifs and would be predicted to alter the relative arrangement of the 7-TM segments from each monomer (Pin *et al.*, 2003; Tateyama *et al.*, 2004). This may in turn alter the orientation of the TM segments within each monomer. Therefore, the mechanism linking ligand binding to receptor activation for these receptors is likely to be more complex (Pin *et al.*, 2003; Yi *et al.*, 2002).

While we have learned a great deal about GPCR structure and the cellular signaling pathways activated by GPCRs over the past 20 years, much remains to be learned about the mechanism of activation of this fascinating family of membrane proteins. A better understanding of the complex process of agonist binding and activation may facilitate the design of more effective and selective pharmaceuticals.

## References

Alewijnse, A. E., Timmerman, H., Jacobs, E. H., Smit, M. J., Roovers, E., Cotecchia, S., and Leurs, R. (2000). The effect of mutations in the dry motif on the constitutive activity and structural instability of the histamine H(2) receptor. *Mol. Pharmacol.* **57,** 890–898.

Altenbach, C., Cai, K., Khorana, H. G., and Hubbell, W. L. (1999a). Structural features and light-dependent changes in the sequence 306–322 extending from helix VII to the palmitoylation sites in rhodopsin: A site-directed spin-labeling study. *Biochemistry* **38,** 7931–7937.

Altenbach, C., Klein-Seetharaman, J., Hwa, J., Khorana, H. G., and Hubbell, W. L. (1999b). Structural features and light-dependent changes in the sequence 59–75 connecting helices I and II in rhodopsin: A site-directed spin-labeling study. *Biochemistry* **38,** 7945–7949.

Altenbach, C., Klein-Seetharaman, J., Cai, K., Khorana, H. G., and Hubbell, W. L. (2001). Structure and rhodopsin: Mapping light-dependent changes in distance between residue 316 in helix 8 and residues in the sequence 60–75, covering the cytoplasmic end of helices TM1 and TM2 and their connection loop CL1. *Biochemistry* **40,** 15493–15500.

Angers, S., Salahpour, A., and Bouvier, M. (2002). Dimerization: An emerging concept for G protein-coupled receptor ontogeny and function. *Annu. Rev. Pharmacol. Toxicol.* **42,** 409–435.

Archer, E., Maigret, B., Escrieut, C., Pradayrol, L., and Fourmy, D. (2003). Rhodopsin crystal: New template yielding realistic models of G-protein-coupled receptors? *Trends Pharmacol. Sci.* **24,** 36–40.

Arnis, S., Fahmy, K., Hofmann, K. P., and Sakmar, T. P. (1994). A conserved carboxylic acid group mediates light-dependent proton uptake and signaling by rhodopsin. *J. Biol. Chem.* **269,** 23879–23881.

Attwood, T. K., and Findlay, J. B. (1994). Fingerprinting G-protein-coupled receptors. *Protein Eng.* **7,** 195–203.

Azzi, M., Charest, P. G., Angers, S., Rousseau, G., Kohout, T., Bouvier, M., and Pineyro, G. (2003). Beta-arrestin-mediated activation of MAPK by inverse agonists reveals distinct active conformations for G protein-coupled receptors. *Proc. Natl. Acad. Sci. USA* **100,** 11406–11411.

Baker, J. G., Hall, I. P., and Hill, S. J. (2003). Agonist actions of "beta-blockers" provide evidence for two agonist activation sites or conformations of the human beta1-adrenoceptor. *Mol. Pharmacol.* **63,** 1312–1321.

Balmforth, A. J., Lee, A. J., Warburton, P., Donnelly, D., and Ball, S. G. (1997). The conformational change responsible for AT1 receptor activation is dependent upon two juxtaposed asparagine residues on transmembrane helices III and VII. *J. Biol. Chem.* **272,** 4245–4251.

Baneres, J. L., and Parello, J. (2003). Structure-based analysis of GPCR function: Evidence for a novel pentameric assembly between the dimeric leukotriene B4 receptor BLT1 and the G-protein. *J. Mol. Biol.* **329,** 815–829.

Ballesteros, J. A., Jensen, A. D., Liapakis, G., Rasmussen, S. G., Shi, L., Gether, U., and Javitch, J. A. (2001a). Activation of the beta 2-adrenergic receptor involves disruption of an ionic lock between the cytoplasmic ends of transmembrane segments 3 and 6. *J. Biol. Chem.* **276,** 29171–29177.

Ballesteros, J. A., Shi, L., and Javitch, J. A. (2001b). Structural mimicry in G protein-coupled receptors: Implications of the high-resolution structure of rhodopsin for structure-function analysis of rhodopsin-like receptors. *Mol. Pharmacol.* **60,** 1–19.

Bulenger, S., Marullo, S., and Bouvier, M. (2005). Emerging role of homo- and heterodimerization in G-protein-coupled receptor biosynthesis and maturation. *Trends Pharmacol. Sci.* **26,** 131–137.

Cerione, R. A., Staniszewski, C., Benovic, J. L., Lefkowitz, R. J., Caron, M. G., Gierschik, P., Somers, R., Spiegel, A. M., Codina, J., and Birnbaumer, L. (1985). Specificity of the functional interactions of the $\beta$-adrenergic receptor and rhodopsin with guanine nucleotide regulatory proteins reconstituted in phospholipid vesicles. *J. Biol. Chem.* **260,** 1493–1500.

Del Carmine, R., Molinari, P., Sbraccia, M., Ambrosio, C., and Costa, T. (2004). "Induced-fit" mechanism for catecholamine binding to the beta2-adrenergic receptor. *Mol. Pharmacol.* **66,** 356–363.

Devi, L. A. (2001). Heterodimerization of G-protein-coupled receptors: Pharmacology, signaling and trafficking. *Trends Pharmacol. Sci.* **22,** 532–537.

Dunham, T. D., and Farrens, D. L. (1999). Conformational changes in rhodopsin. Movement of helix f detected by site-specific chemical labeling and fluorescence spectroscopy. *J. Biol. Chem.* **274,** 1683–1690.

Farrens, D. L., Altenbach, C., Yang, K., Hubbell, W. L., and Khorana, H. G. (1996). Requirement of rigid-body motion of transmembrane helices for light activation of rhodopsin. *Science* **274,** 768–770.

Fredriksson, R., Lagerstrom, M. C., Lundin, L. G., and Schioth, H. B. (2003). The G-protein-coupled receptors in the human genome form five main families. Phylogenetic analysis, paralogon groups, and fingerprints. *Mol. Pharmacol.* **63,** 1256–1272.

Gaborik, Z., Jagadeesh, G., Zhang, M., Spat, A., Catt, K. J., and Hunyady, L. (2003). The role of a conserved region of the second intracellular loop in AT1 angiotensin receptor activation and signaling. *Endocrinology* **144,** 2220–2228.

Gether, U., Ballesteros, J. A., Seifert, R., Sanders-Bush, E., Weinstein, H., and Kobilka, B. K. (1997a). Structural instability of a constitutively active G protein-coupled receptor. Agonist-independent activation due to conformational flexibility. *J. Biol. Chem.* **272,** 2587–2590.

Gether, U., Lin, S., Ghanouni, P., Ballesteros, J. A., Weinstein, H., and Kobilka, B. K. (1997b). Agonists induce conformational changes in transmembrane domains III and VI of the beta2 adrenoceptor. *EMBO J.* **16,** 6737–6747.

Ghanouni, P., Schambye, H., Seifert, R., Lee, T. W., Rasmussen, S. G., Gether, U., and Kobilka, B. K. (2000). The effect of pH on beta(2) adrenoceptor function. Evidence for protonation-dependent activation. *J. Biol. Chem.* **275,** 3121–3127.

Ghanouni, P., Gryczynski, Z., Steenhuis, J. J., Lee, T. W., Farrens, D. L., Lakowicz, J. R., and Kobilka, B. K. (2001a). Functionally different agonists induce distinct conformations in the G protein coupling domain of the beta 2 adrenergic receptor. *J. Biol. Chem.* **276,** 24433–24436.

Ghanouni, P., Steenhuis, J. J., Farrens, D. L., and Kobilka, B. K. (2001b). Agonist-induced conformational changes in the G-protein-coupling domain of the beta 2 adrenergic receptor. *Proc. Natl. Acad. Sci. USA* **98,** 5997–6002.

Greasley, P. J., Fanelli, F., Rossier, O., Abuin, L., and Cotecchia, S. (2002). Mutagenesis and modelling of the alpha(1b)-adrenergic receptor highlight the role of the helix 3/helix 6 interface in receptor activation. *Mol. Pharmacol.* **61,** 1025–1032.

Howard, A. D., McAllister, G., Feighner, S. D., Liu, Q., Nargund, R. P., Van der Ploeg, L. H., and Patchett, A. A. (2001). Orphan G-protein-coupled receptors and natural ligand discovery. *Trends Pharmacol. Sci.* **22,** 132–140.

Hubbell, W. L., Altenbach, C., Hubbell, C. M., and Khorana, H. G. (2003). Rhodopsin structure, dynamics, and activation: A perspective from crystallography, site-directed spin labeling, sulfhydryl reactivity, and disulfide cross-linking. *Adv. Protein Chem.* **63,** 243–290.

Javitch, J. A. (2004). The ants go marching two by two: Oligomeric structure of G-protein-coupled receptors. *Mol. Pharmacol.* **66,** 1077–1082.

Javitch, J. A., Fu, D., Liapakis, G., and Chen, J. (1997). Constitutive activation of the beta2 adrenergic receptor alters the orientation of its sixth membrane-spanning segment. *J. Biol. Chem.* **272,** 18546–18549.

Jensen, A. D., Guarnieri, F., Rasmussen, S. G., Asmar, F., Ballesteros, J. A., and Gether, U. (2001). Agonist-induced conformational changes at the cytoplasmic side of transmembrane 6 in the beta 2 adrenergic receptor mapped by site-selective fluorescent labeling. *J. Biol. Chem.* **276,** 9279–9290.

Ji, T. H., Grossmann, M., and Ji, I. (1998). G protein-coupled receptors. I. Diversity of receptor-ligand interactions. *J. Biol. Chem.* **273,** 17299–17302.

Jongejan, A., Bruysters, M., Ballesteros, J. A., Haaksma, E., Bakker, R. A., Pardo, L., and Leurs, R. (2005). Linking agonist binding to histamine H1 receptor activation. *Nat. Chem. Biol.* **1,** 98–103.

Kenakin, T. (2001). Inverse, protean, and ligand-selective agonism: Matters of receptor conformation. *FASEB J.* **15,** 598–611.

Kenakin, T. (2002). Efficacy at G-protein-coupled receptors. *Nat. Rev. Drug Discov.* **1,** 103–110.

Kenakin, T. (2003). Ligand-selective receptor conformations revisited: The promise and the problem. *Trends Pharmacol. Sci.* **24,** 346–354.

Kim, J. M., Altenbach, C., Thurmond, R. L., Khorana, H. G., and Hubbell, W. L. (1997). Structure and function in rhodopsin: Rhodopsin mutants with a neutral amino acid at E134 have a partially activated conformation in the dark state. *Proc. Natl. Acad. Sci. USA* **94,** 14273–14278.

Knoflach, F., Mutel, V., Jolidon, S., Kew, J. N., Malherbe, P., Vieira, E., Wichmann, J., and Kemp, J. A. (2001). Positive allosteric modulators of metabotropic glutamate 1 receptor: Characterization, mechanism of action, and binding site. *Proc. Natl. Acad. Sci. USA* **98,** 13402–13407.

Kobilka, B. K., Kobilka, T. S., Daniel, K., Regan, J. W., Caron, M. G., and Lefkowitz, R. J. (1988). Chimeric alpha 2-,beta 2-adrenergic receptors: Delineation of domains involved in effector coupling and ligand binding specificity. *Science* **240,** 1310–1316.

Kolakowski, L. F., Jr. (1994). Gcrdb: A G-protein-coupled receptor database. *Receptors Channels* **2,** 1–7.

Krebs, A., Edwards, P. C., Villa, C., Li, J., and Schertler, G. F. (2003). The three-dimensional structure of bovine rhodopsin determined by electron cryomicroscopy. *J. Biol. Chem.* **278,** 50217–50225.

Kudo, M., Osuga, Y., Kobilka, B. K., and Hsueh, A. J. W. (1996). Transmembrane regions V and VI of the human luteinizing hormone receptor are required for constitutive activation by a mutation in the third intracellular loop. *J. Biol. Chem.* **271,** 22470–22478.

Kunishima, N., Shimada, Y., Tsuji, Y., Sato, T., Yamamoto, M., Kumasaka, T., Nakanishi, S., Jingami, H., and Morikawa, K. (2000). Structural basis of glutamate recognition by a dimeric metabotropic glutamate receptor. *Nature* **407,** 971–977.

Leff, P. (1995). The two-state model of receptor activation [see comments]. *Trends Pharmacol. Sci.* **16,** 89–97.

Lefkowitz, R. J., and Shenoy, S. K. (2005). Transduction of receptor signals by beta-arrestins. *Science* **308,** 512–517.

Lefkowitz, R. J., Cotecchia, S., Samama, P., and Costa, T. (1993). Constitutive activity of receptors coupled to guanine nucleotide regulatory proteins. *Trends Pharmacol. Sci.* **14,** 303–307.

Li, J., Edwards, P. C., Burghammer, M., Villa, C., and Schertler, G. F. (2004). Structure of bovine rhodopsin in a trigonal crystal form. *J. Mol. Biol.* **343,** 1409–1438.

Liang, Y., Fotiadis, D., Filipek, S., Saperstein, D. A., Palczewski, K., and Engel, A. (2003). Organization of the G protein-coupled receptors rhodopsin and opsin in native membranes. *J. Biol. Chem.* **278,** 21655–21662.

Liapakis, G., Ballesteros, J. A., Papachristou, S., Chan, W. C., Chen, X., and Javitch, J. A. (2000). The forgotten serine. A critical role for ser-2035.42 in ligands binding to and activation of the beta 2-adrenergic receptor. *J. Biol. Chem.* **275,** 37779–37788.

Liapakis, G., Chan, W. C., Papadokostaki, M., and Javitch, J. A. (2004). Synergistic contributions of the functional groups of epinephrine to its affinity and efficacy at the beta2 adrenergic receptor. *Mol. Pharmacol.* **65,** 1181–1190.

Liggett, S. B., Caron, M. G., Lefkowitz, R. J., and Hnatowich, M. (1991). Coupling of a mutated form of the human beta 2-adrenergic receptor to Gi and Gs. Requirement for multiple cytoplasmic domains in the coupling process. *J. Biol. Chem.* **266,** 4816–4821.

Lin, S. W., and Sakmar, T. P. (1996). Specific tryptophan UV-absorbance changes are probes of the transition of rhodopsin to its active state. *Biochemistry* **35,** 11149–11159.

Luttrell, L. M., and Lefkowitz, R. J. (2002). The role of beta-arrestins in the termination and transduction of G-protein-coupled receptor signals. *J. Cell. Sci.* **115,** 455–465.

McLean, A. J., Zeng, F. Y., Behan, D., Chalmers, D., and Milligan, G. (2002). Generation and analysis of constitutively active and physically destabilized mutants of the human beta(1)-adrenoceptor. *Mol. Pharmacol.* **62,** 747–755.

Mesnier, D., and Baneres, J. L. (2004). Cooperative conformational changes in a G-protein-coupled receptor dimer, the leukotriene b(4) receptor blt1. *J. Biol. Chem.* **279,** 49664–49670.

Mirzadegan, T., Benko, G., Filipek, S., and Palczewski, K. (2003). Sequence analyses of G-protein-coupled receptors: Similarities to rhodopsin. *Biochemistry* **42,** 2759–2767.

Nie, J., and Lewis, D. L. (2001). Structural domains of the CB1 cannabinoid receptor that contribute to constitutive activity and G-protein sequestration. *J. Neurosci.* **21,** 8758–8764.

Nishi, S., Nakabayashi, K., Kobilka, B., and Hsueh, A. J. (2002). The ectodomain of the luteinizing hormone receptor interacts with exoloop 2 to constrain the transmembrane region: Studies using chimeric human and fly receptors. *J. Biol. Chem.* **277,** 3958–3964.

Noda, K., Feng, Y. H., Liu, X. P., Saad, Y., Husain, A., and Karnik, S. S. (1996). The active state of the AT1 angiotensin receptor is generated by angiotensin II induction. *Biochemistry* **35,** 16435–16442.

O'Dowd, B. F., Hnatowich, M., Regan, J. W., Leader, W. M., Caron, M. G., and Lefkowitz, R. J. (1988). Site-directed mutagenesis of the cytoplasmic domains of the human $\beta_2$-adrenergic receptor. Localization of regions involved in G protein-receptor coupling. *J. Biol. Chem.* **263,** 15985–15992.

Okada, T. (2004). X-ray crystallographic studies for ligand-protein interaction changes in rhodopsin. *Biochem. Soc. Trans.* **32,** 738–741.

Okada, T., Le Trong, I., Fox, B. A., Behnke, C. A., Stenkamp, R. E., and Palczewski, K. (2000). X-ray diffraction analysis of three-dimensional crystals of bovine rhodopsin obtained from mixed micelles. *J. Struct. Biol.* **130,** 73–80.

Okada, T., Fujiyoshi, Y., Silow, M., Navarro, J., Landau, E. M., and Shichida, Y. (2002). Functional role of internal water molecules in rhodopsin revealed by X-ray crystallography. *Proc. Natl. Acad. Sci. USA* **99,** 5982–5987.

Okada, T., Sugihara, M., Bondar, A. N., Elstner, M., Entel, P., and Buss, V. (2004). The retinal conformation and its environment in rhodopsin in light of a new 2.2 a crystal structure. *J. Mol. Biol.* **342,** 571–583.

Palanche, T., Ilien, B., Zoffmann, S., Reck, M. P., Bucher, B., Edelstein, S. J., and Galzi, J. L. (2001). The neurokinin a receptor activates calcium and camp responses through distinct conformational states. *J. Biol. Chem.* **276,** 34853–34861.

Palczewski, K., Kumasaka, T., Hori, T., Behnke, C. A., Motoshima, H., Fox, B. A., Le Trong, D. C., Teller, D. C., Okada, T., Stenkamp, R. E., Yamamoto, M., and Miyano, M. (2000). Crystal structure of rhodopsin: A G protein-coupled receptor [see comments]. *Science* **289,** 739–745.

Parnot, C., Miserey-Lenkei, S., Bardin, S., Corvol, P., and Clauser, E. (2002). Lessons from constitutively active mutants of G protein-coupled receptors. *Trends Endocrinol. Metab.* **13,** 336–343.

Petaja-Repo, U. E., Hogue, M., Bhalla, S., Laperriere, A., Morello, J. P., and Bouvier, M. (2002). Ligands act as pharmacological chaperones and increase the efficiency of delta opioid receptor maturation. *EMBO J.* **21,** 1628–1637.

Pin, J. P., Galvez, T., and Prezeau, L. (2003). Evolution, structure, and activation mechanism of family 3/c G-protein-coupled receptors. *Pharmacol. Ther.* **98,** 325–354.

Pin, J. P., Kniazeff, J., Goudet, C., Bessis, A. S., Liu, J., Galvez, T., Acher, F., Rondard, P., and Prezeau, L. (2004). The activation mechanism of class-c G-protein coupled receptors. *Biol. Cell* **96,** 335–342.

Rasmussen, S. G., Jensen, A. D., Liapakis, G., Ghanouni, P., Javitch, J. A., and Gether, U. (1999). Mutation of a highly conserved aspartic acid in the beta2 adrenergic receptor: Constitutive activation, structural instability, and conformational rearrangement of transmembrane segment 6. *Mol. Pharmacol.* **56,** 175–184.

Ray, K., and Northup, J. (2002). Evidence for distinct cation and calcimimetic compound (NPS 568) recognition domains in the transmembrane regions of the human $Ca^{2+}$ receptor. *J. Biol. Chem.* **277,** 18908–18913.

Ridge, K. D., Abdulaev, N. G., Sousa, M., and Palczewski, K. (2003). Phototransduction: Crystal clear. *Trends Biochem. Sci.* **28,** 479–487.

Ruprecht, J. J., Mielke, T., Vogel, R., Villa, C., and Schertler, G. F. (2004). Electron crystallography reveals the structure of metarhodopsin I. *EMBO J.* **23,** 3609–3620.

Sakmar, T. P. (2002). Structure of rhodopsin and the superfamily of seven-helical receptors: The same and not the same. *Curr. Opin. Cell Biol.* **14,** 189–195.

Sakmar, T. P., Franke, R. R., and Khorana, H. G. (1991). The role of the retinylidene schiff base counterion in rhodopsin in determining wavelength absorbance and schiff base pKa. *Proc. Natl. Acad. Sci. USA* **88,** 3079–3083.

Samama, P., Cotecchia, S., Costa, T., and Lefkowitz, R. J. (1993). A mutation-induced activated state of the beta 2-adrenergic receptor. Extending the ternary complex model. *J. Biol. Chem.* **268,** 4625–4636.

Scheer, A., Fanelli, F., Costa, T., De Benedetti, P. G., and Cotecchia, S. (1996). Constitutively active mutants of the alpha 1b-adrenergic receptor: Role of highly conserved polar amino acids in receptor activation. *EMBO J.* **15,** 3566–3578.

Schertler, G. F. (2005). Structure of rhodopsin and the metarhodopsin I photointermediate. *Curr. Opin. Struct. Biol.* **15,** 408–415.

Schertler, G. F., Villa, C., and Henderson, R. (1993). Projection structure of rhodopsin. *Nature* **362,** 770–772.

Schoneberg, T., Liu, J., and Wess, J. (1995). Plasma membrane localization and functional rescue of truncated forms of a G protein-coupled receptor. *J. Biol. Chem.* **270,** 18000–18006.

Shapiro, D. A., Kristiansen, K., Weiner, D. M., Kroeze, W. K., and Roth, B. L. (2002). Evidence for a model of agonist-induced activation of 5-hydroxytryptamine 2a serotonin receptors that involves the disruption of a strong ionic interaction between helices 3 and 6. *J. Biol. Chem.* **277,** 11441–11449.

Sharma, S., and Sharma, S. C. (1997). An update on eicosanoids and inhibitors of cyclooxygenase enzyme systems. *Indian J. Exp. Biol.* **35,** 1025–1031.

Sheikh, S. P., Zvyaga, T. A., Lichtarge, O., Sakmar, T. P., and Bourne, H. R. (1996). Rhodopsin activation blocked by metal-ion-binding sites linking transmembrane helices C and F. *Nature* **383,** 347–350.

Sheikh, S. P., Vilardarga, J. P., Baranski, T. J., Lichtarge, O., Iiri, T., Meng, E. C., Nissenson, R. A., and Bourne, H. R. (1999). Similar structures and shared switch mechanisms of the beta2-adrenoceptor and the parathyroid hormone receptor. Zn(II) bridges between helices III and VI block activation. *J. Biol. Chem.* **274,** 17033–17041.

Shi, L., Liapakis, G., Xu, R., Guarnieri, F., Ballesteros, J. A., and Javitch, J. A. (2002). Beta2 adrenergic receptor activation. Modulation of the proline kink in transmembrane 6 by a rotamer toggle switch. *J. Biol. Chem.* **277,** 40989–40996.

Sprang, S. R. (1997). G protein mechanisms: Insights from structural analysis. *Annu. Rev. Biochem.* **66,** 639–678.

Strader, C. D., Candelore, M. R., Hill, W. S., Dixon, R. A., and Sigal, I. S. (1989a). A single amino acid substitution in the beta-adrenergic receptor promotes partial agonist activity from antagonists. *J. Biol. Chem.* **264,** 16470–16477.

Strader, C. D., Candelore, M. R., Hill, W. S., Sigal, I. S., and Dixon, R. A. (1989b). Identification of two serine residues involved in agonist activation of the beta-adrenergic receptor. *J. Biol. Chem.* **264,** 13572–13578.

Strader, C. D., Sigal, I. S., and Dixon, R. A. (1989c). Structural basis of beta-adrenergic receptor function. *FASEB J.* **3,** 1825–1832.

Swaminath, G., Xiang, Y., Lee, T. W., Steenhuis, J., Parnot, C., and Kobilka, B. K. (2004). Sequential binding of agonists to the {beta}2 adrenoceptor: Kinetic evidence for intermediate conformational states. *J. Biol. Chem.* **279,** 686–691.

Swaminath, G., Deupi, X., Lee, T. W., Zhu, W., Thian, F. S., Kobilka, T. S., and Kobilka, B. (2005). Probing the beta2 adrenoceptor binding site with catechol reveals differences in binding and activation by agonists and partial agonists. *J. Biol. Chem.* **280,** 22165–22171.

Tateyama, M., Abe, H., Nakata, H., Saito, O., and Kubo, Y. (2004). Ligand-induced rearrangement of the dimeric metabotropic glutamate receptor 1alpha. *Nat. Struct. Mol. Biol.* **11,** 637–642.

Teller, D. C., Okada, T., Behnke, C. A., Palczewski, K., and Stenkamp, R. E. (2001). Advances in determination of a high-resolution three-dimensional structure of rhodopsin, a model of G-protein-coupled receptors (GPCRs). *Biochemistry* **40,** 7761–7772.

Tsuchiya, D., Kunishima, N., Kamiya, N., Jingami, H., and Morikawa, K. (2002). Structural views of the ligand-binding cores of a metabotropic glutamate receptor complexed with an antagonist and both glutamate and Gd3+. *Proc. Natl. Acad. Sci. USA* **99,** 2660–2665.

Ward, S. D., Hamdan, F. F., Bloodworth, L. M., and Wess, J. (2002). Conformational changes that occur during M3 muscarinic acetylcholine receptor activation probed by the use of an in situ disulfide cross-linking strategy. *J. Biol. Chem.* **277,** 2247–2257.

Wei, H., Ahn, S., Shenoy, S. K., Karnik, S. S., Hunyady, L., Luttrell, L. M., and Lefkowitz, R. J. (2003). Independent beta-arrestin 2 and G protein-mediated pathways for angiotensin II activation of extracellular signal-regulated kinases 1 and 2. *Proc. Natl. Acad. Sci. USA* **100,** 10782–10787.

Weiss, J. M., Morgan, P. H., Lutz, M. W., and Kenakin, T. P. (1996). The cubic ternary complex receptor-occupancy model. III. Resurrecting efficacy. *J. Theor. Biol.* **181,** 381–397.

Wieland, K., Zuurmond, H. M., Krasel, C., Ijzerman, A. P., and Lohse, M. J. (1996). Involvement of Asn-293 in stereospecific agonist recognition and in activation of the beta 2-adrenergic receptor. *Proc. Natl. Acad. Sci. USA* **93,** 9276–9281.

Wilson, A. L., Guyer, C. A., Cragoe, E. J., and Limbird, L. E. (1990). The hydrophobic tryptic core of the porcine alpha 2-adrenergic receptor retains allosteric modulation of binding by Na+, H+, and 5-amino-substituted amiloride analogs. *J. Biol. Chem.* **265,** 17318–17322.

Yi, C. S., Song, Y. S., Ryu, K. S., Sohn, J., Ji, I., and Ji, T. H. (2002). Common and differential mechanisms of gonadotropin receptors. *Cell. Mol. Life Sci.* **59,** 932–940.

# KINETIC ANALYSIS OF G PROTEIN–COUPLED RECEPTOR SIGNALING USING FLUORESCENCE RESONANCE ENERGY TRANSFER IN LIVING CELLS

By MARTIN J. LOHSE, CARSTEN HOFFMANN, VIACHESLAV O. NIKOLAEV, JEAN-PIERRE VILARDAGA, AND MORITZ BÜNEMANN

Institute of Pharmacology and Toxicology, University of Würzburg, D-97078 Würzburg, Germany



## Abstract

We describe and review methods for the kinetic analysis of G protein–coupled receptor (GPCR) activation and signaling that are based on optical methods. In particular, we describe the use of fluorescence resonance energy transfer (FRET) as a means of analyzing conformational changes within a single protein (for example a receptor) or between subunits of a protein complex (such as a G protein heterotrimer) and finally between distinct proteins (such as a receptor and a G protein). These methods allow the analysis of signaling kinetics in intact cells with proteins that retain their essential functional properties. They have produced a number of unexpected results: fast receptor activation kinetics in the millisecond range, similarly fast kinetics for receptor–G protein interactions, but much slower activation kinetics for G protein activation.

## I. Introduction

In spite of their fundamental importance for the rapid regulation of many biological systems, relatively little is known about the kinetics of G protein–coupled receptor (GPCR) activation and signaling. Most

ADVANCES IN PROTEIN CHEMISTRY, Vol. 74
DOI: 10.1016/S0065-3233(07)74005-6


0065-3233/07 $35.00

pharmacological studies—for example most ligand-binding studies as well as studies investigating second messenger generation—are usually done under equilibrium conditions and thus neglect kinetic aspects. Several temporally resolved methods have been widely used to investigate receptor–G protein coupling, such as the agonist-induced binding of GTP-$\gamma$-S; however, these assays can be done only on membrane preparations which may distort the physiological setting.

The speed of GPCR-induced signals can also be gauged from various types of physiological responses to agonists. Observation of everyday life tells us that cardiac frequency is increased in response to sympathetic stress within seconds, and electrophysiological experiments that link channel opening or closing to GPCR activation provide even shorter, subsecond time intervals required for GPCR signals.

A notable exception to our lack of knowledge about the kinetics of receptor-mediated signals is the system of light perception by rhodopsin. The fascinating sensitivity and speed of this system together with the possibility to exactly trigger its activation with light pulses and finally the ability to investigate rhodopsin via the light absorption by retinal have attracted a number of biophysicists to this system and led to the development of a large number of optical assays that permit the monitoring of different steps of rhodopsin activation by light and of the subsequent signaling steps (Okada *et al.*, 2001).

More recently, a few laboratories have attempted to develop optical methods for the analysis of other receptor systems. In view of the much lower concentrations of the components of such signaling systems in intact cells compared to the rhodopsin system, much more sensitive methods were required for their study. This is why the use of fluorescence has gained foremost importance for these systems. Two applications of fluorescence have gained particular importance in these studies: fluorescence quenching and fluorescence resonance energy transfer (FRET). In *fluorescence quenching*, one uses the—often unwanted—sensitivity of fluorescence to the immediate surrounding of the fluorophore. The fluorescence intensity of many fluorophores is either increased or decreased when its immediate environment becomes more hydrophobic or more hydrophilic. Thus, translocation of a fluorophore from the cytosol to the cell membrane or changes in its orientation within a protein or in the cell membrane can have dramatic consequences in its fluorescence emission. In *FRET*, one uses the nonradiative transfer of energy from a fluorophore with a shorter emission wavelength to another one with a longer emission wavelength, which is in proximity to the first fluorophore ($\sim$<10 nm); this transfer of energy results in less emission at the shorter wavelength and the appearance of emission at the longer wavelength. FRET is very sensitive—with the sixth power—to changes in the distance between the two fluorophores and, thus, is a means to report both conformational changes and protein–protein associations. A variant of FRET is bioluminescence resonance energy transfer (BRET), where energy is transferred from a luminescent moiety to a fluorophore.

Fluorescence-quenching techniques have been used by Kobilka *et al.*, who developed chemical labeling of purified $\beta_2$-adrenergic receptors in order to study ligand-induced conformational changes. Agonist-induced activation of GPCRs appears to cause a relative change of the receptor's transmembrane helices, most notably of helix III and VI (Farrens *et al.*, 1996; Sheik *et al.*, 1996; Ward *et al.*, 2002; Wieland *et al.*, 1996); this process appears to be critical for agonist-induced but presumably not for constitutive activation of receptors (Hannawacker *et al.*, 2002). Based on this concept, Kobilka *et al.* have begun to label purified receptors with fluorophores in order to find conformationally sensitive receptor fluorescence due to fluorescence quenching. These studies confirmed not only the notion of agonist-induced relative movements of helices III and VI but also provided the first method to monitor directly conformational changes in a G protein–coupled receptor (Gether *et al.*, 1995; Ghanouni *et al.*, 2001) other than rhodopsin. This approach also showed that partial agonists caused only partial changes in the receptor fluorescence (reviewed in Gether, 2000), and suggested a sequential mode of activation of receptors involving intermediate conformational states (Swaminath *et al.*, 2004, 2005).

This technology allows the sensitive analysis of small conformational changes and will ultimately allow an understanding of the receptor activation mechanism. These studies have shown two distinct switch mechanisms in the $\beta_2$-adrenergic receptor, both located in the third intracellular loop and the cytoplasmic end of transmembrane helix VI: an ionic lock mechanism that involves an interaction of the cytoplasmic ends of helix VI and helix III (and is reported by a label in position 271 of the receptor), and a "rotamer toggle" switch mechanism that is reported by a label in position 265 (Yao *et al.*, 2006).

In terms of investigating the kinetics of this process, however, these studies have met with difficulties. The observed changes were clearly slower than biological responses to receptor activation, which can occur within seconds. Studies that used carefully reconstituted receptors observed agonist-induced changes of intramolecular receptor fluorescence on the timescale of about 30 s (Swaminath *et al.*, 2004, 2005; Yao *et al.*, 2006). Several possible reasons have been discussed that may explain these kinetic differences, most importantly the fact that purified, reconstituted receptors were used that may lack their natural environment such as the cell membrane but also G proteins and possibly other cellular components (Yao *et al.*, 2006).

The first application of FRET for the analysis of GPCR signaling was done by Heithier *et al.* (1992). These authors chemically labeled $\beta$-adrenergic receptors purified from turkey erythrocytes (a $\beta_1$-like subtype) and $G_o$ protein subunits purified from bovine brain. Both receptors and G proteins were labeled with small fluorophores and then reconstituted in phospholipid vesicles in order to study their interaction by FRET; the authors found affinity constants in the submicromolar range. Even though this was a value for a nonphysiological receptor–G protein pair, this study showed that such an approach can be used to determine affinity constants between receptors and G proteins, and that FRET is a tool to study receptor coupling and signaling.

In order to be able to study receptor signaling kinetics with optical methods in a more native environment, we set out to develop new FRET-based methods that would allow the study of these processes in intact cells. These methods are based on genetically encoded fluorescent tags (reviewed in Giepmans *et al.*, 2006; Miyawaki and Tsien, 2000)—either by using variants of green fluorescent proteins (GFPs) fused to the relevant proteins or by introducing small motifs that can be labeled in intact cells. At the same time, similar technologies based on BRET were developed by Bouvier *et al.* While FRET has the advantage of faster kinetics, BRET has a lower background and appears, thus, more useful for the detection of weak interactions, while its limited emission intensity requires longer periods of light acquisition and, hence, results in much lower temporal resolution.

## II. Assays and Methods

### A. *Principle of the Assays*

FRET requires the presence of two fluorophores, one with a shorter emission wavelength (donor) and another with a longer emission wavelength (acceptor). The fluorophores must be chosen such that there is sufficient overlap of the donor emission spectrum and the acceptor excitation spectrum. When FRET occurs, which requires the proximity of the two fluorophores, excitation of the donor results in transfer of energy to the acceptor and, hence, emission at the wavelength characteristic for the acceptor. FRET can be seen with various kinds of fluorophores, but most recently it has been used in particular with variants of GFPs because this permits FRET in intact cells. The most frequently used pairs of GFPs are the cyan fluorescent protein (CFP) and the yellow fluorescent protein (YFP) variants. The donor CFP is excited at its maximum

(436 nm) and emits with a maximum at 480 nm. If FRET occurs, the acceptor YFP becomes excited and emits with a maximum wavelength at 535 nm; at the same time, emission of CFP at 480 nm decreases. The transfer of energy also decreases the fluorescence lifetime of the donor, which is another measure of FRET; it can be used for fluorescence lifetime imaging (FLIM), which is less prone to artifacts that may result from quenching of one or the other fluorophore during an experiment.

The two fluorophores can be present in a single protein or in two separate proteins. In the first case, FRET can be used to monitor small changes in distance as they occur on changes in conformation. In the latter case, association of the two proteins results in appearance of FRET and their dissociation causes loss of FRET. The principle of these assays is depicted in Fig. 1. On irradiation of the CFP-moiety at 436 nm, there was emission not only at 480 nm, the emission wavelength of CFP, but also at 535 nm, the emission wavelength of YFP. The latter is due to FRET between the CFP- and YFP-moieties.

Several attempts have been undertaken to develop fluorescent labels that are also genetically encoded like the GFPs but do not have the large size of these fluorescent proteins. The only technique of this kind that, to our knowledge, has so far been used in the context of GPCRs is based on a small fluorescein derivative called FlAsH (for *fl*uorescein *ar*senical *h*airpin binder). FlAsH has been developed as a small fluorescent probe that only becomes fluorescent when bound to a specific sequence consisting of minimally six amino acids with two flanking cysteines on both sides (Gaietta *et al.*, 2002; Griffin *et al.*, 1998). FlAsH-binding sequences have been further optimized to give high affinity, 12 amino acid motifs, with 3 amino acids flanking the core CCPGCC sequence on both ends (Adams *et al.*, 2002; Martin *et al.*, 2005). These cysteines bind the FlAsH compound, which becomes fluorescent when bound. The fluorescent properties of FlAsH are comparable to those of YFP, and thus it can "replace" YFP in FRET experiments in combination with CFP (Hoffmann *et al.*, 2005).

We and others have developed a number of assays, all based on FRET, that permit recording of individual steps of GPCR signaling. The principles of these assays are depicted in Fig. 1: (1) agonist binding to receptors, (2) activation of the receptors involving a conformational switch, (3) association of activated receptors with G proteins, and (4) activation of G proteins involving rearrangement of their subunits. The design of the various assays is shown in Fig. 2 that illustrates the proteins involved and the sites of fluorophore attachment in the different assays. The key features of these assays are described below.

Agonist binding to receptors on intact cells has been studied by investigating FRET between a fluorescently labeled agonist and receptors

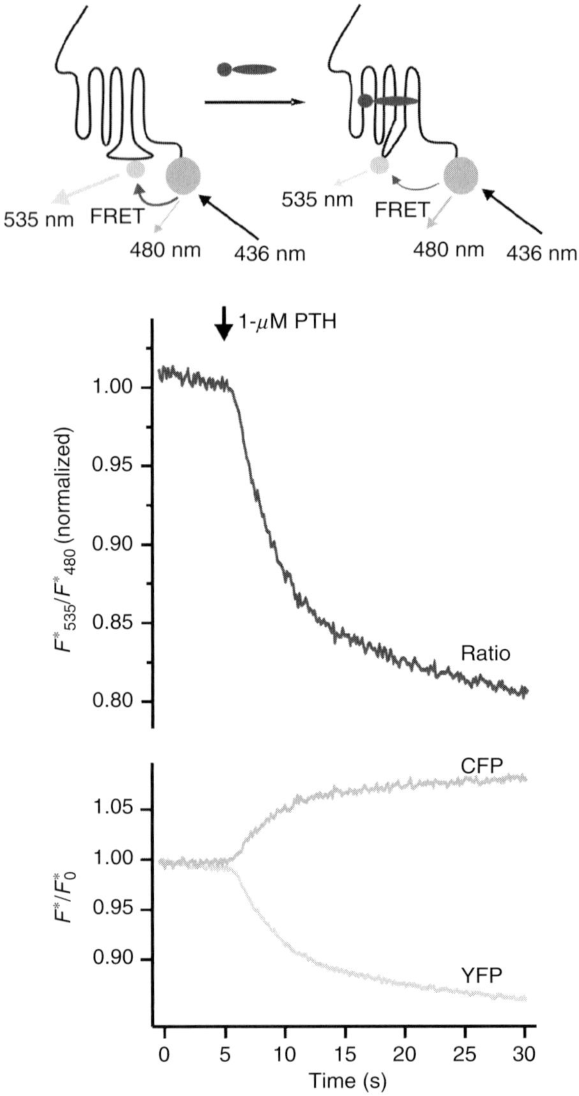


FIG. 1. Time-resolved detection of GPCR activation by single-cell FRET recording. PTH receptors were labeled with CFP and YFP in the third intracellular loop and at the C-terminus, respectively, and were expressed in HEK293 cells (top). Single-cell FRET recordings of CFP and YFP emissions (480 and 535 nm, respectively) on excitation of CFP at 436 nm allow to record a temporally resolved time course of the agonist-induced conformational change of the receptor. The agonist-induced changes are compatible with a movement of the two labels apart, which is reflected by a decrease in FRET between CFP and YFP. This decrease is evident from an increase in the CFP emission, a corresponding decrease in the YFP emission, and, hence, a decrease in the YFP/CFP emission ratio that follows a monoexponential time course (adapted from Vilardaga *et al.*, 2003). (See Color Insert.)

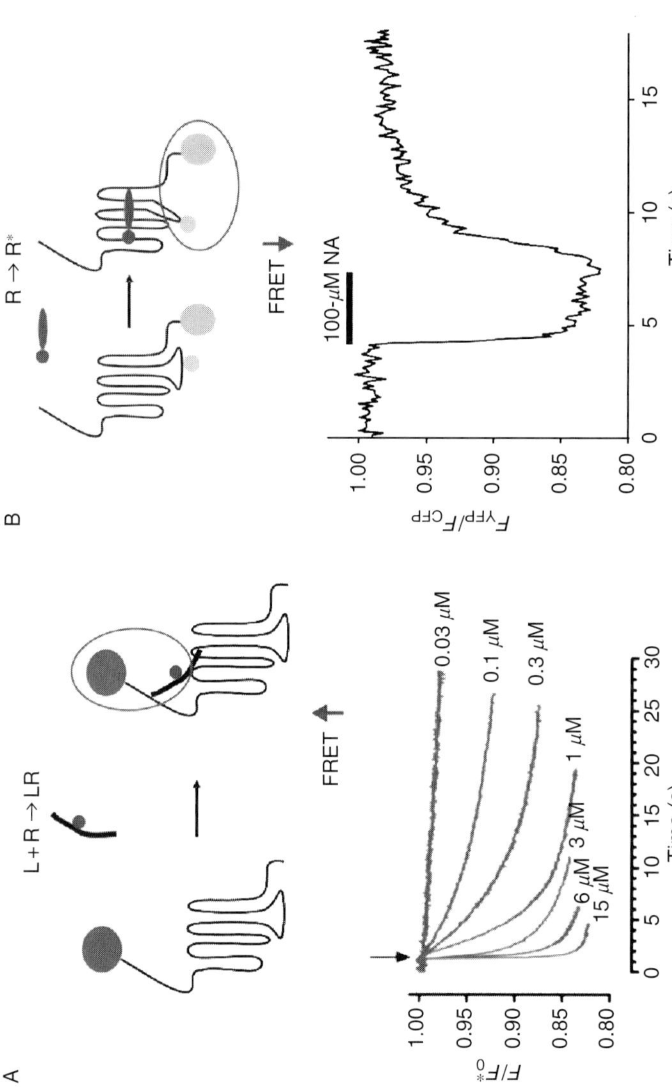


FIG. 2. (*continued*)

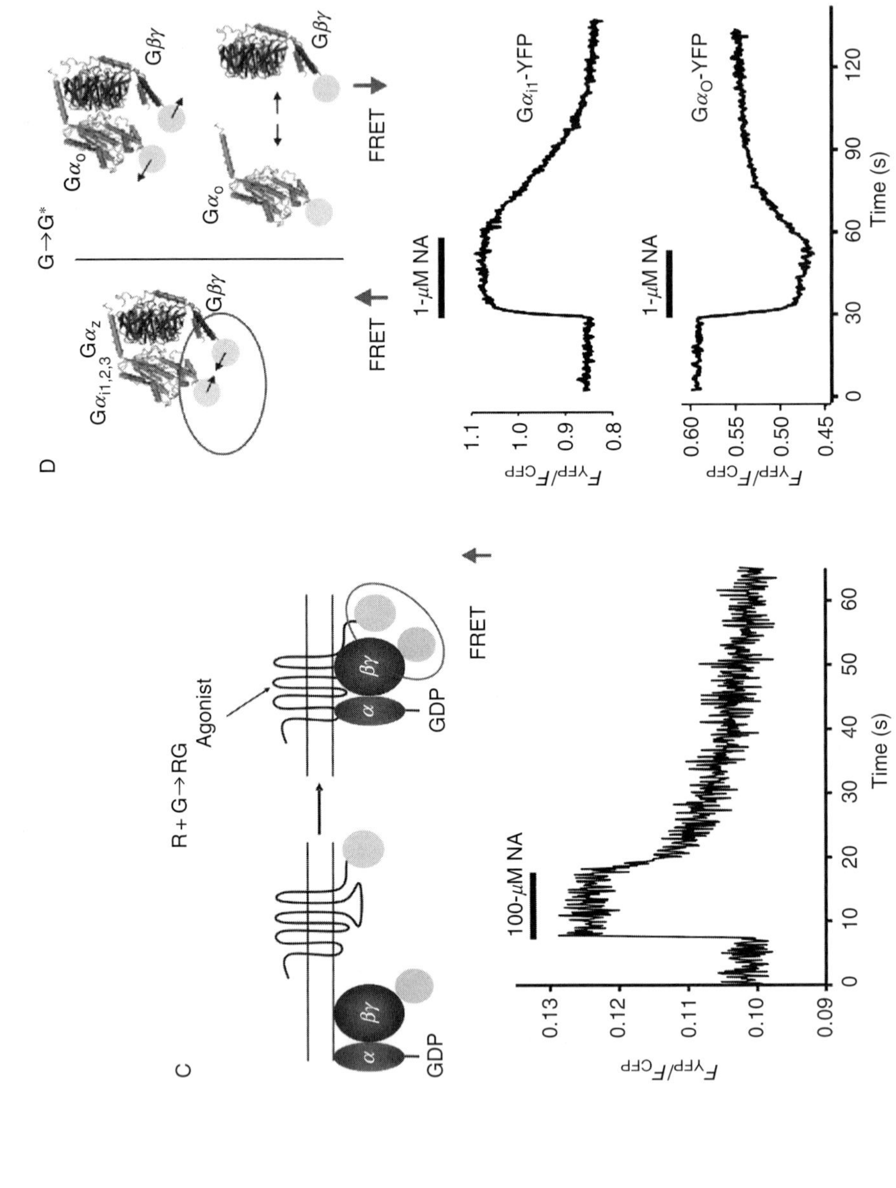
C
R + G→RG
Agonist
GDP
GDP
FRET
100-μM NA
F_YFP/F_CFP
Time (s)
D
G→G*
Gα_z
Gα_i1,2,3
Gα_o
Gβγ
FRET
1-μM NA
Gα_i1-YFP
Gα_O-YFP
F_YFP/F_CFP
Time (s)

tagged with GFP-variants (Castro *et al.*, 2005; Palanche *et al.*, 2001). In these assays, usually a small chemical label needs to be used for the agonist in order to maintain binding and activity. FRET occurs between this label and the GFP-tagged receptor as the agonist approaches the receptor and occupies its binding site. This FRET can be monitored as an emission of the longer wavelength label. More simply, it can be seen from a decrease in the emission of a GFP-tagged receptor when a red fluorescent dye, such as Texas Red (TR) or tetramethyl rhodamine (TMR), comes close to the GFP and takes up some of its energy by FRET. There is some evidence that the initial binding event and the subsequent occupancy of the binding site may be two kinetically separable steps, resulting in biphasic curves for this interaction (see below).

Similar fluorescent assays have also been described for the analysis of second messengers, cAMP and cGMP, and used for the analysis of their synthesis and degradation (Mongillo *et al.*, 2004; Nikolaev *et al.*, 2004, 2006a). These assays have been reviewed (Nikolaev and Lohse, 2006) and will not be discussed here. Electrophysiological methods appear to be best suited to monitor the kinetics of effector activation (Bünemann *et al.*, 1996; Pfaffinger *et al.*, 1985), since the activation of second messenger producing steps—such as adenylyl cyclase—is not as readily monitored. Such activation of G protein–regulated ion channels, for example the G protein–activated inward rectifying $K^+$-channel (GIRK), occurs with subsecond speed and depends on the type and density of the stimulating receptor (Bünemann *et al.*, 2001; Pfaffinger *et al.*, 1985).

---

FIG. 2. Principles of FRET assays for recording initial steps of GPCR-mediated signaling in intact cells: agonist binding (A), receptor activation (B), receptor–G protein interaction (C), and G protein activation (D). (A) The agonist–receptor interaction was measured by FRET between TMR-labeled PTH and GFP-labeled PTH receptors. Illustrated are recordings of the quenching of the GFP emission by FRET in response to application of various TMR-PTH concentrations (modified from Castro *et al.*, 2005). (B) The agonist-induced activation switch of $\alpha_{2A}$-adrenergic receptors was measured similarly as described above (Fig. 1). The third intracellular loop was labeled with FlAsH instead of YFP, and an agonist-induced increase in the distance between helix V and VII was detected by a decrease in FRET (Hoffmann *et al.*, 2005). (C) The interaction between YFP-labeled receptor and CFP-labeled G protein was detected by FRET. On agonist exposure, a fast increase in FRET indicated formation of the ternary complex between ligand, receptor, and G protein subunits. (D) Detection of G protein activation based on changes in FRET between YFP-labeled $\alpha$ and CFP-labeled $\beta\gamma$ subunits revealed differential activation patterns of $G_i$- versus $G_o$ proteins and suggested that $G_i$ protein subunits undergo a subunit rearrangement on activation instead of complete dissociation. (See Color Insert.)

Finally FRET assays can be used to monitor subsequent steps such as binding of $\beta$-arrestins to activated receptors (Bertrand *et al.*, 2002; Krasel *et al.*, 2005; Vilardaga *et al.*, 2003; Violin *et al.*, 2006).

### *B. Construction and Expression of Fluorescent Receptor and G Protein Constructs*

The most critical step in these experiments is the choice of the site for fluorescent labeling. All labels, in particular however the large GFP variants, carry the risk of impairing the function and disposition of the labeled protein, and care must be taken to assure that indeed the labeled protein retains essential properties of the parent protein, such as cell surface expression, ligand binding, and signaling in the case of receptors. The choice of the insertion site is facilitated by structural information, and by knowledge about functional sites in these structures.

In GPCRs themselves, we have generally inserted fluorescent labels at the C-terminus and in the third intracellular loop. C-terminal GFP variants as well as smaller labeling sequences are remarkably well tolerated in most receptors, such that expression, ligand binding, internalization, and even G protein activation are preserved. In some instances, however, they appear to impair cell surface expression. Labeling of the third intracellular loop, in contrast, is more problematic, and may result in significant loss of G protein activation (Hoffmann *et al.*, 2005); however, in other cases at least a significant fraction of G protein signaling can be preserved (Vilardaga *et al.*, 2003, 2005).

For the study of FRET involving G protein subunits, several sites for labeling have been used. In the $\beta$ and $\gamma$ subunits, both C- and N-terminal fusions have been made and shown to preserve dimer and heterotrimer formation. Surprisingly, this is also true for the C-terminal labeling of the $\gamma$ subunit which results in loss of the lipid modification site and, consequently, in reduced membrane targeting (Bünemann *et al.*, 2003). In the $\alpha$ subunits, the site of labeling appears to have a major influence on the results obtained (see below). Insertions of GFPs at position 91 gave rise to $G\alpha$ subunits that were functional with respect to proper receptor coupling as well as effector activation (Bünemann *et al.*, 2003). Insertion of YFP into position 121 of $G\alpha_{i1}$ (Gibson and Gilman, 2006) gave rise to higher expression levels and better preservation of guanine nucleotide exchange, but presumably resulted in a loss of $G_i$-typical behavior on activation (see below). No labeling of G protein subunits with small fluorophores such as FlAsH has been reported to date.

Identically labeled G protein subunits have also been used to study receptor–G protein interactions in intact cells, in combinations with receptors carrying C-terminal GFP-variants as fluorescent tags (Gales *et al.*, 2005; Hein *et al.*, 2005).

For the reasons detailed above, the GFP variants mostly used for such experiments are the YFP and CFP because of good spectral overlap and, hence, FRET. Their cDNAs are fused into the relevant sites, in general without the use of longer linkers. Transient and/or stable expression of these constructs can be achieved in standard cell lines, for example HEK293 and CHO cells, and also in primary cells and transgenic animals. As an alternative to YFP, the small fluorescein-based probe FlAsH has been used in FRET experiments to study the activation of receptors (Hoffmann *et al.*, 2005). FlAsH binds relatively specifically to short tetracysteine motifs, spaced apart by two central amino acids (i.e., CCXXCC); improved specificity of binding has been reported for the sequence CCPGCC (Gaietta *et al.*, 2002), and 12-amino acid sequences with even better specificity have recently been reported but so far not used in FRET experiments. FlAsH is membrane-permeable and is used immediately before an experiment in order to label the expressed protein that contains the binding motif; after application of FlAsH to cells, a careful washing procedure needs to be used in order to remove nonspecifically bound FlAsH (Hoffmann *et al.*, 2005).

### C. *Microscopic FRET Measurements and Imaging*

FRET recordings are best done with fluorescent microscopes studying individual cells, using objectives with high magnification (63× or 100×) and either photodiodes or CCD cameras to record the emission at the donor and the acceptor wavelengths and to determine the ratio of these emissions as a measure of FRET. While other methods to show FRET are also possible (and sometimes less prone to artifacts, see above), only ratiometric recording is suitable to resolve the fast kinetics encountered with receptors and G proteins. Recording from a single cell that is first visualized before the experiment permits recording from a single cell and, thus, visual control of proper expression and targeting of the proteins under investigation—such as membrane targeting of receptors and G protein, and expression at equal levels to permit optimal FRET. Recordings from cell populations are also possible by using objectives with smaller magnification, but such recordings generally have several disadvantages. Such recordings often include cells showing suboptimal expression and targeting (this is particularly important when transiently transfected cells are studied), they therefore generally show smaller signals, and do not allow accurate recordings of kinetic changes because agonists are likely to reach the different cells at slightly different time points after their addition. FRET and BRET recordings can also be done on microtiter plate readers, but good recordings—in particular for FRET—require that the samples can be focused.

For these studies, cells are grown on coverslips and then maintained in a physiological external buffer such as 137-mM NaCl, 5-mM KCl, 1-mM $CaCl_2$, 1-mM $MgCl_2$, 10-mM HEPES, 0.1% BSA, pH 7.4 at room temperature. If FlAsH is used instead of YFP, cells are first stained with FlAsH on the coverslip and then carefully washed as described in detail by Hoffmann *et al.* (2005). The coverslips are then placed on an inverted microscope (e.g., ZeissAxiovert200) equipped with an oil immersion 63× objective and a dual emission photometric system (e.g., Till Photonics). Excitation for standard recording is done at the optimum wavelength for the shorter wavelength fluorophore (=donor), that is in general CFP with an excitation at 436 nm. It is important to be able to switch the excitation light on and off to be able to adapt lighting to the type of experiment and to reduce photobleaching of the fluorophores; this can be achieved by using shutters or tunable light sources such as a polychrome IV (Till Photonics). Illumination times must be long enough to obtain stable recordings, but short enough in order to reduce photobleaching. In general, illumination and recording times of 5–40 ms for each data point are sufficient, and data points are collected (and samples excited) with a frequency between 1 and 75 Hz depending on the speed of the signal to be recorded. Emission at both wavelengths can then be recorded simultaneously using a beam splitter that divides the emissions of the donor and acceptor fluorophores (i.e., in general 480 nm for CFP and 535 nm for YFP or FlAsH). FRET can then be monitored as the emission ratio of YFP to CFP. Examples for filters and beam splitters in the emission path for such experiments are 535 ± 15 nm (YFP) and 480 ± 20 nm (CFP) with a beam splitter DCLP 505 nm. Excitation can be done at 436 ± 10 nm with a beam splitter DCLP 460 nm. The emissions need to be corrected by the respective spillover between the channels (i.e., direct excitation of YFP by 436-nm light and emission of CFP into the 535-nm channel) before a corrected ratio $F^{*}_{535} F^{*}_{480}$ can be calculated. The factors for such corrections need to be determined often, since they vary not only with the type of instrumentation but also with the age of light sources and other specific influences.

A number of controls need to be done to verify that the observed changes are indeed due to FRET (and not, e.g., to fluorescence quenching). This is particularly important when the fluorophores are located on two different proteins and thus may be much affected by changes in the environment of these proteins such as those that may occur on translocation from the cytosol to the cell membrane. There are multiple ways of establishing FRET. One possibility is acceptor photobleaching: if the acceptor is bleached by intense light (at its own excitation maximum), then it can no longer accept energy from the donor and emission by the donor will, thus, increase. The extent of this increase is a measure of FRET efficiency. An alternative measure is the determination of fluorescence lifetime: transfer of energy results in a shortening of fluorescence lifetime of the donor, which is more specific than a simple ratiometric determination. Other controls can be done on a biochemical level: if only one of two fluorescent partners is transfected into a cell, then no changes of its fluorescence should occur when agonist is added.

To determine agonist-induced changes in FRET and to permit the study of rapid kinetics, cells are continuously superfused with buffer and agonists are applied using a computer-assisted solenoid valve controlled rapid superfusion device such as ALA-VM8 (ALA Scientific Instruments). This allows solution exchanges over a time of about 5–10 ms. Signals are detected by avalanche photodiodes and are digitalized using an AD converter (Digidata1322A, Axon Instruments) and stored on PC using Clampex 8.1 software (Axon Instruments).

The kinetics of agonist-induced changes in the corrected FRET ratio can generally be fitted to monoexponential equations of the type: $r(t) = A \times (1 - e^{-t/\tau})$, where $\tau$ is the time constant (s) and $A$ is the magnitude of the signal, or to biexponential equations.

## III. Results and Discussion

### A. *Agonist Binding*

The FRET approaches described above have been used to investigate many steps involved in the activation and signaling of GPCRs. The first step that has been studied in some detail is the ligand-binding step. FRET between a labeled ligand and a labeled receptor can best be investigated by measuring the decrease of emission of shorter wavelength fluorophore in the receptor (e.g., GFP) by the binding of a ligand labeled with a longer wavelength fluorophore. This is because such studies generally require a large excess of ligand over receptor and, hence, the emission from the receptor changes more than that from the ligand. In the case of the small peptidic neurokinin a ligand and their neurokinin NK2 receptors, these signals are compatible with a simple bimolecular binding reaction with observed rate constants $k_{obs}$ of 1.6 and 0.05 $s^{-1}$ (Palanche *et al.*, 2001).

It has been suggested that these two rate constants may reflect two distinct active conformations of the receptor, the fast one linked to an intracellular Ca-signal, the slower one correlating with intracellular cAMP-accumulation (Palanche *et al.*, 2001). In the case of the receptor for parathyroid hormone (PTH), which is an 81-amino acid peptide ligand and is usually studied with its active 34-amino acid N-terminus, the binding reaction appears to be more complex. A first rapid interaction ($\tau \approx 140$ ms)—presumably between the large N-terminus of the receptor and the C-terminus of the ligand—is followed by a second, slower event ($\tau \approx 1$ s) that presumably represents the interaction of the ligand's N-terminus with the transmembrane core of the receptor (Castro *et al.*, 2005). The kinetics of the second binding step correlate with the activation step of the receptor as judged from its conformational change (see below). Whether such biphasic binding reactions are a general feature of GPCRs, and whether they reflect different receptor conformations or separate sequential binding steps remains to be elucidated.

### B. *Receptor Activation*

In order to construct G protein–coupled receptors that showed agonist-sensitive changes in fluorescence, CFP and YFP variants of the GFP have been fused to the receptors into regions that are thought to move during agonist-induced activation. As detailed above, constructs carrying YFP and CFP in the third intracellular loop, that is, adjacent to transmembrane helix VI, and in the C-terminus, respectively, have been found to respond to agonists with a change in FRET (Vilardaga *et al.*, 2003, 2005). A series of analogous constructs have been generated along these lines for other receptors, suggesting that this principle is generally applicable for GPCR sensors. In these experiments, FRET has always been monitored ratiometrically, that is, as the ratio of the (corrected) emission intensities of YFP and CFP (or FlAsH and CFP). In all cases investigated so far, agonist activation results in a loss of FRET as indicated by a reduction in the YFP/CFP-emission ratio. The signal amplitudes that can be obtained in such experiments vary between the different receptor sensors that have been constructed and range from a few percent to about 20% in the best cases. Although such signal amplitudes may appear small, these recordings have a very small noise and this makes them easy to measure.

The kinetics of such FRET recordings of agonist-induced conformational changes in GPCRs are remarkably similar. In all instances where we have studied receptors for small ligands such as adrenaline or adenosine, the signals occurred with half-lives of 30–50 ms. The only exception encountered so far is the PTH receptor, which showed an activation half-life in the range of 1 s (Vilardaga *et al.*, 2003). This correlates well with the second phase of PTH binding as described above. The PTH receptor may be slower than others that have been investigated so far because it has a much larger ligand with a more complex binding process, or because it is a class B receptor (whereas all receptors for small ligands investigated so far are class A receptors) and this class may have a different type of activation.

Activation of GPCRs within 30–50 ms is faster than had previously been thought, but still slower than that of rhodopsin, where a series of steps leads to the active, G protein–coupling metarhodopsin II form within about 1 ms (Okada *et al.*, 2001). It is not clear yet whether these differences are real or just due to technical issues. Technical problems may arise from the facts that in rhodopsin the "ligand" retinal is already bound to its binding pocket in its inactive form, while the receptor ligands used in our studies need to diffuse to the receptor and "find" the binding pocket; that we are observing a large number of receptors on a cell rather than an individual one; that not all receptors may see the agonist at the same time; and that, finally, some of the apparent activation time (up to $\approx$5–10 ms) is required for solution exchange by the perfusion system. On the other hand, it may be that rhodopsin has evolved to have a fast activation switch whereas such switch is not required in other types of GPCRs.

As an alternative to the large GFP-based labels, the small fluorophore FlAsH (see above) has been used to label the third intracellular loop (Hoffmann *et al.*, 2005). This was done in part because it was thought that the large GFP-moieties attached to a receptor might impair the speed of its conformational change. Therefore, the FlAsH-binding motif was introduced into the third intracellular loop, the site thought to move most in GPCR activation. Use of the FlAsH label also improved the ability of the modified receptor to couple to and activate its G proteins, and it also increased, in the specific case of the $A_{2A}$-adenosine receptor studied, the amplitude of the FRET signal. However, surprisingly, the kinetics of the agonist-induced FRET change were exactly the same as in a receptor labeled with CFP and YFP. Thus, at least so far, we have no evidence that the GFP-moieties in our receptor sensors impair the speed of their agonist-induced conformational change or that their replacement with smaller labels would increase that speed.

An interesting issue is the fact that inverse agonists produce effects in the opposite direction from those caused by agonists, that is they result in an increase of FRET, and that partial agonists produce partial effects (Nikolaev *et al.*, 2006b; Vilardaga *et al.*, 2003, 2005). The rate constants for these different types of compounds also appear to vary: the conformational changes induced by partial agonists are considerably slower than those induced by full agonists, and they depend on the degree of "partiality," that is those with the smallest agonist effects induce the slowest changes. Inverse agonist signals were even slower. These observations are in line with the hypothesis that the different compounds may induce distinct conformations of the receptors and that the kinetics of assuming these distinct conformations are different.

### C. *Receptor–G Protein Interaction*

Receptor–G protein interactions have been monitored with optical techniques both via FRET and BRET (Gales *et al.*, 2005; Hein *et al.*, 2005). Even though there are important differences in details of the results obtained, the general conclusions from the two approaches are similar. The interaction between activated receptors and their G proteins can be very rapid and can, at high expression levels, reach the speed of receptor activation itself (i.e., occur with a $\tau$ in the 30- to 50-ms range, Hein *et al.*, 2005).

An important aspect of receptor–G protein interactions has not been solved in these studies—the general principle of coupling. Two different concepts have been postulated long ago, free collision coupling versus coupling within a ''precoupled'' complex. Measuring FRET between a YFP-tagged $\alpha_{2A}$-adrenergic receptor and G$\alpha$i$_1$,CFP-G$\beta\gamma$, Hein *et al.* (2005) observed agonist-*independent* FRET. However, this FRET signal was not significantly different from the signal between G$\alpha$i$_1$,CFP-G$\beta\gamma$ and a membrane-anchored YFP. Therefore, it was concluded that precoupling between $\alpha_{2A}$-adrenergic receptors and G$_i$ proteins may either not exist or occur with a probability below the limit of detection. In contrast, two other studies using similar conditions (identical receptor–G protein combinations) reported the detection of an association of inactive receptors with G proteins (Gales *et al.*, 2005; Nobles *et al.*, 2005). A major difference in these studies is the choice of YFP-labeled reference constructs to determine FRET or BRET above nonspecific or ''bystander'' FRET or BRET that is caused by random interactions. The contrary conclusions of these studies highlight limitations of such studies for the determination of *steady state* protein interactions. In contrast, FRET or BRET measurements designed to detect *dynamics* of protein–protein interactions or conformational changes may give rise to a wealth of information regarding the nature of protein–protein interactions. In this context it is interesting to note that we observed an increased speed of agonist-induced receptor–G protein coupling with increasing levels of G protein expression (Hein *et al.*, 2005)—a finding that is predicted to occur with collision coupling.

### *D. G Protein Activation*

There is a wealth of biochemical data on the activation of G proteins by receptors. Based on this very large body of evidence coming largely from experiments with cell membranes and reconstituted systems with the purified proteins, the now classical G protein cycle has been proposed (Bourne *et al.*, 1990; Gilman, 1987). The core characteristic of this cycle is that the activated receptor causes GTP binding and subsequent dissociation of the heterotrimeric G protein into the active $\alpha$ and $\beta\gamma$ subunits. FRET- and BRET-based assays have recently been developed in order to investigate G protein activation in intact systems and also to assess whether G protein dissociation is indeed a requirement for G protein activation in the intact cell membrane.

The method to utilize FRET for studying G protein activation was initially developed for *Dictyostelium* (Janetopoulos *et al.*, 2001) and was later also used in other systems, including yeast (Yi *et al.*, 2003) and mammalian G proteins (Azpiazu and Gautam, 2004; Bünemann *et al.*, 2003; Frank *et al.*, 2005; Gibson and Gilman, 2006). Surprisingly, in the case of mammalian $G_i$ proteins, Bünemann *et al.* (2003) and Frank *et al.* (2005) observed that G protein activation can actually lead to an increase in FRET between $G\alpha$-YFP and CFP-$G\beta\gamma$ (Fig. 2). This increase in FRET is not compatible with the concept of complete dissociation of $\alpha$ and $\beta\gamma$ subunits during activation. Instead, this observation supports the concept that G protein subunits can signal without complete separation of the subunits (Klein *et al.*, 2000).

An increase in FRET on G protein activation has been found for $G\alpha_{i1,2,3}$- and $G\alpha_z$-containing G proteins. In contrast, the closely related $G_o$ proteins exhibited a decrease in FRET on activation (Fig. 2). This difference was attributed to the region that is most divergent between $G\alpha_i$ and $G\alpha_o$, the so-called $\alpha_B$-$\alpha_C$ helical region, and exchange of this region in a $G\alpha_i$ with the corresponding one from $G\alpha_o$ resulted in a $G_o$-like *decrease* in FRET on activation (Frank *et al.*, 2005). A similar decrease was also seen in a $G\alpha_{i1}$ construct where YFP was fused right into the $\alpha_B$-$\alpha_C$ helical region (amino acid 120; Gibson and Gilman, 2006). This observation is not surprising given the fact that an intact $\alpha_B$-$\alpha_C$ helical region of $G\alpha_{i1}$ appears to be critically important for generation of an agonist-induced increase in FRET (Frank *et al.*, 2005). Taken together, the published results are consistent with the hypothesis that $G\alpha_i$ proteins undergo a subunit rearrangement on activation and not a complete dissociation.

A very recent study (Gales *et al.*, 2006) using BRET to probe the receptor–G protein interaction as well as G protein activation reports also an increase in energy transfer for one of three $G\alpha_{i1}$ constructs, but also for a G$\alpha_s$–G$\beta_2$ system. This raises the possibility that G protein rearrangement rather than dissociation is a more general pattern of G protein activation. This pattern would also help to solve the question why G$\beta\gamma$ subunits from some G proteins activate certain effectors while others do not: if the G protein subunits do not fully dissociate on activation, then a contribution of the $\alpha$ subunit to the G$\beta\gamma$-effects appears plausible. However, additional experimentation and confirmation of these results seem necessary to settle this question.

It is still an open question whether in other G proteins ($G_o$, $G_q$, and $G_s$) $\alpha$ and $\beta\gamma$ subunits separate during activation or also stay in proximity, even though the FRET signals decrease during activation. The decrease in FRET on activation of YFP-G$\alpha_o$ and CFP-$\beta\gamma$ is typically not more than 30% (Frank *et al.*, 2005; Gibson and Gilman, 2006). This suggests that either the major fraction of $G_o$ proteins is not activated even in the presence of saturating receptor stimulation or $\alpha$ and $\beta\gamma$ subunits do not separate completely. This important question needs to be addressed in future studies.

On the level of kinetics it is interesting to note that activation of G proteins by receptors appears to be rather slow. For example, in the case of $\alpha_2$-adrenergic receptors and $G_i$, the comparison of receptor–G protein interaction kinetics, which occur with time constants of 30–50 ms (Hein *et al.*, 2005), with the kinetics of $G_i$ activation, which require 0.5–1 s (Bünemann *et al.*, 2003), reveals that the G proteins may be a time-limiting step in this signaling pathway. It remains to be seen whether the release of GDP from the $\alpha$ subunit, which is induced by the agonist-occupied receptor, is indeed the reason for the slow kinetics of G protein activation, as may be inferred from biochemical experiments in cell membranes and in reconstituted systems.

## IV. Conclusions

The development of FRET-based monitoring of GPCR signaling pathways and protein activity has opened new possibilities to study their temporal (and also spatial) pattern in living cells. It is now possible to study the kinetics of the different steps from agonist binding to G protein activation with millisecond temporal resolution in single cells. Figure 3 shows this for the binding, receptor activation, and cAMP accumulation in the PTH–PTH receptor system. Together with electrophysiological readouts of effectors such as the G protein-activated inwardly rectifying $K^+$ (GIRK) channels, we can now decipher the kinetics of entire signaling cascades. In the future, FRET-based imaging of GPCR signaling offers the possibility

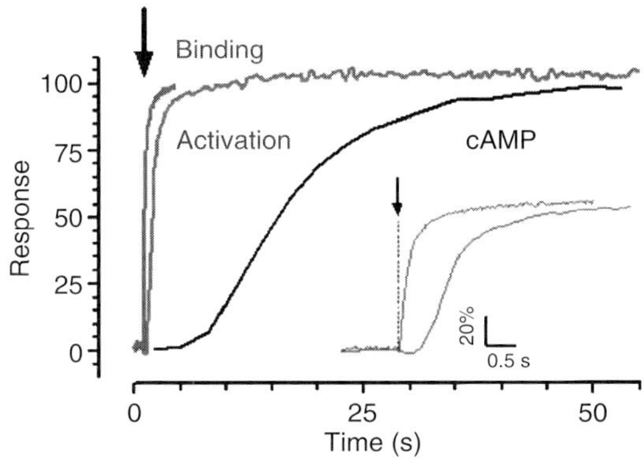


FIG. 3. Kinetics of the different steps in GPCR activation and signaling. Shown are the fast and slow phases of PTH binding to the PTH receptor, and the PTH-induced increase in intracellular cAMP. PTH binding was measured by FRET as indicated in Fig. 2A. cAMP levels were determined by FRET using the epac1-sensor (Nikolaev *et al.*, 2004). Note that the conformational change of the PTH receptor (measured as depicted in Fig. 2B) has the same kinetics as the slow phase of PTH binding [modified from Castro *et al.* (2005)]. (See Color Insert.)

to visualize signals in subcellular compartments and to image their propagation across living cells. Such data will add the dimensions of space and time to the study of GPCR activation and the downstream signaling process.

## Acknowledgments

Work in the authors' laboratory was supported by the Deutsche Forschungsgemeinschaft, the Ernst-Jung-Award, and the Fonds der Chemischen Industrie.

## References

Adams, S. R., Campbell, R. E., Gross, L. A., Martin, B. R., Walkup, G. K., Yao, Y., Llopis, J., and Tsien, R. Y. (2002). New biarsenical ligands and tetracysteine motifs for protein labeling *in vitro* and *in vivo*: Synthesis and biological applications. *J. Am. Chem. Soc.* **124,** 6063–6076.

Azpiazu, I., and Gautam, N. (2004). A fluorescence resonance energy transfer-based sensor indicates that receptor access to a G protein is unrestricted in a living mammalian cell. *J. Biol. Chem.* **279,** 27709–27718.

Bertrand, L., Parent, S., Caron, M., Legault, M., Joly, E., Angers, S., Bouvier, M., Brown, B., Houle, B., and Menard, L. (2002). The BRET2/arrestin assay in stable recombinant cells: A platform to screen for compounds that interact with G protein-coupled receptors (GPCRs). *J. Recept. Signal Transduct. Res.* **22,** 533–541.

Bourne, H. R., Sanders, D. A., and McCormick, F. (1990). The GTPase superfamily: A conserved switch for diverse cell functions. *Nature* **348,** 125–132.

Bünemann, M., Brandts, B., and Pott, L. (1996). Downregulation of muscarinic M2 receptors linked to $K^+$ current in cultured guinea-pig atrial myocytes. *J. Physiol.* **494,** 351–362.

Bünemann, M., Bücheler, M. M., Philipp, M., Lohse, M. J., and Hein, L. (2001). Activation and deactivation kinetics of $\alpha_{2A}$- and $\alpha_{2C}$-adrenergic receptor-activated GIRK currents. *J. Biol. Chem.* **276,** 47512–47517.

Bünemann, M., Frank, M., and Lohse, M. J. (2003). Gi protein activation in intact cells involves subunit rearrangement rather than dissociation. *Proc. Natl. Acad. Sci. USA* **100,** 16077–16082.

Castro, M., Nikolaev, V. O., Palm, D., Lohse, M. J., and Vilardaga, J. P. (2005). Turn-on switch in parathyroid hormone receptor by a two-step PTH binding mechanism. *Proc. Natl. Acad. Sci. USA* **102,** 16084–16089.

Farrens, D. L., Altenbach, C., Yang, K., Hubbell, W. L., and Khorana, H. G. (1996). Requirement of rigid body motion of transmembrane helices for light activation of rhodopsin. *Science* **274,** 768–770.

Frank, M., Thümer, L., Lohse, M. J., and Bünemann, M. (2005). G-protein-activation without subunit dissociation depends on a Gαi-specific region. *J. Biol. Chem.* **280,** 24584–24590.

Gaietta, G., Deerinck, T. J., Adams, S. R., Bouwer, J., Tour, O., Laird, D. W., Sosinsky, G. E., Tsien, G. E., and Ellisman, M. H. (2002). Multicolor and electron microscopic imaging of connexin trafficking. *Science* **296,** 503–507.

Gales, C., Rebois, R. V., Hogue, M., Trieu, P., Breit, A., Hebert, T. E., and Bouvier, M. (2005). Real-time monitoring of receptor and G-protein interactions in living cells. *Nat. Methods* **2,** 177–184.

Gales, C., Van Durm, J. J., Schaak, S., Pontier, S., Percherancier, Y., Audet, M., Paris, H., and Bouvier, M. (2006). Probing the activation-promoted structural rearrangements in preassembled receptor-G protein complexes. *Nat. Struct. Mol. Biol.* **13,** 778–786.

Gether, U. (2000). Uncovering molecular mechanism involved in activation of G protein-coupled receptor. *Endocr. Rev.* **21,** 90–113.

Gether, U., Lin, S. B., and Kobilka, B. K. (1995). Fluorescent labeling of purified $\beta_2$-adrenergic receptor: Evidence for ligand-specific conformational changes. *J. Biol. Chem.* **270,** 28268–28275.

Griffin, B. A., Adams, S. R., and Tsien, R. Y. (1998). Specific covalent labeling of recombinant protein molecules inside live cells. *Science* **281,** 269–272.

Ghanouni, J., Steenhuis, J., Farrens, D. L., and Kobilka, B. K. (2001). Agonist-induced conformational changes in the G-protein-coupling domain of the $\beta_2$ adrenergic receptor. *Proc. Natl. Acad. Sci. USA* **98,** 5997–6002.

Gibson, S. K., and Gilman, A. G. (2006). Giα and Gβ subunits both define selectivity of G protein activation by α2-adrenergic receptors. *Proc. Natl. Acad. Sci. USA* **103,** 212–217.

Giepmans, B. N. G., Adams, S. R., Ellisman, M. H., and Tsien, R. Y. (2006). The fluorescent toolbox for assesssing protein location and function. *Science* **312,** 217–224.

Gilman, A. G. (1987). G-proteins: Transducers of receptor-generated signals. *Ann. Rev. Biochem.* **56,** 615–649.

Hannawacker, A., Krasel, C., and Lohse, M. J. (2002). Mutation of Asn293 to Asp in transmembrane helix V. I. abolishes agonist-induced but not constitutive activity of the $\beta_2$-adrenergic receptor. *Mol. Pharmacol.* **62,** 1431–1437.

Hein, P., Frank, M., Hoffmann, C., Lohse, M. J., and Bünemann, M. (2005). Dynamics of receptor/G protein coupling in living cells. *EMBO J.* **24,** 4106–4114.

Heithier, H., Fröhlich, M., Dees, C., Baumann, M., Haring, M., Gierschik, P., Schiltz, E., Vaz, W. L., Hekman, W. L., and Helmreich, E. J. (1992). Subunit interactions of GTP-binding proteins. *Eur. J. Biochem.* **204,** 1169–1181.

Hoffmann, C., Gaietta, G., Bünemann, M., Adams, S., Oberdorff-Maass, S., Behr, B., Vilardaga, J. P., Tsien, J. P., Ellisman, M. H., and Lohse, M. J. (2005). A FlAsH-based FRET approach to determine G-protein coupled receptor activation in living cells. *Nat. Methods* **2,** 171–176.

Janetopoulos, C., Jin, T., and Devreotes, P. (2001). Receptor-mediated activation of heterotrimeric G-proteins in living cells. *Science* **291,** 2408–2411.

Klein, S., Reuveni, H., and Levitzki, A. (2000). Signal transduction by a nondissociable heterotrimeric yeast G. protein. *Proc. Natl. Acad. Sci. USA* **97,** 3219–3223.

Krasel, C., Bünemann, M., Lorenz, K., and Lohse, M. J. (2005). $\beta$-Arrestin binding to the $\beta_2$-adrenergic receptor requires both receptor phosphorylation and receptor activation. *J. Biol. Chem.* **280,** 9528–9535.

Martin, B. R., Giepmans, B. N., Adams, S. R., and Tsien, R. Y. (2005). Mammalian cell-based optimization of the biarsenical-binding tetracysteine motif for improved fluorescence and affinity. *Nat. Biotechnol.* **23,** 1308–1314.

Miyawaki, A., and Tsien, R. Y. (2000). Monitoring protein conformations and interactions by fluorescence resonance energy transfer between mutants of green fluorescent protein. *Methods Enzymol.* **327,** 472–501.

Mongillo, M., Evellin, S., Lissandron, V., Terrin, A., Hannawacker, A., Lohse, M. J., Pozzan, M., Houslay, M., and Zaccolo, M. (2004). FRET-based analysis of cAMP dynamics in live cells reveals distinct functions of compartmentalized phosphodiesterases in cardiac myocytes. *Circ. Res.* **95,** 67–75.

Nikolaev, V. O., and Lohse, M. J. (2006). Monitoring of cAMP synthesis and degradation in living cells. *Physiology* **21,** 86–92.

Nikolaev, V. O., Bünemann, M., Hein, L., Hannawacker, A., and Lohse, M. J. (2004). Novel single chain cAMP sensors for receptor-induced signal propagation. *J. Biol. Chem.* **279,** 37215–37218.

Nikolaev, V. O., Gambaryan, S., and Lohse, M. J. (2006a). Fluorescent sensors for rapid monitoring of intracellular cGMP. *Nat. Methods* **3,** 23–25.

Nikolaev, V. O., Hoffmann, C., Bünemann, M., Lohse, M. J., and Vilardaga, J. P. (2006b). Molecular basis of partial agonism at the neurotransmitter $\alpha_{2A}$-adrenergic receptor and $G_i$-protein heterotrimer. *J. Biol. Chem.* **281,** 24506–24511.

Nobles, M., Benians, A., and Tinker, A. (2005). Heterotrimeric G proteins precouple with G protein-coupled receptors in living cells. *Proc. Natl. Acad. Sci. USA* **102,** 18706–18711.

Okada, T., Ernst, O. P., Palczewski, K., and Hofmann, K. P. (2001). Activation of rhodopsin: New insights from structural and biochemical studies. *Trends Biochem. Sci.* **26,** 318–324.

Palanche, T., Ilien, B., Zoffmann, S., Reck, M.-P., Bucher, B., Edelstein, S. J., and Galzi, J.-L. (2001). The neurokinin A. receptor activates calcium and cAMP responses through distinct conformational states. *J. Biol. Chem.* **276,** 34853–34861.

Pfaffinger, P. J., Martin, J. M., Hunter, D. D., Nathanson, N. M., and Hille, B. (1985). GTP-binding proteins couple cardiac muscarinic receptors to a K channel. *Nature* **317,** 536–538.

Sheik, S. P., Zvyaga, T. A., Lichtarge, O., Sakmar, T. O., and Bourne, H. R. (1996). Rhodopsin activation blocked by metal-ion-binding sites linking transmembrane helices C and F. *Nature* **383,** 347–350.

Swaminath, G., Xiang, Y., Lee, T. W., Steenhuis, J., Parnot, C., and Kobilka, B. K. (2004). Sequential binding of agonists to the $\beta_2$-adrenoceptor. Kinetic evidence for intermediate conformational states. *J. Biol. Chem.* **279,** 686–691.

Swaminath, G., Deupi, X., Lee, T. W., Zhu, W., Thian, F. S., Kobilka, T. S., and Kobilka, B. K. (2005). Probing the $\beta_2$-adrenoceptor binding site with catechol reveals differences in binding and activation by agonists and partial agonists. *J. Biol. Chem.* **280,** 22165–22171.

Vilardaga, J. P., Bünemann, M., Krasel, C., Castro, M., and Lohse, M. J. (2003). Measurement of the millisecond activation switch of G-protein-coupled receptors in living cells. *Nat. Biotechnol.* **21,** 807–812.

Vilardaga, J. P., Steinmeyer, R., Harms, G. S., and Lohse, M. J. (2005). Molecular basis of inverse agonism in a G protein-coupled receptor. *Nat. Chem. Biol.* **1,** 25–28.

Violin, J. D., Ren, X. R., and Lefkowitz, R. J. (2006). G-protein-coupled receptor kinase specificity for $\beta$-arrestin recruitment to the $\beta_2$-adrenergic receptor revealed by fluorescence resonance energy transfer. *J. Biol. Chem.* **281,** 20577–20588.

Ward, S. D. C., Hamdan, F. F., Bloodworth, L. M., and Wess, J. (2002). Conformational changes that occur during $M_3$ muscarinic acetylcholine receptor activation probed by the use of an in situ disulfide cross-linking strategy. *J. Biol. Chem.* **277,** 2247–2257.

Wieland, K., Zuurmond, H. M., Andexinger, S., IJzerman, A. P., and Lohse, M. J. (1996). Stereospecificity of agonist binding to $\beta_2$-adrenergic receptors involves Asn-293. *Proc. Natl. Acad. Sci. USA* **93,** 9276–9281.

Yao, K., Parnot, C., Deupi, X., Ratnala, V. R., Swaminath, G., Farrens, D., and Kobilka, B. K. (2006). Coupling ligand structure to specific conformational switches in the $\beta_2$-adrenoceptor. *Nat. Chem. Biol.* **2,** 417–422.

Yi, T. M., Kitano, H., and Simon, M. I. (2003). A quantitative characterization of the yeast heterotrimeric G protein cycle. *Proc. Natl. Acad. Sci. USA* **100,** 10764–10769.

# REGULATION OF RHO GUANINE NUCLEOTIDE EXCHANGE FACTORS BY G PROTEINS

By PAUL C. STERNWEIS,* ANGELA M. CARTER,* ZHE CHEN,[†] SHAHAB M. DANESH,* YING-FAN HSIUNG,* AND WILLIAM D. SINGER*

*Department of Pharmacology, University of Texas Southwestern Medical Center, Dallas, Texas 75390;
[†]Department of Biochemistry, University of Texas Southwestern Medical Center, Dallas, Texas 75390



## ABSTRACT

Monomeric Rho GTPases regulate cellular dynamics through remodeling of the cytoskeleton, modulation of immediate signaling pathways, and longer-term regulation of gene transcription. One family of guanine nucleotide exchange factors for Rho proteins (RhoGEFs) provides a direct pathway for regulation of RhoA by cell surface receptors coupled to heterotrimeric G proteins. Some of these RhoGEFs also contain RGS domains that can attenuate signaling by the $G_{12}$ and $G_{13}$ proteins. The regulation provided by these RhoGEFs is defined by their selective regulation by specific G proteins, phosphorylation by kinases, and potential localization with signaling partners. Evidence of their physiological importance is



*ADVANCES IN PROTEIN CHEMISTRY, Vol. 74*
DOI: 10.1016/S0065-3233(07)74006-8


0065-3233/07 $35.00

derived from gene knockouts in *Drosophila* and mice. Current understanding of the basic regulatory mechanisms of these RhoGEFs is discussed. An overview of identified interactions with other signaling proteins suggests the growing spectrum of their involvement in numerous signaling pathways.

## I. Introduction

Sensing the extracellular environment and integrating appropriate responses are essential processes for the survival of unicellular organisms and coordination of function in multicellular entities. In eukaryotes, two families of G proteins often mediate this sensing and early integration of signals. The heterotrimeric G proteins interact directly with the largest group of membrane receptors, the G protein–coupled receptors (GPCRs). These receptors regulate numerous physiological processes such as vision, cardiovascular function, chemotaxis, olfaction, neural transmission, and many endocrine processes. The Ras superfamily of monomeric GTPases is composed of several distinct families, which modulate such diverse functions as cell growth, intracellular vesicular traffic, nuclear transport, and cellular migration. Regulation of several of these GTPases was first shown in response to the tyrosine kinase family of growth factors. It is now clear that many of the Ras-like GTPases can also be regulated by GPCRs and other classes of cell surface receptors. The most direct pathway for regulation by GPCRs is the modulation of Rho GTPases by the $G_{12}$ class of heterotrimeric G proteins via the p115 family of Rho-dependent guanine nucleotide exchange factors (RhoGEFs). The mechanisms and physiological impact of this and closely related regulatory modules are the focus of this discussion.

### A. *Intrinsic Mechanisms of G Proteins*

Both classes of G proteins specifically bind guanine nucleotides and are regulated by a GTP cycle (Fig. 1). Classically, the inactive state of the G protein contains GDP. Formation of the active state is facilitated by guanine nucleotide exchange factors (GEFs), which increase the rate of dissociation of GDP, and thus facilitate binding of the activating ligand, GTP. In the case of the heterotrimeric G proteins, the GEFs are GPCRs, integral membrane proteins with extracellular and intracellular domains, which are functionally linked through seven transmembrane $\alpha$ helices. In contrast, many different families of GEFs regulate nucleotide exchange on the monomeric GTPases and have cellular locations that include various membranes, cytoplasm, and the nucleus. Once activated, the G proteins can slowly

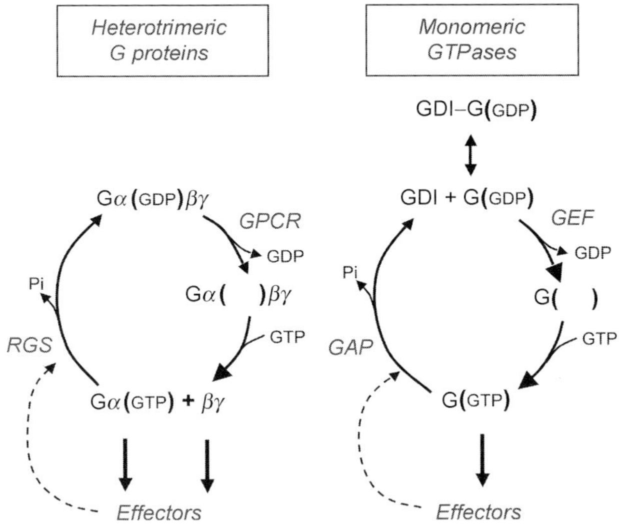


FIG. 1. G protein cycles. Regulation of both the heterotrimeric G proteins and the monomeric GTPases involves a guanine nucleotide cycle. The active state of the G proteins has bound GTP and is determined by the balance between two major regulatory mechanisms determining guanine nucleotide exchange (activation) and hydrolysis of GTP (inactivation). See text for further details.

inactivate themselves via intrinsic GTPase activities that dictate half-lives ranging from a few seconds to hours. Theoretically, this gives the G proteins a capability to act as timed switches for providing predetermined quanta of regulation. However, the abundance of proteins that stimulate hydrolysis of GTP on the G proteins and the rapidity of signaling events suggests that this intrinsic activity is rarely used as a timer in its own right. In the case of the monomeric GTPases, the proteins that stimulate GTPase activity are known as GAPs (GTPase-activating proteins). Hydrolysis of GTP bound to the heterotrimeric G proteins is stimulated by RGS proteins (regulators of G protein signaling).

The oligomeric structure of the heterotrimeric G proteins increases the complexity and diversity of their signaling mechanisms. The intrinsic GTP cycle is directly associated with a cycle of subunit dissociation in which the $\alpha$ subunit (which contains the nucleotide-binding site) can dissociate from a stable $\beta\gamma$ dimer (Fig. 1). In the inactive GDP-bound state, high affinity between the $\alpha$ and $\beta\gamma$ subunits favors association. When GTP is bound, the affinity between subunits is lowered and dissociation is promoted. This allows both $\alpha$ and $\beta\gamma$ subunits to act as coordinate or diverse regulators of downstream targets. Reassociation of the subunits inhibits downstream activities. While there is evidence for and against complete dissociation of subunits during activation cycles, it is clear that large rearrangement of subunit interfaces is required for downstream regulation. The $\alpha$ subunits of heterotrimeric G proteins are most diverse and have definitive unique functions. Therefore, heterotrimeric G proteins are named by their $\alpha$ subunit. Specificity of $\beta\gamma$ subunits for downstream function is less clear. Nevertheless, the inactive heterotrimer is the most efficient substrate for activation by GPCRs and both $\alpha$ and $\beta\gamma$ subunits will specify the potency and efficacy of this interaction. Further details are summarized in numerous reviews including (Cabrera-Vera *et al.*, 2003; Gilman, 1987; Hepler, 1999; Ross and Wilkie, 2000; Sternweis, 1996; Wettschureck and Offermanns, 2005).

The Rho subfamily of monomeric GTPases is composed of several proteins including the well-characterized Rho (A, B, and C), Rac (1 and 2), and Cdc42 proteins. Their roles in cellular regulation include modulation of cytoskeletal structure, motility, cell division, gene transcription, vesicular transport, and various enzymatic activities. One of the first and better-known themes of these proteins is their control of actin-based structures. Thus, Hall and colleagues first noted the ability of RhoA, Rac, and Cdc42 to differentially regulate formation of stress fibers, lamellipodia, and more protrusive filopodia, respectively (Etienne-Manneville and Hall, 2002). Progress in the enumeration of several mechanisms for these events and their roles in such processes as wound healing, transendothelial migration, cell polarization, and phagocytosis have been reviewed extensively (Brown *et al.*, 2006; Burridge and Wennerberg, 2004; Etienne-Manneville and Hall, 2002; Karnoub *et al.*, 2004; Wittchen *et al.*, 2005).

The involvement of the Rho proteins in a broad spectrum of physiological functions and the identification of numerous targets for downstream regulation indicate a need for tight control of their activities. In addition, the Rho proteins act at various sites in the cell, a property facilitated by association with GDI (GDP dissociation inhibitor) proteins. The Rho proteins are modified by isoprenylation at their C-termini; association with GDIs shields these hydrophobic groups for transport through the cytoplasm. Activation of Rho is mediated by a large family of GEFs (about 70 in the human genome) that are characterized by tandem Dbl homology (DH) and pleckstrin homology (PH) domains (Rossman *et al.*, 2005). The DH domain provides the site for catalyzing nucleotide exchange on and thus activation of Rho proteins; the C-terminal PH domains are less homologous and appear more diverse with functions ranging from facilitation of interaction with Rho to localization of the GEF through association with specific polyphosphoinositides. These GEFs also contain a variety of other signaling domains that mediate interaction with numerous inputs and multiple mechanisms for regulation of their exchange activity.

Once activated, the regulatory tenure of the Rho proteins is determined by another large family of GAPs; based on genome sequence analysis, this may number up to 80 members (Moon and Zheng, 2003). These proteins provide an array of inputs to selectively limit the action of Rho proteins by stimulating their GTPase activity or, in some cases, also provide the means for regulation of downstream pathways.

### *B. Regulation of Rho Proteins by Heterotrimeric G Proteins*

The heterotrimeric G proteins are divided into four families ($G_s$, $G_i$, $G_q$, and $G_{12}$). All have been implicated in the regulation of Rho GTPases, mostly through indirect pathways involving transactivation of growth factor pathways or subsequent to generation of second messengers like cAMP, calcium, and phosphatidylinositol 3,4,5-trisphosphate ($PIP_3$). Some of these pathways have been reviewed (Bhattacharya *et al.*, 2004). The capability of a subclass of GPCRs including receptors for thrombin, lysophosphatidic acid, and endothelin to stimulate the RhoA subclass of GTPases became well established in the 1990s (Sah *et al.*, 2000). Much of the evidence was initially derived from use of botulinum C3 exoenzyme, a specific inhibitor of Rho proteins, and subsequently confirmed with more direct measurements of activated RhoA. The use of constitutively activated G protein $\alpha$ subunits and gene knockouts implicated the heterotrimeric $G_{12}$ subfamily ($G_{12}$ and $G_{13}$) in many physiological processes involving Rho, including cell proliferation, cellular morphology, embryological development, neurite growth, and cellular interactions (Riobo and Manning, 2005). Overexpression of the $G_{12}$ and $G_{13}$ $\alpha$ subunits can also cause activation of RhoA and such downstream events as formation of actin stress fibers, stimulation of Rho-dependent kinases, and modulation of gene transcription. A requirement for Rho has been shown for many of these physiological effects of $G_{12/13}$. A direct mechanism for the regulation of RhoA by the $G_{12}$ and $G_{13}$ proteins occurs via a subfamily of RhoGEFs. Current understanding of this pathway and the status of its role in GPCR regulation via Rho proteins are reviewed below.

## II. RGS-RhoGEFs

### *A. Discovery and Relationships*

The first member of a new subfamily of RhoGEFs was identified simultaneously by affinity chromatography of COS cell lysates with RhoA in the absence of nucleotide (Hart *et al.*, 1996) and by expression cloning in 3T3 cells with transformation as the selection phenotype (Whitehead *et al.*, 1996). The two respective homologues, p115-RhoGEF (human) and Lsc (mouse), were shown to be expressed most abundantly in hematopoietic tissues but could be observed in other tissues and were expressed broadly in cell lines of diverse origins. The oncogenic potential of these proteins, defined by the expression cloning of Lsc and demonstrated by expression of truncated versions containing the DH-PH domains of both proteins, corresponds to findings with several members of the broader RhoGEF family. An interesting feature of p115-RhoGEF (Hart *et al.*, 1996) and Lsc (Glaven *et al.*, 1996) is their selective catalysis of guanine nucleotide exchange on RhoA, but not Rac and Cdc42. While the ability of dominant negative forms of Rac and Cdc42 to block the focus-forming activity of overexpressed Lsc (Whitehead *et al.*, 1996) seems to argue for broader selectivity, this inhibition was likely due to broader impacts of the dominant negative proteins on signaling by monomeric GTPases. Subsequent studies of p115-RhoGEF identified an RGS domain and established the functional link of this protein with the $G_{12}$ and $G_{13}$ proteins through direct characterization *in vitro* (Hart *et al.*, 1998; Kozasa *et al.*, 1998) and by expression studies in cultured cells (Mao *et al.*, 1998). This relationship was also supported by the finding that a homologous protein in *Drosophila melanogaster*, DRhoGEF2, was essential for gastrulation and mutants were phenotypically similar to defects in concertina, a G protein $\alpha$ subunit most closely related to the mammalian $G_{12}\alpha$ family (Barrett *et al.*, 1997; Hacker and Perrimon, 1998).

Two additional mammalian proteins with strong homology to p115-RhoGEF in their DH-PH domains and regions of similarity to the RGS domains are leukemia-associated Rho-GEF (LARG) (KIAA0382) and PDZ-RhoGEF (KIAA0380, GTRAP48). LARG was identified in a patient with acute myeloid leukemia as a coding sequence fused to the MLL gene by a chromosome rearrangement (Kourlas *et al.*, 2000). Both LARG and PDZ-RhoGEF (identified by homology searches) were expressed and shown to stimulate RhoA *in vivo* and could be shown to specifically interact with $G_{12}\alpha$ and $G_{13}\alpha$ through their RGS domains (Fukuhara *et al.*, 1999, 2000). Like p115-RhoGEF, these exchange factors were specific for RhoA and did not stimulate either Rac or Cdc42 (Reuther *et al.*, 2001; Rumenapp *et al.*, 1999; Taya *et al.*, 2001; Togashi *et al.*, 2000). In addition to their RGS, DH, and PH domains, which are characteristic of Lsc/p115-RhoGEF, LARG and PDZ-RhoGEF also contain a PDZ domain near their N-termini; this domain may play a role in their spectrum of regulation (see below). In contrast to the localization of p115-RhoGEF/Lsc, LARG appears to be distributed broadly through tissues (Kourlas *et al.*, 2000). PDZ-RhoGEF is also expressed broadly but much higher levels of mRNA were found in

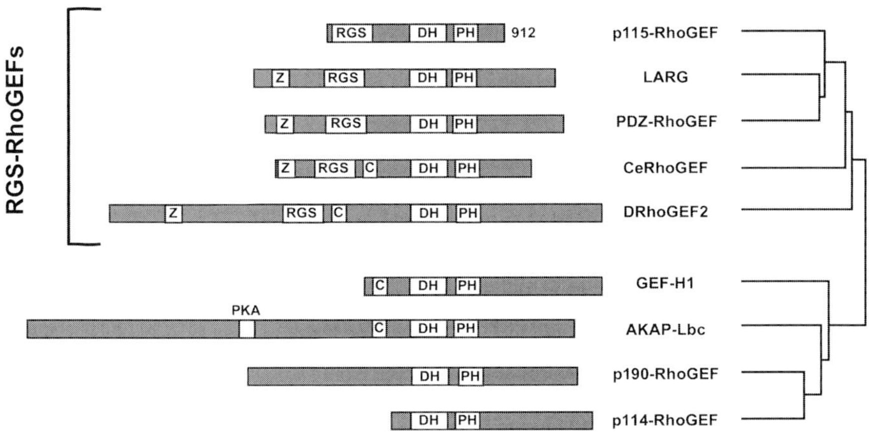


FIG. 2. RGS-RhoGEFs. A schematic representation of the RhoGEFs and their known structural elements. The sizes of the proteins range from 912 amino acid in p115-RhoGEF to 2817 amino acid of AKAP-Lbc. The dendrogram at the right was determined from homology comparisons of the DH and PH domains of the proteins. See text for descriptions and references of the RhoGEFs. Alternate names include: p115-RhoGEF, Lsc (mouse)/Arhgef1; PDZ-RhoGEF, GTRAP48/KIAA0380/Arhgef11; LARG, KIAA0382/Arhgef12; AKAP-Lbc, AKAP13; p114-RhoGEF, KIAA0521/Arhgef18; p190-RhoGEF, Rip2. Domains: RGS, *r*egulator of *G* protein *s*ignaling; DH, *d*bl *h*omology; PH, *p*leckstrin *h*omology; Z, PDZ (*P*SD-95/SAP90-*D*iscs-larg-*Z*O-1); C, C1 homology domain; PKA, binding site for *p*rotein *k*inase *A*.

brain and testes (Fukuhara *et al.*, 1999; Jackson *et al.*, 2001). A schematic representation of structure and relationships among members of this family of RhoGEFs is shown in Fig. 2.

Recently, a homologous member of the family was described in *Caenorhabditis elegans*. In this organism, GPA-12, a homologue of the mammalian $G_{12}$ family, caused developmental growth arrest when overexpressed (van der Linden *et al.*, 2003). A potential target of GPA-12 is the CeRhoGEF, which can stimulate activation of RhoA when expressed in COS cells and can bind to activated GPA-12 based on coimmunoprecipitation (Yau *et al.*, 2003). In this study, treatment of *C. elegans* with RNAi to either GPA-12 or CeRhoGEF produced similar phenotypes.

Members of a closely related group of RhoGEFs may also be directly regulated by G proteins (Fig. 2). The most analogous member is AKAP-Lbc. The tandem DH-PH domains of Lbc were originally isolated as an oncogene by transformation of NIH-3T3 cells and shown to be specific for stimulation of the RhoA subclass and not Rac or Cdc42 (Toksoz and Williams, 1994; Zheng *et al.*, 1995). Subsequently, it was demonstrated that the gene for Lbc could produce various splice variants of which the longest was AKAP-Lbc, which contained a protein kinase A-anchoring domain, and could be shown to synergistically activate Rho-GTP when coexpressed with an activated form of $G_{12}\alpha$ (Diviani *et al.*, 2001). At this time, there is not a characterized RGS domain for this protein. p114-RhoGEF was shown to synergistically stimulate reporter genes for RhoA when expressed in cells that were stimulated with LPA (Niu *et al.*, 2003). Synergy could also be obtained by coexpression with G protein $\beta\gamma$ subunits but not activated $\alpha$ subunits; attenuation of the LPA response by expression of the transducin $\alpha$ subunit supports a mechanism for hormone regulation via $\beta\gamma$ subunits. Coimmunoprecipitation of overexpressed p114-RhoGEF and $\beta\gamma$ subunits suggest that such regulation could occur through direct interaction of the proteins.

Two additional related proteins have not been identified as targets of G protein regulation at this time. p190-RhoGEF was identified in a two-hybrid screen with RhoA (Gebbink *et al.*, 1997) and evidence suggests that it can be regulated by binding to and phosphorylation by focal adhesion kinase (FAK) (Zhai *et al.*, 2003). GEF-H1 was originally described as a RhoGEF that could bind to microtubules (Ren *et al.*, 1998). Several studies have elaborated on mechanisms that indicate suppression of GEF-H1 activity when associated with microtubules and on potential regulation of the exchange factor by disruption of microtubules, the action of PAK kinases, and levels of expression. See Birukova *et al.* (2006), Callow *et al.* (2005), Chang *et al.* (2006), and Krendel *et al.* (2002), for examples and references therein for more complete characterization. In contrast to the selectivity of p190-RhoGEF for RhoA, p114-RhoGEF and GEF-H1 can also act on Rac, but not Cdc42 (Gebbink *et al.*, 1997; Ren *et al.*, 1998). The following discussion will be limited to properties and function of the RGS-RhoGEFs and AKAP-Lbc.

### *B. The RGS Domain and GTPase Activity*

The RGS domain in the N-terminus of p115-RhoGEF was first shown to have weak homology with known classical members of the RGS family and to potently increase the intrinsic GTPase activity of the $\alpha$ subunits of $G_{12}$ and $G_{13}$, but not $G_s\alpha$, $G_{i1}\alpha$, or $G_q\alpha$ (Kozasa *et al.*, 1998). This unique property sets the proteins apart from other RGS proteins that act on the $G_i$ and $G_q$ subfamilies with various selectivities (Hepler, 1999; Ross and Wilkie, 2000). The RGS region was also required to observe the physical interaction of activated $G_{13}\alpha$ with p115-RhoGEF by coimmunoprecipitation from cell lysates (Hart *et al.*, 1998). Subsequent characterization of LARG (Fukuhara *et al.*, 2000) and PDZ-RhoGEF (Fukuhara *et al.*, 1999) showed a similar requirement of the RGS (LH) domain for interaction with $G_{12}\alpha$ or $G_{13}\alpha$, and established this as a common feature of this subfamily of homologous RhoGEFs. Therefore, they have been called the RGS-RhoGEFs.

RGS proteins are related by homology in a region called the "RGS box." While quite diverse in sequence, these regions of about 120 amino acids form similar structures consisting of an 8 or 9 helix bundle with 2 subdomains (de Alba *et al.*, 1999; Slep *et al.*, 2001; Spink *et al.*, 2000; Tesmer *et al.*, 1997). These minimal domains can sometimes be expressed alone with retention of their GAP activity (Ross and Wilkie, 2000). The RhoGEF-RGS domains (rgRGS) constitute a distinct subgroup and can be characterized by their requirement for regions that are both N- and C-terminal to their homologous "RGS box" sequences. Stable expression of the rgRGS domain of p115-RhoGEF required about 60 additional C-terminal amino acids; structure determination revealed that this C-terminal region formed three to four additional $\alpha$ helices that are intimately associated with the conserved helical bundle through hydrophobic interactions and presumably help stabilize the overall structure (Chen *et al.*, 2001). At the same time, a similar structure was elucidated for the rgRGS (LH) domain of PDZ-RhoGEF (Longenecker *et al.*, 2001). Neither structure contained residues N-terminal to the conserved RGS box but functional measurements with deletion mutants established a requirement for N-terminal residues for GAP activity (Wells *et al.*, 2002a). This requirement was localized to specific residues in an N-terminal acidic region (Chen *et al.*, 2003), which were subsequently shown in structural studies to interact with a basic groove on the surface of $G_{13}\alpha$ (Chen *et al.*, 2005). Overall, the functional rgRGS domains then consist of ~210 residues as shown in Fig. 3.

The structure of a complex of the p115 rgRGS bound to an activated $G_{13/i1}$ chimeric $\alpha$ subunit revealed the functional nature of this interaction (Chen *et al.*, 2005). The chimeric protein contained the structural core of $G_{i1}\alpha$ but the regulatory switch regions and $\alpha$ helical domain of $G_{13}\alpha$. While the chimeric protein expresses in bacteria with high abundance like $G_{i1}\alpha$, it interacts functionally like $G_{13}\alpha$ to include specific interactions with the RGS domains of RhoGEFs *in vitro* and activation of Rho-dependent reporter genes by transient expression. The structure shows that two distinct surfaces of the rgRGS interact with $G_{13}\alpha$. An interactive surface within the "RGS box" is unique from but roughly corresponds to the position of the interactive surfaces of RGS4 (Tesmer *et al.*, 1997) and RGS9 (Slep *et al.*, 2001) when bound to either $G_{i1}\alpha$ or $G_t\alpha$ (transducin), respectively; this confirmed the interactive surface predicted from prior mutational analysis (Chen *et al.*, 2003). However, rather than interacting with the switches of $G\alpha$, the "RGS box" of p115-RhoGEF occupies putative sites

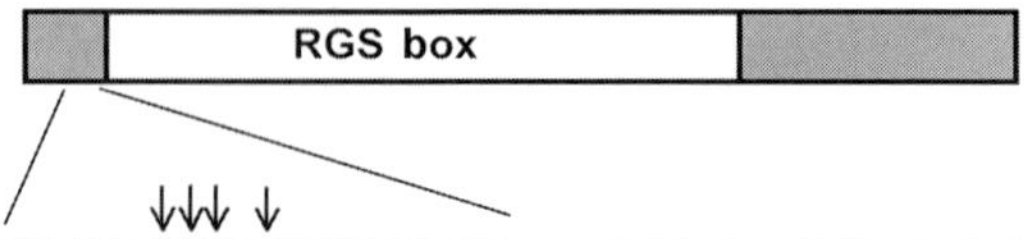


| Sequence | Protein | Residues |
|---|---|---|
| VSI**I**GA**EDEDFE**N**E**LETNSE | p115-RhoGEF | (21-40) |
| PHIIGAEDDDFGTEHEQING | LARG | (348-367) |
| PLIIGP-EEDYDPGYFNNES | PDZ-RhoGEF | (289-307) |
| NMDIDSDEEENDVVYLPESQ | CeRhoGEF | (180-199) |
| KEIISMEDENSDLDEPFIDE | DRhoGEF2 | (902-921) |

FIG. 3. The acidic regions of rgRGS domains. The rgRGS domains consist of a core RGS box, which is homologous in sequence and structure to other RGS domains, a C-terminal extension that helps stabilize the core structure, and an N-terminal acidic region that forms a second binding motif to the $\alpha$ subunit. Amino acids in bold-underline were shown to have contacts with $G_{13}\alpha$: mutations of residues indicated with an arrow could attenuate GAP activity. This acidic region plays a fundamental role in GAP activity, and differences among the RGS-RhoGEFs appear to determine their ability to act as GAPs. See text for details.

for binding of $G_{13}\alpha$ to effectors. A distinct difference is the generally negatively charged surface of the rgRGS, which contrasts with the positively charged interactive surfaces of RGS4 and RGS9. The respective interactions of the RGS domains with the more positively charged surface of $G_{13}\alpha$ and the negatively charged surfaces of the $G_i$ proteins provide a very strong basis for the absolute specificity of these RGS subclasses for specific G proteins (Chen *et al.*, 2005).

The second interface is an acidic region, which is N-terminal to the conserved RGS box. This region binds to a complementary groove between the $\alpha$ helical domain and the switches of $G_{13}\alpha$. The contacts formed here are closest to the catalytic site and presumably drive the increase in catalytic rate (Chen *et al.*, 2005). This is supported by the demonstration that peptides mimicking this region could accelerate the intrinsic GTPase activity of $G_{13}\alpha$, albeit at robust concentrations. A more detailed discussion of putative catalytic mechanisms is found in the papers by Sprang *et al.* (this volume). The absence of readily defined structure in the region connecting the RGS and acidic domains suggests that this basically serves as an unordered tether.

Insightful observations include mutagenesis analysis which indicates that both the N-terminal acidic region and the "RGS box" are required for high-affinity binding (Bhattacharyya and Wedegaertner, 2003a; Chen *et al.*, 2003) of the rgRGS domain to activated $G_{13}\alpha$ and for efficient stimulation of GTPase activity (Chen *et al.*, 2003). Thus, a model for effective function of the rgRGS domains will include both subdomains. An appealing hypothesis is an ordered interaction. Association of the "RGS box" region with $G_{13}\alpha$ would provide initial specific recognition; the tethered acidic region could then interact effectively with the $G_{13}\alpha$ by virtue of its localized high affinity.

Interestingly, not all of the rgRGS domains act as GAPs. While the rgRGS domain of LARG functions like p115-RhoGEF (Suzuki *et al.*, 2003), the rgRGS domain of PDZ-RhoGEF/GTRAP48 binds to activated $G_{13}\alpha$, but does not stimulate its intrinsic GTPase activity (Jackson *et al.*, 2001; Wells *et al.*, 2002a). This discrepancy is now explained by structural requirements in the acidic region. LARG is highly homologous to p115-RhoGEF in this region, especially the placement of a glutamic acid four residues upstream from the hydrophobic phenylalanine (Fig. 3). In the structure of the rgRGS domain of p115-RhoGEF, Phe31 associates with a hydrophobic pocket in $G_{13}\alpha$ and this interaction subserves high-affinity interaction and GAP activity (Chen *et al.*, 2003). The salt bridge formed between Glu27 and Arg200 of the $G_{13}\alpha$-Switch I is proposed to facilitate a catalytic role for Arg200 in hydrolysis and even conservative mutations of Glu27 markedly inhibited activity of the rgRGS (Chen *et al.*, 2005). Inspection of the acidic region in PDZ-RhoGEF shows that this domain lacks the acidic residue in this critical position and the remaining acidic residues are simply not in the correct register. By this criterion, putative GAP activity of the rgRGS domains from CeRhoGEF and DRhoGEF2 becomes questionable. While both possess acidic residues in the correct position, they lack an aromatic residue in the position equivalent to Phe31 (Fig. 3). It will be of interest to see whether the respective G protein $\alpha$ subunits for these domains will structurally accommodate this difference to allow facilitation of their GTPase activities. Finally, a putative RGS domain has been proposed for proto-Lbc (Dutt *et al.*, 2004); if this region of low homology is an interactive region for $G_{12}\alpha$ or $G_{13}\alpha$, it lacks the structure that appears required for GAP activity.

The absence or presence of GAP activity in different RhoGEFs adds complexity to the question of its physiological significance. The simplest interpretation of GAP activity is that it will terminate the active state of the G protein and, therefore, attenuate signaling. Indeed, this is readily accomplished by expression of the rgRGS domain alone and suggests that regulation of PDZ-RhoGEF by $G_{13}\alpha$ should be more efficacious. In contrast, it has been proposed that the ability of the PLC$\beta$ proteins to activate the GTPase activity of $G_q\alpha$ subunits actually facilitates functional coupling by maintaining a receptor-$G_q$-PLC complex through rapid cycling (Ross and Wilkie, 2000). In this scenario, extended activation of PLC by $G_q$ is proposed to promote dissociation and diffusion of activated components; further signaling after inactivation would then be delayed by the need to reassemble signaling complexes. It will be of interest to test this hypothesis and determine the role of GAP activity in the function of the RGS-RhoGEFs.

### *C. The DH and PH Domains: Structure and Relative Activities*

Structures of complexes of RhoA with the DH-PH domains of LARG (Kristelly *et al.*, 2004) and PDZ-RhoGEF (Derewenda *et al.*, 2004) provide a view of downstream catalysis by the RhoGEFs. The core of the DH domains is an elongated bundle of six major $\alpha$ helices that is similar to structures for other DH domains described for various members of the broader RhoGEF family. Numerous interactions with the bound RhoA GTPase, especially its switch regions, provide explanations for catalysis of nucleotide dissociation and the specificity of the RGS-RhoGEFs for RhoA. Details can be found in the respective papers. Two observations that set these exchange factors apart from some of the other subfamilies of RhoGEFs are an extended region at the N-terminus of the core DH domain and the direct interaction of the PH domain with the Rho GTPase. The N-terminal extension is conserved in a subfamily of RhoGEFs that also includes Lfc, Net1, and intersectin. Kristelly *et al.* (2004), who have also determined the structure of the DH-PH domain of LARG in the absence of RhoA, noted that this $\alpha$N1/$\alpha$N2 extension shifted about 2–3 Å toward bound RhoA making contact with Switch I of the GTPase. Apparent contribution of this interaction to catalytic rate was shown by an 80% attenuation of activity when the N-terminal region was deleted or mutated.

Contact of the PH domain with RhoA also suggests a potential role in catalysis. With LARG, this appears to be minor; in the structural study, the DH domain alone was almost as active (75%) as in the presence of its tandem PH domain (Kristelly *et al.*, 2004) and an earlier study indicated a modest threefold increase with inclusion of the PH domain (Reuther *et al.*, 2001). This contrasts with observations using the DH-PH domains from PDZ-RhoGEF, where the tandem DH-PH construct was 50 times more active than the DH domain alone (Derewenda *et al.*, 2004). The latter observation is similar to p115-RhoGEF where removal of the PH domain in several constructs which retained the DH domain consistently lowered exchange activity by >90% (Wells *et al.*, 2001). Overall, it would appear that the PH domain increases the exchange activity of the RGS-RhoGEFs, either through stabilization of the DH domain or direct impact on the substrate GTPase. Whether significant differences in the extent of this action by specific RhoGEFs represent divergence of physiological mechanism or simple differences in stability or design of the expressed domains remains to be seen. In both cases, such action of the extended N-termini of the DH domain and the PH domains in the context of the full length RhoGEFs remains to be confirmed.

## III. Mechanisms of Regulation

### A. *Direct Regulation of GEF Activity by $G_{12}$ and $G_{13}$*

Studies of p115-RhoGEF *in vitro* first demonstrated that $G_{13}\alpha$ could directly stimulate its exchange activity for RhoA (Hart *et al.*, 1998). In this study, regulation by $G_{13}\alpha$ required the rgRGS domain. All other $\alpha$ subunits were ineffective including $G_{12}\alpha$. However, activated $G_{12}\alpha$ could block the stimulation driven by $G_{13}\alpha$. This is consistent with the interaction of both G protein $\alpha$ subunits with the rgRGS domain (Kozasa *et al.*, 1998), but raises questions about the regulatory mechanism. This selectivity for stimulation of exchange activity in p115-RhoGEF was also observed *in vivo* in that coexpression of the GEF with an activated form of $G_{13}\alpha$ ($G_{13}\alpha$ QL) produced synergistic stimulation of a Rho reporter gene, whereas combinations of the GEF with activated $G_{12}\alpha$ or $G_q\alpha$ were additive or less than additive (Mao *et al.*, 1998). This synergism with $G_{13}\alpha$ observed *in vivo* was also dependent on the presence of the rgRGS domain. Thus, both G protein $\alpha$ subunits interact functionally with the RGS domain to evolve GAP activity but differentially regulate stimulation of RhoA. This indicates that a simple mechanism for activation, in which autoinhibition via the RGS domain was relieved by its interaction with the G protein $\alpha$ subunits, is not viable. Indeed, subsequent studies, which produced only lower basal activities in p115-RhoGEF on deletion of several regions including the rgRGS domain, provide no evidence for the existence of an autoinhibitory domain (Wells *et al.*, 2001).

An alternative hypothesis is that there is a second site for regulation of RhoA exchange activity by $G_{13}\alpha$, which is not functionally interactive with $G_{12}\alpha$. Logically, such a site would exist in the DH-PH domain for more direct action on the catalytic mechanism. Two observations support this idea. First, activated $G_{13}\alpha$ could bind directly to a protein containing just the DH-PH domains (Wells *et al.*, 2002a). Second, activation of RhoGEF activity can be uncoupled from high-affinity association with the rgRGS domain. Thus, point mutations in the rgRGS region that attenuate binding of activated $G_{13}\alpha$ to this domain had no effect on the potency of the $\alpha$ subunit to activate exchange activity for RhoA (Chen *et al.*, 2003). Bhattacharyya and Wedegaertner (2003a) showed that similar mutations attenuated high-affinity binding as assessed by co-immunoprecipitation (co-IP) with activated $G_{13}\alpha$, but these mutations did not affect the shift in localization of p115-RhoGEF to particulate fractions in the presence of expressed activated $G_{13}\alpha$. The major impediment to this two-site hypothesis is the consistent failure of $G_{13}\alpha$ to regulate forms of the RhoGEF that lack the rgRGS domain. One potential explanation is that the rgRGS domain provides a required structural role for regulation of exchange activity in the DH domain that is independent of its ability to bind G protein $\alpha$ subunits. If so, this structural requirement is highly conserved as the rgRGS domain of the PDZ-RhoGEF restored activation of Rho exchange activity by $G_{13}\alpha$ in p115-RhoGEF lacking its own rgRGS domain (Wells *et al.*, 2002a).

Regardless, this model does not explain the ability of $G_{12}\alpha$ to inhibit the action of $G_{13}\alpha$ unless occupation of one site excludes the other or $G_{12}\alpha$ interacts nonproductively with an independent second site in the DH-PH domains. An alternate model that could explain the observation is one in which the G protein $\alpha$ subunit binds to one site but with two mutually compatible surfaces either for interaction with the RGS domain or for activation of RhoA exchange activity. Thus, $G_{12}\alpha$ would not interact productively with the surface stimulating exchange but would effectively block interaction by $G_{13}\alpha$ by binding through the rgRGS surface. One observation that must be included in any mechanism is the loss-of-function by mutation of Lys204 to alanine in $G_{13}\alpha$. While this mutation leaves the basic properties of the $\alpha$ subunit intact, it causes loss of both functional interaction with the rgRGS domain and stimulation of the Rho exchange activity (Nakamura *et al.*, 2004). This suggests that the two activities may share a common interface that disallows divergent signaling through separate surfaces.

While the molecular mechanism for direct regulation of RhoA exchange activity by the G proteins remains a mystery, the domains of the RGS-RhoGEF required for this regulation are limited. Based on structural and deletion analysis discussed above, two regions that might be used to mediate increased functional activity of the DH domain are the PH domain and the N-terminal extension of the DH domain (Derewenda *et al.*, 2004; Kristelly *et al.*, 2004; Reuther *et al.*, 2001; Wells *et al.*, 2001). However, Wells *et al.* (2002a) showed that the PH domain of p115-RhoGEF could be deleted without affecting the fold-stimulation obtained with $G_{13}\alpha$, even though basal exchange activities were markedly reduced. This study also eliminated the C-terminal region and part of the linker region between the rgRGS and DH domains as participants in this function.

The RhoA exchange activity of LARG is also modestly stimulated by G proteins *in vitro* (Suzuki *et al.*, 2003). In this case, both $G_{13}\alpha$ and $G_{12}\alpha$ were effective, although requirements differed (Section III.E.). Surprisingly, the exchange activity of PDZ-RhoGEF was not stimulated by $G_{13}\alpha$ *in vitro* (Wells *et al.*, 2002a). This discrepancy is not due to the lack of GAP activity in the RGS domain of the PDZ-RhoGEF as this domain can facilitate regulation of exchange activity when substituted into p115-RhoGEF (Wells *et al.*, 2002a). While these results may indicate a significant difference in the mechanisms of regulation of the GEFs, it is also possible that the purified PDZ-RhoGEF has lost a functional component.

Coexpression of the G proteins and RhoGEFs, along with subsequent measurement of downstream activities such as the formation of Rho-GTP or stimulation of Rho-dependent transcription via the SRE.L promoter element, has been used to show regulatory capability. This is often interpreted to indicate direct regulation because of the ability of the components to bind to each other. The selective stimulation of p115-RhoGEF by $G_{13}\alpha$ was discussed earlier (Mao *et al.*, 1998). Suzuki *et al.* (2003) demonstrated significant synergism with the SRE.L reporter when either activated $G_{12}\alpha$ or $G_{13}\alpha$ was expressed in HeLa cells with either LARG or PDZ-RhoGEF. Similar synergism for expression of LARG with $G_{12}\alpha$ or $G_{13}\alpha$ was observed in NIH-3T3 cells by direct examination of Rho-GTP formation (Booden *et al.*, 2002). Coexpression of AKAP-Lbc with activated $G_{12}\alpha$ produced more than additive stimulation of Rho-GTP; this was coupled with evidence of direct interaction via coimmunoprecipitation (Diviani *et al.*, 2001). In the case of Lbc, there is not a readily defined binding domain, although subsequent reiteration of these experiments with proto-Lbc narrowed potential sites and a region near the C-terminus with low homology to the ''RGS box'' sequences has been proposed (see above) (Dutt *et al.*, 2004).

### B. *Indirect Regulation of RGS-RhoGEF Activity*

The modest direct stimulation of p115-RhoGEF with $G_{13}\alpha$ (less than fivefold) and the lack of such regulation with PDZ-RhoGEF *in vitro* indicate that another mechanism for regulation by GPCRs is operative. This regulation is likely to occur at the plasma membrane as the site of both cell surface receptors and the substrate, RhoA. The Rho proteins are posttranslationally modified at their C-termini with geranylgeranyl groups. Therefore, they partition readily into membranes, but can be transported or stored in cytosol in complexes with RhoGDI proteins. While the abundant RhoA–RhoGDI complexes in cytosol would seem an abundant source for substrate, interaction with the GDI occludes interfaces on RhoA required for interaction with the RhoGEFs (Ghosh *et al.*, 2003; Nakamura *et al.*, 2004). Therefore, the exchange factors presumably act on the population of free RhoA that is partitioned into membranes either through passive or facilitated dynamic equilibrium with its GDI complex in cytosol. An additional aspect to this reaction is the observation that isoprenylated RhoA is a much better substrate for p115-RhoGEF than the nonprenylated form when assayed *in vitro* (Wells *et al.*, 2002b). While this supports the assumption that the RhoGEF acts on RhoA in the membrane, it raises questions about mechanism and may suggest that the GEF extracts the RhoA C-terminus from the membrane either prior to or during activation.

Two of the RGS-RhoGEFs, p115-RhoGEF and LARG, are located largely in cytosol in resting cells where they are presumably inactive (Bhattacharyya and Wedegaertner, 2000; Hirotani *et al.*, 2002; Togashi *et al.*, 2000; Wells *et al.*, 2001). In contrast, PDZ-RhoGEF is reported to localize significantly to actin structures in the cell and near the plasma membrane (Banerjee and Wedegaertner, 2004; Hirotani *et al.*, 2002; Togashi *et al.*, 2000). Therefore, a second mechanism for regulation of nucleotide exchange on RhoA would be simple translocation of the RGS-RhoGEFs to the membrane for proximal action on available substrate, free RhoA. This could be readily achieved by binding of activated $G_{13}\alpha$ to the rgRGS domains of the RhoGEFs. Evidence for such translocation has been shown with p115-RhoGEF. Expression of an activated form of $G_{13}\alpha$ ($G_{13}\alpha$ QL), which localizes to plasma membranes, was shown to cause redistribution of p115-RhoGEF from cytosol to particulate fractions (cell fractionation) and the plasma membrane (immunolocalization with microscopy) (Bhattacharyya and Wedegaertner, 2000). More interestingly, significant translocation of p115-RhoGEF to membranes could be obtained on stimulation of NIH-3T3 cells with lysophosphatidic acid (presumably acting through LPA receptors) (Wells *et al.*, 2001) or stimulation of overexpressed thromboxane $A_2$ receptors in HEK293 cells (Bhattacharyya and Wedegaertner, 2003b). If p115-RhoGEF was overexpressed, it partitioned into both soluble and particulate fractions and was not altered by hormone stimulation (Wells *et al.*, 2001). The translocation of p115-RhoGEF to membranes by expression of activated $G_{13}\alpha$ QL was dependent on the presence of the rgRGS domain but not the acidic region required for GAP activity and high-affinity binding (Bhattacharyya and Wedegaertner, 2003b). Translocation experiments also suggested a role for the PH domain in that constructs missing this region had impaired membrane recruitment by $G_{13}\alpha$ QL (Bhattacharyya and Wedegaertner, 2003b). The nature of this requirement remains to be determined.

Studies of Lsc in neutrophils also suggest that localization may play a role in activation of Rho (Francis *et al.*, 2006). Immunostaining showed that Lsc was largely distributed uniformly on the plasma membrane. After 5 min of stimulation with *N*-formyl-methionyl-leucyl-phenylalanine (fMLP), the Lsc became enriched in the leading and trailing edges of the cells, sites of active cytoskeletal reorganization in these stimulated cells. While simple translocation to the plasma membranes does not appear to be a viable mechanism for activation in this case, localization within the membrane to direct activity to specific sites is indicated.

### C. *Role of the C-Terminus in Oligomerization and Regulation of* In Vivo *Activity*

An inhibitory role for the C-terminus of the RGS-RhoGEFs was noted in the first description of p115-RhoGEF. Expression in NIH-3T3 cells of a construct of the exchange factor that lacked the C-terminal 111 amino acids produced a greater than tenfold increase in transforming activity as measured by the formation of foci (Hart *et al.*, 1996). Subsequently, p115-RhoGEF missing the C-terminus was shown to be a much more potent stimulator of RhoA *in vivo* than the wild-type GEF, a result which contrasted with the loss of exchange activity observed when the protein missing the C-terminus was assayed *in vitro* (Wells *et al.*, 2001). Deletion of the C-terminus of the mouse orthologue, Lsc, produced the same gain-of-function (Eisenhaure *et al.*, 2003). A working hypothesis is that the C-terminus mediates an inhibitory constraint on the RhoGEF through interaction with other factors found in the cell, but lacking in the *in vitro* assessments. A function of $G_{12}$ or $G_{13}$ *in vivo* may be the removal of this constraint through induction of allosteric changes in the RhoGEF or direct action on the associated factor. This inhibitory role of the C-terminus on activities measured *in vivo* was also observed for LARG and PDZ-RhoGEF (Chikumi *et al.*, 2004). In this study, deletion of the C-terminus of PDZ-RhoGEF had little effect on exchange activity *in vitro*; however, both LARG and PDZ-RhoGEF lacking their C-termini showed a much greater capacity than the wild-type proteins to activate RhoA *in vivo* and a significant increase in the focus-forming assay. Finally, proto-Lbc was much less active in a focus-forming assay than ONCO-Lbc, which lacks the C-terminal domain (Sterpetti *et al.*, 1999). In this study, the exchange activities of the two forms of the protein were similar when measured *in vitro*. In contrast, a subsequent report showed significant increases in activities for ONCO-Lbc when measured either *in vitro* or with a Rho-reporter assay *in vivo* (Dutt *et al.*, 2004).

The C-termini of these proteins also contain an oligomerization domain. Chikumi *et al.* (2004) used immunoprecipitation of differentially tagged overexpressed proteins to provide evidence that three RhoGEFs (LARG, PDZ-RhoGEF, and p115-RhoGEF) could all form homo-oligomers, which were dependent on their C-termini. In addition, evidence was presented for the formation of hetero-oligomers between LARG and PDZ-RhoGEF, but not between p115-RhoGEF and the other exchange factors. It is tempting to speculate that this oligomerization has a regulatory role related to the inhibitory impact of the C-terminal domains *in vivo*. However, evidence indicates this is not the case. In an earlier study, Eisenhaure *et al.* (2003) provided evidence for oligomerization of Lsc, but went on to produce mutations in the C-terminus that could impair homo-oligomerization without producing a gain-of-function characteristic of C-terminal deletion. Stimulation of the exchange factor activity by $G_{13}\alpha$ is also not affected by oligomerization as mutants of p115-RhoGEF lacking the C-terminus are effectively activated by $G_{13}\alpha$ *in vitro* (Wells *et al.*, 2002a). At this time, a physiological role for oligomerization of these classical RGS-RhoGEFs is not known, and the mechanism by which the C-termini attenuate exchange activity on RhoA *in vivo* remains to be discerned.

There is a clearer role for oligomerization of AKAP-Lbc. Deletion of two leucine-zipper motifs within the C-terminal region of this protein prevents oligomerization, and this correlates with an increase in basal activity *in vivo* (Baisamy *et al.*, 2005). Furthermore, a robust inhibitory impact of forskolin (presumably via cAMP/PKA) on the activity of wild-type AKAP-Lbc was absent in protein missing the oligomerization region. Recapitulation of similar behavior with AKAP-Lbc containing point mutations that disrupt oligomerization strongly indicates that this function is important for downregulation by PKA. It is also possible that disruption of oligomerization by association with activated G proteins could provide a mechanism for activation of this protein.

### D. *Specificity of Regulation by G Proteins*

Initial characterization of the RGS-RhoGEFs identified their interactions with $G_{12}\alpha$ and $G_{13}\alpha$. The selective activation of p115-RhoGEF by $G_{13}\alpha$ was demonstrated by experiments both *in vivo* (Mao *et al.*, 1998) and *in vitro* (Hart *et al.*, 1998) (see above). In contrast, AKAP-Lbc showed a synergistic activation of Rho-GTP when coexpressed with $G_{12}\alpha$ but not $G_{13}\alpha$ or other G protein $\alpha$ subunits (Diviani *et al.*, 2001). This selectivity is supported by the ability of dominant negative forms of proto-Lbc to block stimulation of a Rho reporter by activated $G_{12}\alpha$ but not activated $G_{13}\alpha$(Dutt *et al.*, 2004). LARG and PDZ-RhoGEF appear less selective in that both can be activated by both G protein $\alpha$ subunits (Booden *et al.*, 2002; Suzuki *et al.*, 2003).

Several reports have provided evidence that selected RhoGEFs can also be regulated by $G_q$. A comparison study utilizing immunoprecipitation of expressed proteins showed that the RGS domain of LARG but not p115-RhoGEF could bind $G_q\alpha$ which was activated with $AlF_4^-$ (Booden *et al.*, 2002). The authors went on to show that the LARG rgRGS could partially attenuate the stimulation of the SRE.L reporter by overexpressed $G_q\alpha$ (QL) or activation of stably expressed m1-muscarinic receptors in NIH-3T3 cells. Finally, coexpression of $G_q\alpha$ and LARG produced synergistic stimulation of Rho-GTP that was dependent on the LARG rgRGS region.

A second report used a genetic approach to produce evidence for a $G_q$ pathway (Vogt *et al.*, 2003). Mouse embryonic fibroblasts (MEFs) from mutant mice and treatment with pertussis toxin were used to provide cells devoid of the $\alpha$ subunits of $G_{12}$ and $G_{13}$ and blocked for signaling through the $G_i$ family, respectively. In these cells, thrombin, LPA, and bradykinin still mediated increases in Rho-GTP and this could be blocked by expression of LARG lacking its DH-PH domains. This suggests that the $G_q$-mediated pathways are blocked through a dominant negative interaction of the mutated LARG with $G_q\alpha$. An interesting observation in these cells was a 100-fold higher requirement for thrombin for stimulation of Rho-GTP in the $G_{12/13}$ knockout cells (Vogt *et al.*, 2003). This suggests that the $G_{12}$ family is the primary stimulus in wild-type cells.

A third report provides evidence for the interaction of $G_q$ with the Lbc oncogene, although any regulation is unclear. Thus, stably expressed GPCRs that sense calcium could activate the Rho reporter gene in HEK293 cells in an apparent $G_q$-dependent fashion; receptor-dependent stimulation of this pathway was also reduced by expression of the Lbc oncogene lacking either its DH or PH domain (Pi *et al.*, 2002). The simplest interpretation is that the inactive Lbc is tying up $G_q$ through interaction with a binding domain. Sagi *et al.* (2001) used overexpression in COS-7 cells and immunoprecipitation to also demonstrate interaction of proto-Lbc with $G_q\alpha$, but any regulatory impact is unknown. While activated $G_q\alpha$ could stimulate production of luciferase via SRE.L, it did not activate Rho-GTP, either alone or synergistically with proto-Lbc. Overall, the study appeared to show that $G_q\alpha$ enhanced the ability of Rho to stimulate SRE.L but not Rho itself. The authors also showed interaction of proto-Lbc with the $\alpha$ subunits of $G_{12}\alpha$ and $G_{13}\alpha$ but not synergism in function. This contrasts with two reports above suggesting selective regulation of proto-Lbc by $G_{12}\alpha$ (Diviani *et al.*, 2001; Dutt *et al.*, 2004). It remains to be seen if such differences evolve from the use of different cell lines and different constructs (e.g., full-length AKAP-Lbc vs proto-Lbc).

A summary of the current interactions of G proteins with the RhoGEFs is shown in Fig. 4. Regulation by the $G_{12}$ family is well documented with a spectrum where $G_{12}$ and $G_{13}$ share potential regulation of the PDZ-containing RhoGEFs but diverge in their potential spectra of action in differentially engaging either AKAP-Lbc or p115-RhoGEF, respectively. The identification of $G_q$ as a regulator widens markedly the spectrum of potential inputs into selected RhoGEFs. The regulation of two of the RhoGEFs is complicated by the issue of their ability to attenuate the active

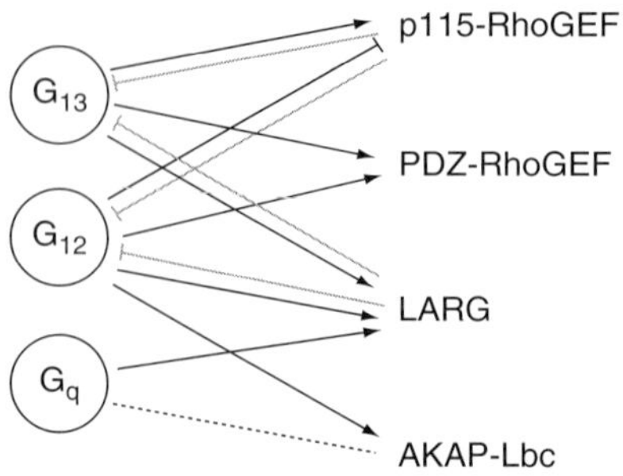


FIG. 4. Potential selectivity for regulation of the RhoGEFs by G proteins. Black arrows indicate that the specified G protein can stimulate the exchange activity of the RhoGEF. A crossbar indicates that function can be blocked; thus, $G_{12}\alpha$ can block regulation of p115-RhoGEF, and the LARG and p115-RhoGEF RGS domains act as GAPs to attenuate functioning of the G proteins. A dotted line indicates evidence for potential but undetermined regulation.

state of their stimulating molecules. It will be of interest to sort out the physiological impacts of these differences in cellular regulation. Although no regulatory impact of $\beta\gamma$ subunits has been noted for the RhoGEFs described here, the interaction of $\beta\gamma$ subunits with p114-RhoGEF (see above) may eventually add another layer of complexity to these signaling pathways.

### E. *Regulation of RGS-RhoGEFs by Phosphorylation*

One indication that phosphorylation may be involved in the modulation of RhoA by heterotrimeric G protein pathways was the reduction of longer-term but not early stimulation of RhoA by thrombin when HEK293 cells were treated with inhibitors of tyrosine kinase activity (Chikumi *et al.*, 2002). Interestingly, the inhibitors could also attenuate stimulation of RhoA by expression of activated $G_{13}\alpha$, but not overexpression of PDZ-RhoGEF; this suggested an effect of phosphorylation at the G protein–RhoGEF interface. A potential site for this regulation was suggested by the observation that thrombin induced phosphorylation of tyrosine in both LARG and PDZ-RhoGEF, probably in their C-termini (Chikumi *et al.*, 2002). Longer-term stimulation of RhoA could also be inhibited by a dominant negative form of FAK, which is also phosphorylated in response to thrombin. While dominant negative FAK could also inhibit phosphorylation of PDZ-RhoGEF by overexpressed FAK, it remains unclear whether the phosphorylation of the RhoGEF plays a role in its activity or whether any effect of FAK on Rho utilizes the RhoGEF. A potential positive effect on RhoA must also be balanced against the potential downregulation of RhoA activity by p190RhoGAP, which can also be activated by tyrosine phosphorylation. Indeed, FAK appears to modulate such inhibition in pulmonary endothelial cells in response to thrombin (Holinstat *et al.*, 2006). Thus, this modulation of RhoA via tyrosine kinases may ultimately depend on the net effect of regulation on both stimulatory and inhibitory pathways or localized effects within the cell.

Suzuki *et al.* (2003) reported more direct evidence for regulation of the RhoGEFs by phosphorylation. These studies showed that LARG, which lacked its PDZ domain (ΔPDZ-LARG), could be activated *in vitro* by $G_{13}\alpha$, but not $G_{12}\alpha$. However, phosphorylation of LARG by the Tec kinase (*t*yrosine kinase *e*xpressed in hepatocellular *c*arcinoma) could induce regulation of Rho exchange activity by $G_{12}\alpha$ without modulating that already observed with $G_{13}\alpha$. *In vivo*, a modest activation of ΔPDZ-LARG by $G_{12}\alpha$ and $G_{13}\alpha$ could be greatly enhanced by expression of the Tec tyrosine kinase, and this effect was dependent on the kinase activity of Tec (Suzuki *et al.*, 2003). Overall, the data indicate that G protein regulation of LARG can be facilitated by phosphorylation of a tyrosine(s), which remains unknown. The *in vitro* data suggest that regulation by $G_{12}\alpha$ may require such phosphorylation, which is presumably lost in the purified protein.

Whereas p115-RhoGEF does not appear to be a substrate for tyrosine phosphorylation, it can be phosphorylated in a protein kinase C (PKC)-dependent fashion in endothelial cells stimulated with thrombin (Holinstat *et al.*, 2003). In this scenario, activation of Rho by thrombin appeared to be dependent on stimulation of both $G_{12}$ or $G_{13}$ (blocked by the rgRGS domain of p115-RhoGEF) and PKCα (blocked by the PKC inhibitor, Gö6976, or expression of a dominant negative form of the protein). Evidence that PKCα and p115-RhoGEF could be immunoprecipitated with each other after stimulation of cells with thrombin suggests direct regulation, but leaves mechanism a mystery. These studies in human umbilical vein endothelial cells suggest that p115-RhoGEF may be a convergence point for coordinated regulation of Rho by both the $G_{12/13}$ and $G_q$ (PKC) pathways.

### F. *Function of the PDZ Domain*

The presence of PDZ (*P*SD-95/SAP90-*D*iscs-larg-*Z*O-1) domains in the RhoGEFs predicted their likely association with other proteins. Several groups identified plexin B as a target through two-hybrid analysis with either the PDZ domain of RhoGEFs (Perrot *et al.*, 2002; Swiercz *et al.*, 2002) or the C-terminal cytoplasmic domains of plexins B1 and B2 (Aurandt *et al.*, 2002; Driessens *et al.*, 2002; Hirotani *et al.*, 2002) as bait. With some variance, groups confirmed interaction of the PDZ domain with the C-terminus of the plexins (B1, B2, and B3) using pull-downs *in vitro* or immunoprecipitations from cell lysates; furthermore, mutation of the C-terminal residues, TDL, greatly impaired the observed associations. Specificity of the interaction was demonstrated by the failure of other plexins with different C-termini (Aurandt *et al.*, 2002; Perrot *et al.*, 2002; Swiercz *et al.*, 2002) or RhoGEFs lacking PDZ domains (Perrot *et al.*, 2002; Swiercz *et al.*, 2002) to bind the respective counterparts. The selectivity of the RhoGEFs for the C-terminal TDL residues of plexins classifies their PDZ domains as having a Type 1 PDZ-binding motif (Songyang *et al.*, 1997). However, a broader spectrum of interaction may be indicated by other interactions (see below).

The plexins are receptors for semaphorins, and selective interactions between members of these two protein families regulate intercellular communication affecting adhesion, motility, and guidance of axonal pathways. Plexin B1 and B2 receptors are activated by semaphorin 4D, which can lead to stimulation of various kinase pathways and Rho proteins (Kruger *et al.*, 2005). Therefore, association of the PDZ domains of LARG and PDZ-RhoGEF with the C-terminus of plexin B proteins suggests a pathway for activation of RhoA by semaphorin 4D. Initial experiments demonstrated that overexpression of plexin B1 and LARG in HEK293 cells could produce synergistic activation of RhoA by stimulation with semaphorin 4D; this stimulation was dependent on the plexin C-terminus (Swiercz *et al.*, 2002). Furthermore, expression of the RhoGEF lacking the DH domain could prevent semaphorin-induced collapse of growth cones in primary neurons (Swiercz *et al.*, 2002). Experiments showing similar synergism and use of the PDZ domain to block neurite retraction were done at the same time in NIH-3T3 and PC12 cells, respectively (Perrot *et al.*, 2002).

Subsequent studies have identified Rnd1 (Oinuma *et al.*, 2003), another member of the Rho GTPase subfamily, and the ErbB-2 tyrosine kinase (Swiercz *et al.*, 2004) in the regulation of RhoA by plexins. Additionally, an apparent requirement for interaction of plexin B with RhoGEFs for potentiation of angiogenesis by semaphorins (Basile *et al.*, 2004) and activation of MAPK pathways (Aurandt *et al.*, 2006) by plexin B1 was identified. Yet, direct evidence for the mechanism by which plexins achieve this activation of RhoA has been elusive. A simple mechanism would be localization of the RhoGEF to membranes on stimulation of the pathway as discussed in Section III.B. Evidence for this comes from a study (Oinuma *et al.*, 2003) with Rnd1, which can bind directly to plexin B1.

Expression of Rnd1 along with plexin B1 in COS-7 cells enhanced the contraction of the cells and the stimulation of RhoA-GTP when exposed to semaphorin 4D. An increased binding of PDZ-RhoGEF to plexin B1 was also observed in the presence of Rnd1. Such increased binding, especially induced by ligand, would localize the RhoGEF to the plasma membrane and its potential substrate RhoA, as discussed above. This model implies an underlying mechanism for preventing interaction of the plexin C-terminus with the RhoGEFs in the resting state.

Identification of ErbB-2 tyrosine kinase in plexin B1 signaling suggests a potential role for phosphorylation (Swiercz *et al.*, 2004). Semaphorin 4D stimulates phosphorylation of both the kinase and plexin B1. This raises the interesting possibility that the PDZ-RhoGEFs could also be targets of the kinase and that phosphorylation would facilitate their activation as discussed in the previous section. The activation mechanism remains to be determined but could also include tonic stimulation by the $G_{12}$- or $G_{13}$ proteins.

The PDZ domain may also play a direct role in the regulation of the RhoGEFs by heterotrimeric G proteins. Yamada *et al.* (2005) used immunoprecipitation to show interaction between the PDZ domains of the RhoGEFs and the C-termini of the $LPA_1$ and $LPA_2$ receptors. These receptors contain class 1 consensus motifs, which were required for binding of the PDZ domains. No association was observed with the $LPA_3$ receptor, which does not contain the motif. Subsequent studies showed that activation of RhoA, mediated by overexpressed $LPA_1$ and $LPA_2$ receptors in HEK293 cells, was dependent on the C-terminal residues of the receptors and that this stimulation could be attenuated by expression of the PDZ domains (Yamada *et al.*, 2005). These observations lead to the prediction that efficacious activation of RhoA by these receptors requires direct association with the RhoGEFs. Such association would promote efficient regulation of the RhoGEFs by the activated $G_{12}$ or $G_{13}$ proteins. It will be important to address whether the RhoGEF is associated with the receptor prior to stimulation with agonists. The coordinate association of the RhoGEF with both the G protein and the PDZ-binding motif on receptor activation is an appealing hypothesis. Binding through the PDZ motif would help stabilize a receptor-G protein-RhoGEF complex during successive cycles of activation and facilitate robust and rapid responses. This parallels stabilization of complexes proposed for kinetic coupling via the RGS domain (Section II.B.). Potential interaction with the PDZ domain also indicates a selectivity of these receptors for the RhoGEFs with PDZ domains. A coordinate hypothesis is that signaling through p115-RhoGEF or receptors without such motifs will be less efficient or that there are other mechanisms for facilitating interactions with selectivity for other isoforms.

Interaction of the PDZ domains of RhoGEFs with plexins or GPCRs is thought to facilitate signaling. Two reports of interaction of the PDZ domains with the insulin-like growth factor-1 (IGF-1) receptor (Taya *et al.*, 2001) or a microtubule-associated protein (MAP) (Longhurst *et al.*, 2006) either agree with a facilitating function or suggest the opposite, respectively. After identification in a two-hybrid screen, the interaction of LARG with the IGF-1 receptor was demonstrated by coimmunoprecipitations from cells and pull-downs *in vitro* (Taya *et al.*, 2001). This was shown to be dependent on interaction of the PDZ domain of LARG with the C-terminal 20 residues of the IGF-1 receptor, although the latter did not contain a classic consensus PDZ-binding motif. In MDCK cells, which express high levels of IGF-1 receptors, exogenous and overexpressed LARG showed PDZ domain-dependent localization to lateral membranes as opposed to its cytoplasmic distribution in most cells. It was hypothesized that the localization is due to the IGF-1 receptors but association with other membrane factors in the associated membranes of these confluent cells was not ruled out. Functional interaction was strongly suggested by the stimulation of Rho-GTP and downstream pathways by IGF-1 and its attenuation by overexpression of the PDZ domain (Taya *et al.*, 2001). However, the mechanism by which stimulation is achieved is unclear especially since LARG was not phosphorylated by stimulation with IGF-1 (Section III.E.) and the stimulation of RhoA was also inhibited by overexpression of the RGS domain of LARG. The latter result indicates that stimulation of $G_{12}\alpha$ or $G_{13}\alpha$ is also required.

A recent study reports an interaction of the PDZ domain of PDZ-RhoGEF with the light chain 2 (LC2) of MAP1. The interaction occurs through the C-terminus of LC2, which has a class 2 PDZ-binding motif (Longhurst *et al.*, 2006). This further indicates a diverse selectivity for these PDZ domains. Overexpression of the RhoGEF lacking the PDZ domain in HeLa cells produced enhanced activation of RhoA as compared with the wild-type protein, rounding of the cells, and enrichment of the RhoGEF to cell surface structures. The simplest interpretation is that localization of the overexpressed RhoGEF through interaction with LC2 prevents unregulated activity, similar to the inhibitory action of the RhoGEF C-termini. Whether this interaction indicates a positive regulatory function through localization of the endogenous PDZ-RhoGEF remains unclear (Section IV.C.).

Overall, the potential interaction of the PDZ-RhoGEFs with multiple cellular proteins predicts a diversity of regulatory possibilities but also potential competition for the RhoGEFs during signaling. One prediction is that the participation of a specific RhoGEF in any single pathway may depend largely on the cell being examined and the expression of proteins that associate with the RhoGEF being studied. It will be of interest to determine whether there is preferential binding of the PDZ domains to specific proteins and how these interactions are affected by the activity of these proteins. Also, how will this association through the PDZ domains coordinate with interaction through other domains in the RhoGEF.

### G. *Expression of PDZ-RhoGEFs*

A common mechanism for modulating the function of regulatory pathways is controlled expression of one or more of the pathway components. This could be especially useful for the RGS-RhoGEFs where such expression could determine the coordinated regulation of RhoA by $G_{12/13}$ versus regulation of other downstream pathways by the G proteins. As noted previously and in the next section, potential selective functions of specific RhoGEFs may determine differential cellular behavior.

The expression of LARG can be upregulated by stimulation of cultured aortic vascular smooth muscle cells with angiotensin II. Thus, a transient threefold increase in the mRNA for LARG is followed by a similar increase in protein expression when these cells are treated with the agonist (Ying *et al.*, 2006). This effect appears to be dependent on activation of the PI-3 kinase pathway. In these cells, the Rho pathway can increase the contractile response. The authors use relaxation of contracted aortic rings by an inhibitor of Rho kinase to show a small change in sensitivity to the inhibitor after treatment of the tissue with angiotensin II for 3 h. No change was observed in the contractile response, and it is not clear that any change was due to increases in expression of LARG.

Upregulation of Lsc in the immune system is discussed in Section IV. It is anticipated that future efforts will reveal many other inputs for regulation of expression of the RGS-RhoGEFs and the impact of this modulation on a variety of signaling paradigms.

## IV. Physiological Function of the RGS-RhoGEFs

The physiological significance of the RGS-RhoGEFs has been implicated by the knockout of specific genes in model organisms and mice. Yet the full spectrum of their physiological impact remains to be elucidated. Their impact on Rho signaling provides one major indication. The interaction of the RGS-RhoGEFs with other signaling proteins suggests a much broader spectrum of regulation. Finally, the multiplicity of genes in mammalian organisms raises the possibility of redundant roles or specialized signaling either in selected cells or with specific stimuli. Current evidence for potential roles and specificity is summarized in this section.

### A. *Pathway Specificity*

Three RGS-RhoGEFs, which can be regulated by the $G_{12/13}$ proteins, have been identified in mammalian cells. Moreover, several GPCRs can regulate the activation of RhoA via these G proteins. While it has often been assumed that the RGS-RhoGEFs mediate this regulation, it has rarely been demonstrated. Thus, the actual role of endogenous RGS-RhoGEFs in hormonal regulation and any selectivity for specific isoforms of the RhoGEFs remains largely unexplored.

One study that addressed this issue used an RNAi strategy to examine the activation of RhoA by thrombin and LPA receptors (Wang *et al.*, 2004). In HEK293 cells, a modest activation of RhoA by thrombin was abolished by the knockdown of LARG, but not by knockdowns of p115-RhoGEF or PDZ-RhoGEF. The selectivity for LARG was mimicked in PC3 cells, where it was required to mediate cell rounding induced by thrombin. In contrast, the ability of LPA to cause rounding in these cells was attenuated by the knockdown of PDZ-RhoGEF rather than LARG or p115-RhoGEF. These results demonstrate that the PDZ-RhoGEFs are necessary for mediation of RhoA by these $G_{12/13}$-dependent pathways. In addition, the selectivity for use of the RhoGEFs by specific receptors may suggest the potential for unique signaling by these stimuli. It will be of interest to determine whether the receptor selectivity demonstrated here is universal or cell specific. The selectivity of the LPA response is somewhat surprising in view of the interaction of the PDZ domains from both LARG and PDZ-RhoGEF with the $LPA_1$ and $LPA_2$ receptors (Yamada *et al.*, 2005); this may indicate that other factors are involved in these processes.

Results from *in vitro* experiments and overexpression in cells (Section III) indicate that selectivity in pathways regulating RhoA by G proteins can be expected from the G protein utilized, participation of other RhoGEFs related to the RGS-RhoGEF family, and other pathways that may regulate the signaling capabilities of the RhoGEFs. Thus, future experiments are expected to elucidate more complicated usage of the proteins that may be influenced by both the composition of individual cells and their interaction with other cellular stimuli.

### B. *Regulation in Hematopoietic Cells*

The relatively high expression of p115-RhoGEF/Lsc in hematopoietic tissues (Hart *et al.*, 1996; Whitehead *et al.*, 1996) suggested a specialized function for this protein. Indeed, the locus for this protein in mice can be knocked out without apparent gross effects on mouse development, survival, and reproduction (Francis *et al.*, 2006; Girkontaite *et al.*, 2001).

This has allowed evaluation of roles for Lsc in development of the immune system and regulatory pathways in cells of hematopoietic origin. Initial characterization noted defects in the homeostasis of B cells in the marginal zone of the spleen and impaired immune response (Girkontaite *et al.*, 2001). In wild-type mice, Lsc is upregulated during lymphocyte development. When Lsc was knocked out, specific observations included defective polymerization of actin in lymphocytes in response to thromboxane A2 ($TXA_2$) and LPA, and impaired responses to challenges with antigens that activate B cells via both T-cell independent and costimulatory mechanisms. The authors noted abnormal motility of the B cells *in vitro*, with decreased basal migration but increased motility in response to serum, which could contribute to their abnormal homeostasis *in vivo*. Further studies examined the absence of Lsc in BALB/c mice (Harenberg *et al.*, 2005). As opposed to the normal numbers of thymocytes observed in $Lsc^{-/-}$ mice with an inbred 129/Sv background (Girkontaite *et al.*, 2001), $Lsc^{-/-}$ BALB/c mice had increased thymi and overall numbers of thymocytes (Harenberg *et al.*, 2005). Here the authors showed directly a defect in the ability of agonists for $TXA_2$-receptors to stimulate Rho-GTP as well as polymerization of actin and proposed that the increased number of thymocytes could be due to decreased apoptosis via stimulation of Rho-dependent pathways by $TXA_2$ (Harenberg *et al.*, 2005).

A second group examined neutrophils from $Lsc^{-/-}$ mice (Francis *et al.*, 2006). While confirming the general normality of the mutant mice and problems with lymphocyte homeostasis, these studies focused on identifying defects in peripheral leukocytes. The $Lsc^{-/-}$ mice had a twofold increase in circulating neutrophils and lymphocytes but normal platelet counts and red blood cells. In functional studies, the $Lsc^{-/-}$ neutrophils showed a reduced ability to stimulate formation of Rho-GTP and abnormal pseudopod development in response to fMLP. An increased motility and lower adherence of the neutrophils when stimulated with fMLP (Francis *et al.*, 2006) are similar to the behavior of B cells from $Lsc^{-/-}$ mice when stimulated with serum (Girkontaite *et al.*, 2001). While these studies imply a role for a $G_{12/13}$-Lsc-RhoA pathway in the phenotypes of these cells, the molecular details are not yet known.

Overall, the defects in lymphocyte homeostasis and immune response may also include defective responses to the lysolipids, LPA and S1P, which regulate Rho pathways in numerous cells, including lymphocytes (Lin and Boyce, 2006). Rubtsov *et al.* (2005) found a decrease in marginal zone B cells in $Lsc^{-/-}$ mice and these MZB cells had an enhanced migratory response to S1P. In contrast to the observations in neutrophils, the MZB cells also showed an increased adherence in the absence of Lsc that could prevent appropriate migration in response to S1P. This would correspond to a model in which S1P provides a migratory signal for B cells in the spleen. A key observation is the requirement for the $S1P_1$ (edg1) receptor for retention of B cells in the marginal zone (Cinamon *et al.*, 2004). The contrasting observations for the same ligand to both increase migration and increase adherence may very well reflect effects on multiple receptors and divergent signaling pathways, which require further exploration. More detailed hypotheses can be found in the references of this section and an additional commentary (Gunn and Kelsoe, 2004).

### C. *Interactions with Other Proteins*

The interaction of the RhoGEFs with multiple signaling proteins through their PDZ domains has been discussed (Section III.F.). Interactions of proteins with other domains of the RhoGEFs are summarized below and in Fig. 5.

The discovery of DRhoGEF2 and demonstration of its requirement for gastrulation during early morphogenesis in *Drosophila* provided early evidence for a direct linkage between the heterotrimeric G proteins and regulation of Rho proteins (Cta and Rho1 in *Drosophila,* respectively) (Barrett *et al.*, 1997; Hacker and Perrimon, 1998). In these studies,

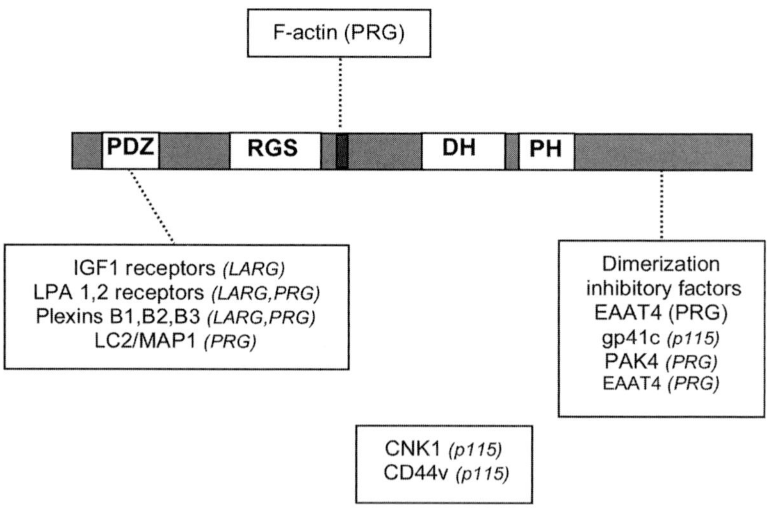


FIG. 5. Identified interactions of the PDZ-RhoGEFs with other signaling proteins or domains. Interactions with specific domains are indicated. Interactions with specific RhoGEFs are indicated in parentheses: p115, p115-RhoGEF; PRG, PDZ-RhoGEF. See text for details.

deletion of DRhoGEF2 resulted in a failure of cells to coordinate the changes in shape needed for gastrulation, a phenotype similar to deletion of Cta and Rho1. These studies also suggested the selectivity of the process for Rho1 in that *Drosophila* Rac or Cdc42 were not required. A more recent study showed that DRhoGEF2 is localized to the apical surface of epithelial cells throughout embryogenesis (Padash *et al.*, 2005). This could help explain the more limited defects of deletion of the exchange factor as opposed to Rho1. The authors hypothesize a selected role for DRhoGEF2 in organizing contractile actomyosin via Rho1 for selected purposes including apical surface constriction to provide the shape changes needed for organized gastrulation.

A second study has identified an interaction of the DRhoGEF2 with EB-1, a protein that associates with the tips of growing microtubules (Rogers *et al.*, 2004). Subsequent studies in *Drosophila* S2 cells demonstrated an association of DRhoGEF2 with the ends of microtubules that was dependent on EB-1 and the ability of activated Cta, Rho1, or overexpressed DRhoGEF2 to induce cellular contraction. The authors hypothesize that DRhoGEF2 is delivered to the cell periphery through microtubule growth, where it can be activated and retained in the presence of an appropriate stimulus such as the FOG pathway at the apical surface. Both this and the previous paper indicate that, in this system, the endogenous RhoGEF is not simply distributed throughout the cell but is localized to effect selected cytoskeletal regulation, perhaps in a directed fashion.

The association of RGS-RhoGEFs with microtubules is reflected in the mammalian PDZ-RhoGEF. Early studies indicated that this protein could localize significantly near the plasma membrane (Hirotani *et al.*, 2002; Togashi *et al.*, 2000) and in some cases appeared to colocalize with polymerized actin. Subsequent studies with overexpressed PDZ-RhoGEF provided detailed localization of the exchange factor with cytoplasmic and peripheral F-actin; this localization was lost if cortical actin was disrupted with latrunculin B (Banerjee and Wedegaertner, 2004). Further experiments demonstrated binding of the PDZ-RhoGEF to actin by immunoprecipitation and sedimentation. The authors went on to demonstrate that this binding was mediated by a region of 25 amino acids (561–585) located between the RGS and DH domains (Banerjee and Wedegaertner, 2004). This region appeared to be both required and sufficient for the association observed. Expression of the PDZ-RhoGEF that lacked this actin-binding region produced modest increases in cellular responses compared to the wild-type enzyme.

A second study provides a more complex picture. Here the authors provide evidence for interaction of the rat PDZ-RhoGEF with the LC2 of MAP1, but through its PDZ domain (Longhurst *et al.*, 2006).

Overexpression of wild-type protein produced stress fibers, whereas expression of the RhoGEF that could not bind to LC2 induced reorganization of cortical actin and cell rounding. Interestingly, disruption of microtubules with nocodazole allowed the overexpressed wild-type protein to produce the same rounded phenotype. This presumably reflects uncontrolled increased activation of RhoA at the cell surface. The authors suggest a model in which the PDZ-RhoGEF can shuttle between sites for binding on microtubules and membranes, presumably in some regulated equilibrium that can be shifted to stress-specific cytoskeletal rearrangements as needed. Interestingly, the reorganization of actin caused by expression of the RhoGEF lacking its PDZ domain produced actin-rich swellings at the membrane surface that were also enriched in the GEF (Longhurst *et al.*, 2006). This could be consistent with continued binding to polymerized actin via a second region (Banerjee and Wedegaertner, 2004), but with a different more active orientation. It will be of interest to get a clearer picture on the behavior of the endogenous protein and the relative ability of the RhoGEFs to activate RhoA in different locales.

The cytoplasmic domain of the *Human immunodeficiency virus 1* transmembrane protein, gp41, is believed to have a role in replication of the virus. A yeast-two-hybrid screen for proteins that interact with the cytoplasmic domain of the protein (gp41C) identified an interaction with the C-terminal 54 amino acids of p115-RhoGEF (Zhang *et al.*, 1999). Binding *in vivo* was indicated by the colocalization of the overexpressed proteins in HeLa cells and expression of gp41C could inhibit stimulation of reporter genes by overexpressed p115-RhoGEF. A mechanism for such inhibition is not clear, but may fit with other data that identified the C-terminal regions of RGS-RhoGEFs as autoinhibitory domains *in vivo* (Section III.C.). A putative leucine-zipper domain of gp41C could be identified as the interactive region, and a point mutation that disrupted binding of gp41C to p115-RhoGEF also attenuated viral reproduction when placed in the context of the HIV virion and infection of T-cell lines (Zhang *et al.*, 1999). This suggested a role for p115-RhoGEF in HIV reproduction but other effects of the mutation were not ruled out.

The potential importance of this interaction is reinforced by a second study (Wang *et al.*, 2000) which showed that overexpression of activated $G_{13}\alpha$ or p115-RhoGEF could attenuate HIV replication in HEK293 or Jurkat T cells. This attenuation was also observed with activated RhoA, and a series of switch domain mutants was used to suggest that the pathway leading to suppression of HIV replication was unique from those leading to stimulation of gene transcription and stress fiber formation (Wang *et al.*, 2000). An overall model for T cells suggests that interaction of gp41C with p115-RhoGEF may provide a mechanism for enhancing HIV replication by suppressing activation of RhoA and its inhibitory activity. A caveat to this model is that disruption of gp41C-p115-RhoGEF interaction did not affect viral reproduction in the HEK293 cells (Zhang *et al.*, 1999).

The p21-activated kinase 4 (PAK4) is another protein that interacts with the C-terminal region of PDZ-RhoGEF (Barac *et al.*, 2004). The initial discovery of this interaction by two-hybrid analysis in yeast was followed by demonstration of coprecipitation when the C-terminus of the RhoGEF was coexpressed with the kinase domain of PAK4 in HEK293 cells. Expression experiments with the whole proteins and immunoprecipitations showed that the interaction of PAK4 was selective for PDZ-RhoGEF and showed only minor interaction with p115-RhoGEF and none with LARG. Interestingly, PAK4 could phosphorylate the C-terminus of PDZ-RhoGEF, and this appears to occur with overexpressed activated proteins *in vivo*. From a functional standpoint, overexpression of activated PAK4, but not dominant negative PAK4, could decrease the stimulation of RhoA-GTP by LPA or expression of activated $G_{13}\alpha$ in Swiss 3T3 cells (Barac *et al.*, 2004). It was hypothesized that PAK4, which also interacts with Cdc42, could mediate regulation of RhoA by Cdc42 as a means of cross talk among the larger family of Rho GTPases. However, the mechanism remains unclear as phosphorylation of PDZ-RhoGEF had no effect on activity *in vitro*. It would also be of interest to know whether PDZ-RhoGEF was the primary mediator of RhoA stimulation by LPA or activated $G_{13}\alpha$ in these cells. The potential for this regulation of hormone responses is supported by the recent demonstration that PAK4 could also interact with another RhoGEF, GEF-H1, and attenuate its ability to induce stress fiber formation through phosphorylation of the GEF (Callow *et al.*, 2005).

One mechanism for directing the action of RhoGEFs and activated RhoA to selected pathways would be interactions with specific scaffolds. This is proposed for the function of CNK1 (Jaffe *et al.*, 2005). Based on pull-downs of overexpressed proteins, CNK1 appears to bind to p115-RhoGEF and rhophilin, as well as MLK3 and MKK7, two kinases that can lead to phosphorylation of JNK. Expression of CNK1 does not attenuate the formation of RhoA-GTP in response to p115-RhoGEF, but blocks several downstream effects by some undefined mechanism, perhaps by sequestering the Rho-GTP from interaction with appropriate sites. In contrast, the overexpressed CNK1 caused a modest enhancement of phosphorylation of JNK induced by the p115-RhoGEF. The ability of CNK1 to bind to kinases in this pathway and p115-RhoGEF suggests that this might provide an ordering of components for efficient stimulation of JNK via hormonal stimulation of RhoA (Jaffe *et al.*, 2005). This is supported by a decrease in the modest stimulation of JNK by activated $G_{12}\alpha$ when CNK1 was knocked down by RNAi in HeLa cells.

The theme of organization through scaffolds is also suggested in the putative interaction of p115-RhoGEF with CD44v (Singleton and Bourguignon, 2002), a transmembrane glycoprotein that can bind hyaluronin (HA), a glycosaminoglycan found in extracellular matrix. Immunoprecipitations from lysates of a breast cancer cell line of either CD44v or p115-RhoGEF could coprecipitate the other respective protein. HA could stimulate activation of Rho-dependent pathways in these and other cells. Previous studies had demonstrated an interaction of CD44 with the Rho-activated kinase, ROK (Bourguignon *et al.*, 2003). While it is not clear that these interactions are direct, a hypothesis can be developed that colocalization of the RhoGEF and ROK to the C-terminus of CD44 would facilitate efficient and perhaps selective regulation of this kinase on stimulation with HA (Singleton and Bourguignon, 2002).

The C-terminal region of Lbc-RhoGEF can associate with $\alpha$-catulin, an $\alpha$-catenin-related protein (Park *et al.*, 2002). The interaction identified by a two-hybrid screen could be verified by complementary pull-downs of overexpressed tagged proteins and use of the tagged proteins to coimmunoprecipitate endogenous proteins. Potential functional relevance is supported by a significant twofold enhancement of activation of Rho-GTP or Rho reporter constructs when $\alpha$-catulin was coexpressed with Lbc; this stimulation was attenuated in constructs that lacked the association domains (Park *et al.*, 2002). At this time, the physiological relevance or mechanism for this activity is not known. Speculation that $\alpha$-catulin may interface with cytoskeletal organization in a fashion similar to the catenins would immediately rationalize the interaction with a RhoGEF. If $\alpha$-catulin can interface with components of the catenin–cadherin machinery, it is of interest to note that a regulator of Lbc, $G_{12}\alpha$, can also functionally interact with the cadherins (Meigs *et al.*, 2001).

A final intriguing role for RhoREFs is suggested by the interaction of GTRAP48 (rat form of PDZ-RhoGEF), with the glutamate transporter, EAAT4 (excitatory amino acid transporter 4) (Jackson *et al.*, 2001). This interaction was first found in a two-hybrid screen with the C-terminus of EAAT4 and identified an interaction with the C-terminal region of the RhoGEF. This interaction was confirmed by immunoprecipitation of overexpressed proteins, and characterization of the RhoGEF provided regulatory information noted in previous sections. The physiological relevance of this interaction is not yet clear. Exogenous expression of the RhoGEF in cells stably expressing EAAT4 resulted in an increase in glutamate transport; subsequent analysis suggested that this increase was due to stabilization of the transporters at the cellular surface through interaction with the RhoGEF. It will be of interest to find out whether the $G_{12}$ family has a role in regulating distribution of these transporters in glutaminergic synapses and plasticity.

In general, the interaction of the RGS-RhoGEFs with a variety of cellular proteins suggests that their function in the context of complexes with other proteins may be the rule. Thus, their specific signaling may be directed by localization and their juxtaposition with downstream signaling components or other regulatory pathways. The interaction of several proteins with the C-terminal regions of the RGS-RhoGEFs raises immediate questions about their impact on the inhibitory function of this region or oligomerization of the RhoGEFs. Will such interactions merely localize the RhoGEF, relieve inhibition of RhoA exchange activity by the C-termini, or enforce inhibition of the RhoGEF as hypothesized for some unknown inhibitor (Section III.C.)? Will such interactions be dependent on dimerization of the RhoGEFs, occur only with monomers, or be independent of the state of oligomerization of the RhoGEFs? Further elucidation of the mechanisms, integration, and physiological impact of these interactions should be highly insightful to this area of cellular function.

## Acknowledgments

The authors thank past and current members of the Sternweis research team for their hard work and contribution to some of the concepts expressed in this chapter and support from the National Institutes of Health (GM31954), the Welch Foundation (I-1262), and the Alfred and Mabel Gilman Chair in Molecular Pharmacology.

## References

Aurandt, J., Vikis, H. G., Gutkind, J. S., Ahn, N., and Guan, K. L. (2002). The semaphorin receptor plexin-B1 signals through a direct interaction with the Rho-specific nucleotide exchange factor, LARG. *Proc. Natl. Acad. Sci. USA* **99,** 12085–12090.

Aurandt, J., Li, W., and Guan, K. L. (2006). Semaphorin 4D activates the MAPK pathway downstream of plexin-B1. *Biochem. J.* **394,** 459–464.

Baisamy, L., Jurisch, N., and Diviani, D. (2005). Leucine zipper-mediated homo-oligomerization regulates the Rho-GEF activity of AKAP-Lbc. *J. Biol. Chem.* **280,** 15405–15412.

Banerjee, J., and Wedegaertner, P. B. (2004). Identification of a novel sequence in PDZ-RhoGEF that mediates interaction with the actin cytoskeleton. *Mol. Biol. Cell.* **15,** 1760–1775.

Barac, A., Basile, J., Vazquez-Prado, J., Gao, Y., Zheng, Y., and Gutkind, J. S. (2004). Direct interaction of p21-activated kinase 4 with PDZ-RhoGEF, a G protein-linked Rho guanine exchange factor. *J. Biol. Chem.* **279,** 6182–6189.

Barrett, K., Leptin, M., and Settleman, J. (1997). The Rho GTPase and a putative RhoGEF mediate a signaling pathway for the cell shape changes in *Drosophila* gastrulation. *Cell* **91,** 905–915.

Basile, J. R., Barac, A., Zhu, T., Guan, K. L., and Gutkind, J. S. (2004). Class IV semaphorins promote angiogenesis by stimulating Rho-initiated pathways through plexin-B. *Cancer Res.* **64,** 5212–5224.

Bhattacharyya, R., and Wedegaertner, P. B. (2000). G alpha 13 requires palmitoylation for plasma membrane localization, Rho-dependent signaling, and promotion of p115-RhoGEF membrane binding. *J. Biol. Chem.* **275,** 14992–14999.

Bhattacharyya, R., and Wedegaertner, P. B. (2003a). Mutation of an N-terminal acidic-rich region of p115-RhoGEF dissociates alpha13 binding and alpha13-promoted plasma membrane recruitment. *FEBS Lett.* **540,** 211–216.

Bhattacharyya, R., and Wedegaertner, P. B. (2003b). Characterization of G alpha 13-dependent plasma membrane recruitment of p115RhoGEF. *Biochem. J.* **371,** 709–720.

Bhattacharya, M., Babwah, A. V., and Ferguson, S. S. (2004). Small GTP-binding protein-coupled receptors. *Biochem. Soc. Trans.* **32,** 1040–1044.

Birukova, A. A., Adyshev, D., Gorshkov, B., Bokoch, G. M., Birukov, K. G., and Verin, A. D. (2006). GEF-H1 is involved in agonist-induced human pulmonary endothelial barrier dysfunction. *Am. J. Physiol. Lung Cell Mol. Physiol.* **290,** L540–L548.

Booden, M. A., Siderovski, D. P., and Der, C. J. (2002). Leukemia-associated Rho guanine nucleotide exchange factor promotes G alpha q-coupled activation of RhoA. *Mol. Cell. Biol.* **22,** 4053–4061.

Bourguignon, L. Y., Singleton, P. A., Zhu, H., and Diedrich, F. (2003). Hyaluronan-mediated CD44 interaction with RhoGEF and Rho kinase promotes Grb2-associated binder-1 phosphorylation and phosphatidylinositol 3-kinase signaling leading to cytokine (macrophage-colony stimulating factor) production and breast tumor progression. *J. Biol. Chem.* **278,** 29420–29434.

Brown, J. H., Del Re, D. P., and Sussman, M. A. (2006). The Rac and Rho hall of fame: A decade of hypertrophic signaling hits. *Circ. Res.* **98,** 730–742.

Burridge, K., and Wennerberg, K. (2004). Rho and Rac take center stage. *Cell* **116,** 167–179.

Cabrera-Vera, T. M., Vanhauwe, J., Thomas, T. O., Medkova, M., Preininger, A., Mazzoni, M. R., and Hamm, H. E. (2003). Insights into G protein structure, function, and regulation. *Endocr. Rev.* **24,** 765–781.

Callow, M. G., Zozulya, S., Gishizky, M. L., Jallal, B., and Smeal, T. (2005). PAK4 mediates morpho'al changes through the regulation of GEF-H1. *J. Cell Sci.* **118,** 1861–1872.

Chang, Y. C., Lee, H. H., Chen, Y. J., Bokoch, G. M., and Chang, Z. F. (2006). Contribution of guanine exchange factor H1 in phorbol ester-induced apoptosis. *Cell Death Differ.* **13**, 2023–2032.

Chen, Z., Wells, C. D., Sternweis, P. C., and Sprang, S. R. (2001). Structure of the rgRGS domain of p115RhoGEF. *Nat. Struct. Biol.* **8,** 805–809.

Chen, Z., Singer, W. D., Wells, C. D., Sprang, S. R., and Sternweis, P. C. (2003). Mapping the G alpha 13 binding interface of the rgRGS domain of p115RhoGEF. *J. Biol. Chem.* **278,** 9912–9919.

Chen, Z., Singer, W. D., Sternweis, P. C., and Sprang, S. R. (2005). Structure of the p115RhoGEF rgRGS domain-G alpha 13/i1 chimera complex suggests convergent evolution of a GTPase activator. *Nat. Struct. Mol. Biol.* **12,** 191–197.

Chikumi, H., Fukuhara, S., and Gutkind, J. S. (2002). Regulation of G protein-linked guanine nucleotide exchange factors for Rho, PDZ-RhoGEF, and LARG by tyrosine phosphorylation: Evidence of a role for focal adhesion kinase. *J. Biol. Chem.* **277,** 12463–12473.

Chikumi, H., Barac, A., Behbahani, B., Gao, Y., Teramoto, H., Zheng, Y., and Gutkind, J. S. (2004). Homo- and hetero-oligomerization of PDZ-RhoGEF, LARG and p115RhoGEF by their C-terminal region regulates their *in vivo* Rho GEF activity and transforming potential. *Oncogene* **23,** 233–240.

Cinamon, G., Matloubian, M., Lesneski, M. J., Xu, Y., Low, C., Lu, T., Proia, R. L., and Cyster, J. G. (2004). Sphingosine 1-phosphate receptor 1 promotes B cell localization in the splenic marginal zone. *Nat. Immunol.* **5,** 713–720.

de Alba, E., De Vries, L., Farquhar, M. G., and Tjandra, N. (1999). Solution structure of human GAIP (G alpha interacting protein): A regulator of G protein signaling. *J. Mol. Biol.* **291,** 927–939.

Derewenda, U., Oleksy, A., Stevenson, A. S., Korczynska, J., Dauter, Z., Somlyo, A. P., Otlewski, J., Somlyo, J., and Derewenda, Z. S. (2004). The crystal structure of RhoA in complex with the DH/PH fragment of PDZRhoGEF, an activator of the $Ca^{(2+)}$ sensitization pathway in smooth muscle. *Structure* **12,** 1955–1965.

Diviani, D., Soderling, J., and Scott, J. D. (2001). AKAP-Lbc anchors protein kinase A and nucleates G alpha 12-selective Rho-mediated stress fiber formation. *J. Biol. Chem.* **276,** 44247–44257.

Driessens, M. H., Olivo, C., Nagata, K., Inagaki, M., and Collard, J. G. (2002). B plexins activate Rho through PDZ-RhoGEF. *FEBS Lett.* **529,** 168–172.

Dutt, P., Nguyen, N., and Toksoz, D. (2004). Role of Lbc RhoGEF in G alpha12/13-induced signals to Rho GTPase. *Cell. Signal.* **16,** 201–209.

Eisenhaure, T. M., Francis, S. A., Willison, L. D., Coughlin, S. R., and Lerner, D. J. (2003). The Rho guanine nucleotide exchange factor Lsc homo-oligomerizes and is negatively regulated through domains in its carboxyl terminus that are absent in novel splenic isoforms. *J. Biol. Chem.* **278,** 30975–30984.

Etienne-Manneville, S., and Hall, A. (2002). Rho GTPases in cell biology. *Nature* **420,** 629–635.

Francis, S. A., Shen, X., Young, J. B., Kaul, P., and Lerner, D. J. (2006). Rho GEF Lsc is required for normal polarization, migration, and adhesion of formyl-peptide-stimulated neutrophils. *Blood* **107,** 1627–1635.

Fukuhara, S., Murga, C., Zohar, M., Igishi, T., and Gutkind, J. S. (1999). A novel PDZ domain containing guanine nucleotide exchange factor links heterotrimeric G proteins to Rho. *J. Biol. Chem.* **274,** 5868–5879.

Fukuhara, S., Chikumi, H., and Gutkind, J. S. (2000). Leukemia-associated Rho guanine nucleotide exchange factor (LARG) links heterotrimeric G proteins of the G (12) family to Rho. *FEBS Lett.* **485,** 183–188.

Gebbink, M. F., Kranenburg, O., Poland, M., van Horck, F. P., Houssa, B., and Moolenaar, W. H. (1997). Identification of a novel, putative Rho-specific GDP/GTP exchange factor and a RhoA-binding protein: Control of neuronal morphology. *J. Cell Biol.* **137,** 1603–1613.

Ghosh, S., Moore, S., Bell, R. M., and Dush, M. (2003). Functional analysis of a phosphatidic acid binding domain in human Raf-1 kinase: Mutations in the phosphatidate binding domain lead to tail and trunk abnormalities in developing zebrafish embryos. *J. Biol. Chem.* **278,** 45690–45696.

Gilman, A. G. (1987). G proteins: Transducers of receptor-generated signals. *Annu. Rev. Biochem.* **56,** 615–649.

Girkontaite, I., Missy, K., Sakk, V., Harenberg, A., Tedford, K., Potzel, T., Pfeffer, K., and Fischer, K. D. (2001). Lsc is required for marginal zone B cells, regulation of lymphocyte motility and immune responses. *Nat. Immunol.* **2,** 855–862.

Glaven, J. A., Whitehead, I. P., Nomanbhoy, T., Kay, R., and Cerione, R. A. (1996). Lfc and Lsc oncoproteins represent two new guanine nucleotide exchange factors for the Rho GTP-binding protein. *J. Biol. Chem.* **271,** 27374–27381.

Gunn, M. D., and Kelsoe, G. (2004). Should we B-leavin' now? *Nat. Immunol.* **5,** 703–704.

Hacker, U., and Perrimon, N. (1998). DRhoGEF2 encodes a member of the Dbl family of oncogenes and controls cell shape changes during gastrulation in *Drosophila. Genes Dev.* **12,** 274–284.

Harenberg, A., Girkontaite, I., Giehl, K., and Fischer, K. D. (2005). The Lsc RhoGEF mediates signaling from thromboxane A2 to actin polymerization and apoptosis in thymocytes. *Eur. J. Immunol.* **35,** 1977–1986.

Hart, M. J., Sharma, S., elMasry, N., Qiu, R. G., McCabe, P., Polakis, P., and Bollag, G. (1996). Identification of a novel guanine nucleotide exchange factor for the Rho GTPase. *J. Biol. Chem.* **271,** 25452–25458.

Hart, M. J., Jiang, X., Kozasa, T., Roscoe, W., Singer, W. D., Gilman, A. G., Sternweis, P. C., and Bollag, G. (1998). Direct stimulation of the guanine nucleotide exchange activity of p115 RhoGEF by G alpha13. *Science* **280,** 2112–2114.

Hepler, J. R. (1999). Emerging roles for RGS proteins in cell signalling. *Trends Pharmacol. Sci.* **20,** 376–382.

Hirotani, M., Ohoka, Y., Yamamoto, T., Nirasawa, H., Furuyama, T., Kogo, M., Matsuya, T., and Inagaki, S. (2002). Interaction of plexin-B1 with PDZ domain-containing Rho guanine nucleotide exchange factors. *Biochem. Biophys. Res. Commun.* **297,** 32–37.

Holinstat, M., Mehta, D., Kozasa, T., Minshall, R. D., and Malik, A. B. (2003). Protein kinase C alpha-induced p115RhoGEF phosphorylation signals endothelial cytoskeletal rearrangement. *J. Biol. Chem.* **278,** 28793–28798.

Holinstat, M., Knezevic, N., Broman, M., Samarel, A. M., Malik, A. B., and Mehta, D. (2006). Suppression of RhoA activity by focal adhesion kinase-induced activation of p190RhoGAP: Role in regulation of endothelial permeability. *J. Biol. Chem.* **281,** 2296–2305.

Jackson, M., Song, W., Liu, M. Y., Jin, L., Dykes-Hoberg, M., Lin, C. I., Bowers, W. J., Federoff, H. J., Sternweis, H. J., and Rothstein, J. D. (2001). Modulation of the neuronal glutamate transporter EAAT4 by two interacting proteins. *Nature* **410,** 89–93.

Jaffe, A. B., Hall, A., and Schmidt, A. (2005). Association of CNK1 with Rho guanine nucleotide exchange factors controls signaling specificity downstream of Rho. *Curr. Biol.* **15,** 405–412.

Karnoub, A. E., Symons, M., Campbell, S. L., and Der, C. J. (2004). Molecular basis for Rho GTPase signaling specificity. *Breast Cancer Res. Treat.* **84,** 61–71.

Kourlas, P. J., Strout, M. P., Becknell, B., Veronese, M. L., Croce, C. M., Theil, K. S., Krahe, R., Ruutu, R., Knuutila, S., Bloomfield, C. D., and Caligiuri, M. A. (2000). Identification of a gene at 11q23 encoding a guanine nucleotide exchange factor: Evidence for its fusion with MLL in acute myeloid leukemia. *Proc. Natl. Acad. Sci. USA* **97,** 2145–2150.

Kozasa, T., Jiang, X., Hart, M. J., Sternweis, P. M., Singer, W. D., Gilman, A. G., Bollag, G., and Sternweis, P. C. (1998). p115 RhoGEF, a GTPase activating protein for G alpha12 and G alpha13. *Science* **280,** 2109–2111.

Krendel, M., Zenke, F. T., and Bokoch, G. M. (2002). Nucleotide exchange factor GEF-H1 mediates cross-talk between microtubules and the actin cytoskeleton. *Nat. Cell Biol.* **4,** 294–301.

Kristelly, R., Gao, G., and Tesmer, J. J. (2004). Structural determinants of RhoA binding and nucleotide exchange in leukemia-associated Rho guanine-nucleotide exchange factor. *J. Biol. Chem.* **279,** 47352–47362.

Kruger, R. P., Aurandt, J., and Guan, K. L. (2005). Semaphorins command cells to move. *Nat. Rev. Mol. Cell Biol.* **6,** 789–800.

Lin, D. A., and Boyce, J. A. (2006). Lysophospholipids as mediators of immunity. *Adv. Immunol.* **89,** 141–167.

Longenecker, K. L., Lewis, M. E., Chikumi, H., Gutkind, J. S., and Derewenda, Z. S. (2001). Structure of the RGS-like domain from PDZ-RhoGEF: Linking heterotrimeric g protein-coupled signaling to Rho GTPases. *Structure* **9,** 559–569.

Longhurst, D. M., Watanabe, M., Rothstein, J. D., and Jackson, M. (2006). Interaction of PDZRhoGEF with microtubule-associated protein 1 light chains: Link between microtubules, actin cytoskeleton, and neuronal polarity. *J. Biol. Chem.* **281,** 12030–12040.

Mao, J., Yuan, H., Xie, W., and Wu, D. (1998). Guanine nucleotide exchange factor GEF115 specifically mediates activation of Rho and serum response factor by the G protein alpha subunit G alpha13. *Proc. Natl. Acad. Sci. USA* **95,** 12973–12976.

Meigs, T. E., Fields, T. A., McKee, D. D., and Casey, P. J. (2001). Interaction of G alpha 12 and G alpha 13 with the cytoplasmic domain of cadherin provides a mechanism for beta-catenin release. *Proc. Natl. Acad. Sci. USA* **98,** 519–524.

Moon, S. Y., and Zheng, Y. (2003). Rho GTPase-activating proteins in cell regulation. *Trends Cell Biol.* **13,** 13–22.

Nakamura, S., Kreutz, B., Tanabe, S., Suzuki, N., and Kozasa, T. (2004). Critical role of lysine 204 in switch I region of G alpha13 for regulation of p115RhoGEF and leukemia-associated RhoGEF. *Mol. Pharmacol.* **66,** 1029–1034.

Niu, J., Profirovic, J., Pan, H., Vaiskunaite, R., and Voyno-Yasenetskaya, T. (2003). G Protein betagamma subunits stimulate p114RhoGEF, a guanine nucleotide exchange factor for RhoA and Rac1: Regulation of cell shape and reactive oxygen species production. *Circ. Res.* **93,** 848–856.

Oinuma, I., Katoh, H., Harada, A., and Negishi, M. (2003). Direct interaction of Rnd1 with Plexin-B1 regulates PDZ-RhoGEF-mediated Rho activation by Plexin-B1 and induces cell contraction in COS-7 cells. *J. Biol. Chem.* **278,** 25671–25677.

Padash, B. M., Rogers, S., and Hacker, U. (2005). DRhoGEF2 regulates actin organization and contractility in the *Drosophila* blastoderm embryo. *J. Cell Biol.* **168,** 575–585.

Park, B., Nguyen, N. T., Dutt, P., Merdek, K. D., Bashar, M., Sterpetti, P., Tosolini, A., Testa, J. R., and Toksoz, D. (2002). Association of Lbc Rho guanine nucleotide exchange factor with alpha-catenin-related protein, alpha-catulin/CTNNAL1, supports serum response factor activation. *J. Biol. Chem.* **277,** 45361–45370.

Perrot, V., Vazquez-Prado, J., and Gutkind, J. S. (2002). Plexin B regulates Rho through the guanine nucleotide exchange factors leukemia-associated Rho GEF (LARG) and PDZ-RhoGEF. *J. Biol. Chem.* **277,** 43115–43120.

Pi, M., Spurney, R. F., Tu, Q., Hinson, T., and Quarles, L. D. (2002). Calcium-sensing receptor activation of rho involves filamin and rho-guanine nucleotide exchange factor. *Endocrinology* **143,** 3830–3838.

Ren, Y., Li, R., Zheng, Y., and Busch, H. (1998). Cloning and characterization of GEF-H1, a microtubule-associated guanine nucleotide exchange factor for Rac and Rho GTPases. *J. Biol. Chem.* **273,** 34954–34960.

Reuther, G. W., Lambert, Q. T., Booden, M. A., Wennerberg, K., Becknell, B., Marcucci, J., Sondek, J., Caligiuri, M. A., and Der, C. J. (2001). Leukemia-associated Rho guanine nucleotide exchange factor, a Dbl family protein found mutated in leukemia, causes transformation by activation of RhoA. *J. Biol. Chem.* **276,** 27145–27151.

Riobo, N. A., and Manning, D. R. (2005). Receptors coupled to heterotrimeric G proteins of the G12 family. *Trends Pharmacol. Sci.* **26,** 146–154.

Rogers, S. L., Wiedemann, U., Hacker, U., Turck, C., and Vale, R. D. (2004). *Drosophila* RhoGEF2 associates with microtubule plus ends in an EB1-dependent manner. *Curr. Biol.* **14,** 1827–1833.

Ross, E. M., and Wilkie, T. M. (2000). GTPase-activating proteins for heterotrimeric G proteins: Regulators of G protein signaling (RGS) and RGS-like proteins. *Annu. Rev. Biochem.* **69,** 795–827.

Rossman, K. L., Der, C. J., and Sondek, J. (2005). GEF means go: Turning on RHO GTPases with guanine nucleotide-exchange factors. *Nat. Rev. Mol. Cell Biol.* **6,** 167–180.

Rubtsov, A., Strauch, P., Digiacomo, A., Hu, J., Pelanda, R., and Torres, R. M. (2005). Lsc regulates marginal-zone B cell migration and adhesion and is required for the IgM T-dependent antibody response. *Immunity* **23,** 527–538.

Rumenapp, U., Blomquist, A., Schworer, G., Schablowski, H., Psoma, A., and Jakobs, K. H. (1999). Rho-specific binding and guanine nucleotide exchange catalysis by KIAA0380, a dbl family member. *FEBS Lett.* **459,** 313–318.

Sagi, S. A., Seasholtz, T. M., Kobiashvili, M., Wilson, B. A., Toksoz, D., and Brown, J. H. (2001). Physical and functional interactions of G alpha q with Rho and its exchange factors. *J. Biol. Chem.* **276,** 15445–15452.

Sah, V. P., Seasholtz, T. M., Sagi, S. A., and Brown, J. H. (2000). The role of Rho in G protein-coupled receptor signal transduction. *Annu. Rev. Pharmacol. Toxicol.* **40,** 459–489.

Singleton, P. A., and Bourguignon, L. Y. (2002). CD44v10 interaction with Rho-kinase (ROK) activates inositol 1,4,5-triphosphate (IP3) receptor-mediated $Ca^{2+}$ signaling during hyaluronan (HA)-induced endothelial cell migration. *Cell Motil. Cytoskeleton.* **53,** 293–316.

Slep, K. C., Kercher, M. A., He, W., Cowan, C. W., Wensel, T. G., and Sigler, P. B. (2001). Structural determinants for regulation of phosphodiesterase by a G protein at 2.0 A. *Nature* **409,** 1071–1077.

Songyang, Z., Fanning, A. S., Fu, C., Xu, J., Marfatia, S. M., Chishti, A. H., Crompton, A., Chan, J. M., Anderson, J. M., and Cantley, L. C. (1997). Recognition of unique carboxyl-terminal motifs by distinct PDZ domains. *Science* **275,** 73–77.

Spink, K. E., Polakis, P., and Weis, W. I. (2000). Structural basis of the Axin-adenomatous polyposis coli interaction. *EMBO J.* **19,** 2270–2279.

Sternweis, P. C. (1996). G proteins in signal transduction. *In* ''Signal Transduction'' (C.-H. Heldin and M. Purton, Eds.), pp. 287–301. Chapman & Hall, London.

Sterpetti, P., Hack, A. A., Bashar, M. P., Park, B., Cheng, S. D., Knoll, J. H., Urano, T., Feig, L. A., and Toksoz, D. (1999). Activation of the Lbc Rho exchange factor proto-oncogene by truncation of an extended C terminus that regulates transformation and targeting. *Mol. Cell Biol.* **19,** 1334–1345.

Suzuki, N., Nakamura, S., Mano, H., and Kozasa, T. (2003). G alpha 12 activates Rho GTPase through tyrosine-phosphorylated leukemia-associated RhoGEF. *Proc. Natl. Acad. Sci. USA* **100,** 733–738.

Swiercz, J. M., Kuner, R., Behrens, J., and Offermanns, S. (2002). Plexin-B1 directly interacts with PDZ-RhoGEF/LARG to regulate RhoA and growth cone morphology. *Neuron* **35,** 51–63.

Swiercz, J. M., Kuner, R., and Offermanns, S. (2004). Plexin-B1/RhoGEF-mediated RhoA activation involves the receptor tyrosine kinase ErbB-2. *J. Cell Biol.* **165,** 869–880.

Taya, S., Inagaki, N., Sengiku, H., Makino, H., Iwamatsu, A., Urakawa, I., Nagao, K., Kataoka, S., and Kaibuchi, K. (2001). Direct interaction of insulin-like growth factor-1 receptor with leukemia-associated RhoGEF. *J. Cell Biol.* **155,** 809–820.

Tesmer, J. J., Berman, D. M., Gilman, A. G., and Sprang, S. R. (1997). Structure of RGS4 bound to AlF4-activated G(i alpha1): Stabilization of the transition state for GTP hydrolysis. *Cell* **89,** 251–261.

Togashi, H., Nagata, K., Takagishi, M., Saitoh, N., and Inagaki, M. (2000). Functions of a rho-specific guanine nucleotide exchange factor in neurite retraction. Possible role of a proline-rich motif of KIAA0380 in localization. *J. Biol. Chem.* **275,** 29570–29578.

Toksoz, D., and Williams, D. A. (1994). Novel human oncogene lbc detected by transfection with distinct homology regions to signal transduction products. *Oncogene* **9,** 621–628.

van der Linden, A. M., Moorman, C., Cuppen, E., Korswagen, H. C., and Plasterk, R. H. (2003). Hyperactivation of the G12-mediated signaling pathway in *Caenorhabditis elegans* induces a developmental growth arrest via protein kinase. C. *Curr. Biol.* **13,** 516–521.

Vogt, S., Grosse, R., Schultz, G., and Offermanns, S. (2003). Receptor-dependent RhoA activation in G12/G13-deficient cells: Genetic evidence for an involvement of Gq/G11. *J. Biol. Chem.* **278,** 28743–28749.

Wang, L., Zhang, H., Solski, P. A., Hart, M. J., Der, C. J., and Su, L. (2000). Modulation of HIV-1 replication by a novel RhoA effector activity. *J. Immunol.* **164,** 5369–5374.

Wang, Q., Liu, M., Kozasa, T., Rothstein, J. D., Sternweis, P. C., and Neubig, R. R. (2004). Thrombin and lysophosphatidic acid receptors utilize distinct rhoGEFs in prostate cancer cells. *J. Biol. Chem.* **279,** 28831–28834.

Wells, C. D., Gutowski, S., Bollag, G., and Sternweis, P. C. (2001). Identification of potential mechanisms for regulation of p115 RhoGEF through analysis of endogenous and mutant forms of the exchange factor. *J. Biol. Chem.* **276,** 28897–28905.

Wells, C. D., Liu, M. Y., Jackson, M., Gutowski, S., Sternweis, P. M., Rothstein, J. D., Kozasa, T., and Sternweis, P. C. (2002a). Mechanisms for reversible regulation between G13 and Rho exchange factors. *J. Biol. Chem.* **277,** 1174–1181.

Wells, C., Jiang, X., Gutowski, S., and Sternweis, P. C. (2002b). Functional characterization of p115 RhoGEF. *Methods Enzymol.* **345,** 371–382.

Wettschureck, N., and Offermanns, S. (2005). Mammalian G proteins and their cell type specific functions. *Physiol. Rev.* **85,** 1159–1204.

Whitehead, I. P., Khosravi-Far, R., Kirk, H., Trigo-Gonzalez, G., Der, C. J., and Kay, R. (1996). Expression cloning of lsc, a novel oncogene with structural similarities to the Dbl family of guanine nucleotide exchange factors. *J. Biol. Chem.* **271,** 18643–18650.

Wittchen, E. S., van Buul, J. D., Burridge, K., and Worthylake, R. A. (2005). Trading spaces: Rap, Rac, and Rho as architects of transendothelial migration. *Curr. Opin. Hematol.* **12,** 14–21.

Yamada, T., Ohoka, Y., Kogo, M., and Inagaki, S. (2005). Physical and functional interactions of the lysophosphatidic acid receptors with PDZ domain-containing Rho guanine nucleotide exchange factors (RhoGEFs). *J. Biol. Chem.* **280,** 19358–19363.

Yau, D. M., Yokoyama, N., Goshima, Y., Siddiqui, Z. K., Siddiqui, S. S., and Kozasa, T. (2003). Identification and molecular characterization of the G alpha12-Rho guanine nucleotide exchange factor pathway in *Caenorhabditis elegans*. *Proc. Natl. Acad. Sci. USA* **100,** 14748–14753.

Ying, Z., Jin, L., Palmer, T., and Webb, R. C. (2006). Angiotensin II up-regulates the leukemia-associated Rho guanine nucleotide exchange factor (RhoGEF), a regulator of G protein signaling domain-containing RhoGEF, in vascular smooth muscle cells. *Mol. Pharmacol.* **69,** 932–940.

Zhai, J., Lin, H., Nie, Z., Wu, J., Canete-Soler, R., Schlaepfer, W. W., and Schlaepfer, D. D. (2003). Direct interaction of focal adhesion kinase with p190RhoGEF. *J. Biol. Chem.* **278,** 24865–24873.

Zhang, H., Wang, L., Kao, S., Whitehead, I. P., Hart, M. J., Liu, B., Duus, K., Burridge, C. J., Der, C. J., and Su, L. (1999). Functional interaction between the cytoplasmic leucine-zipper domain of HIV-1 gp41 and p115-RhoGEF. *Curr. Biol.* **9,** 1271–1274.

Zheng, Y., Olson, M. F., Hall, A., Cerione, R. A., and Toksoz, D. (1995). Direct involvement of the small GTP-binding protein Rho in lbc oncogene function. *J. Biol. Chem.* **270,** 9031–9034.

# AUTHOR INDEX

## A



## B

## C

## D



## E

## F



## G

## H

## I



## J

## K

## L

## M

## N

## O



## P

## T

## U



## V

## W



## X

## Y



## Z

# SUBJECT INDEX

## A



## B



## C

## H



## I



## J



## L

## R



## S

## T



## U



## W



## Y

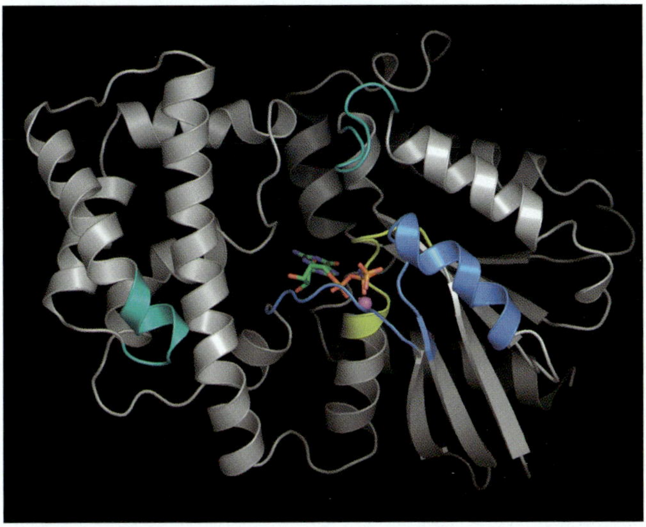

SPRANG *ET AL.*, CHAPTER 1, FIG. 1. (See Legend in Text.)

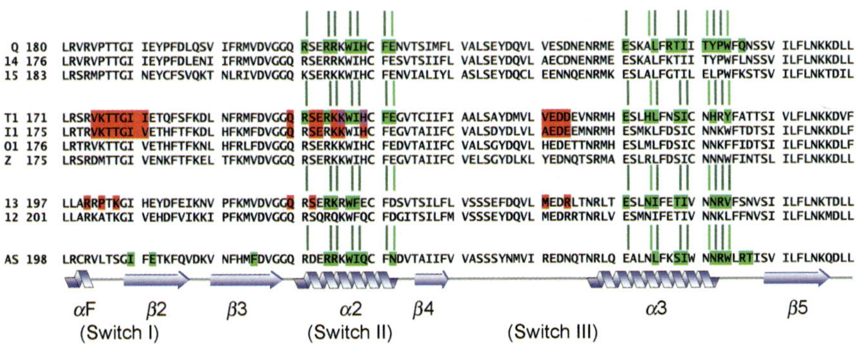


SPRANG *ET AL.*, CHAPTER 1, FIG. 2. (See Legend in Text.)

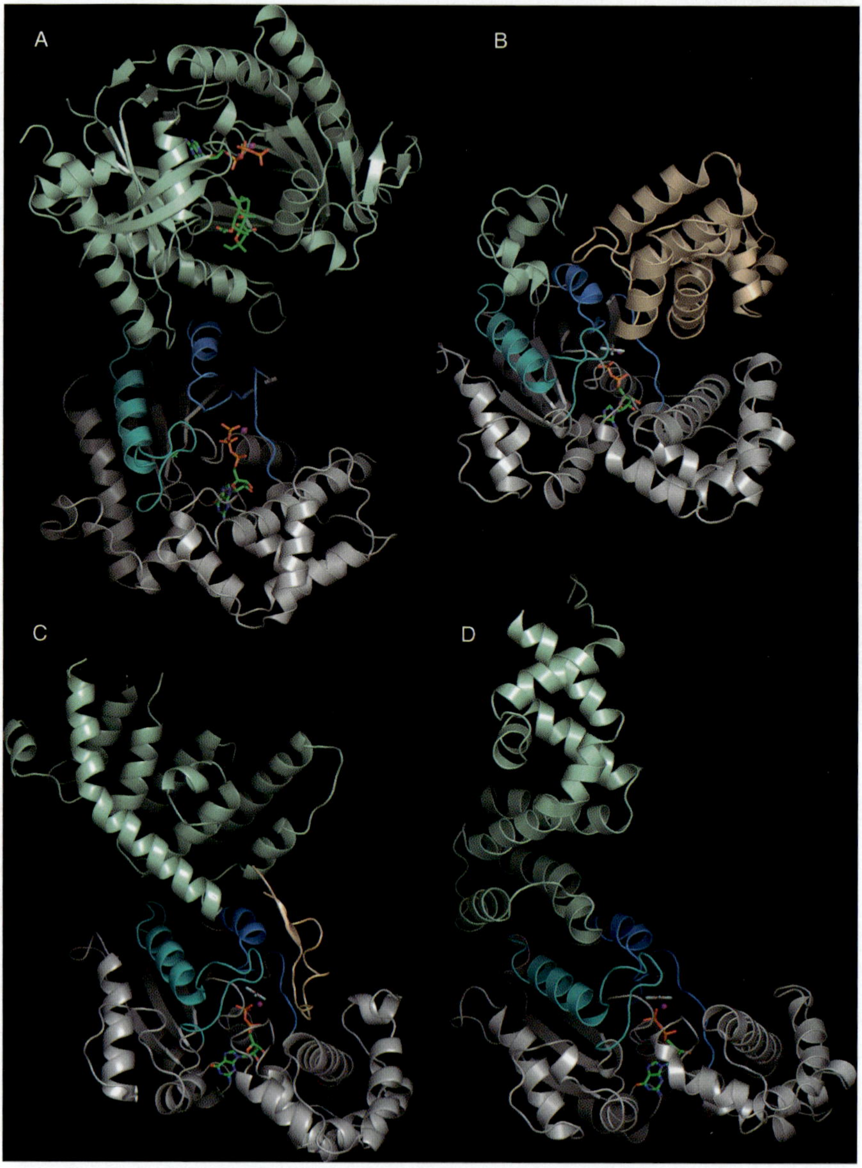


SPRANG *ET AL.*, CHAPTER 1, FIG. 3. (See Legend in Text.)

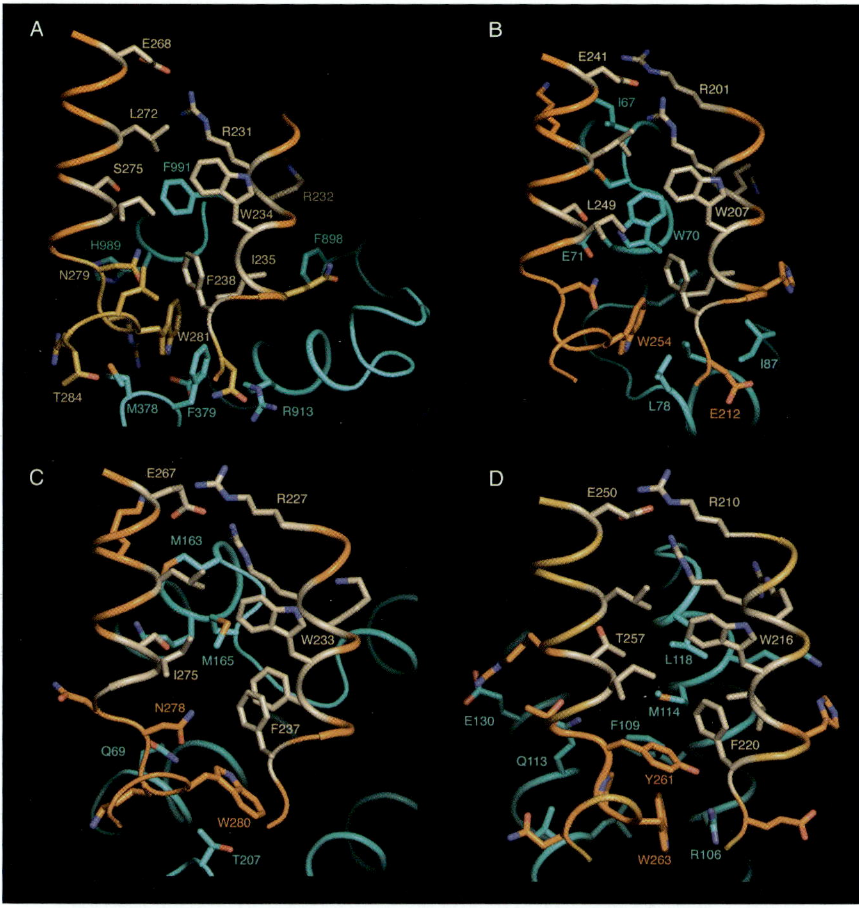


Sprang *et al.*, Chapter 1, Fig. 4. (See Legend in Text.)

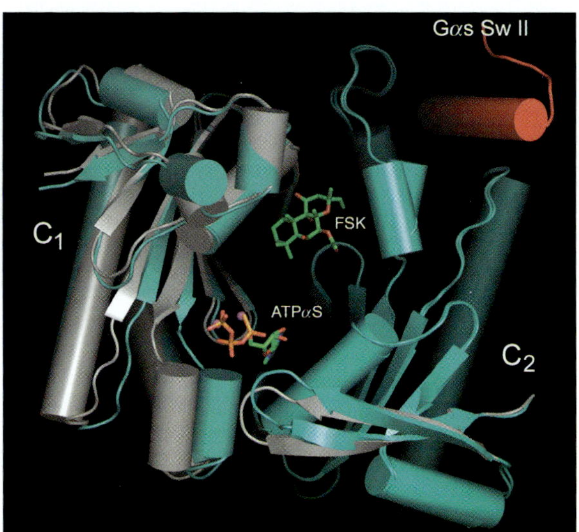


Sprang *et al.*, Chapter 1, Fig. 5. (See Legend in Text.)

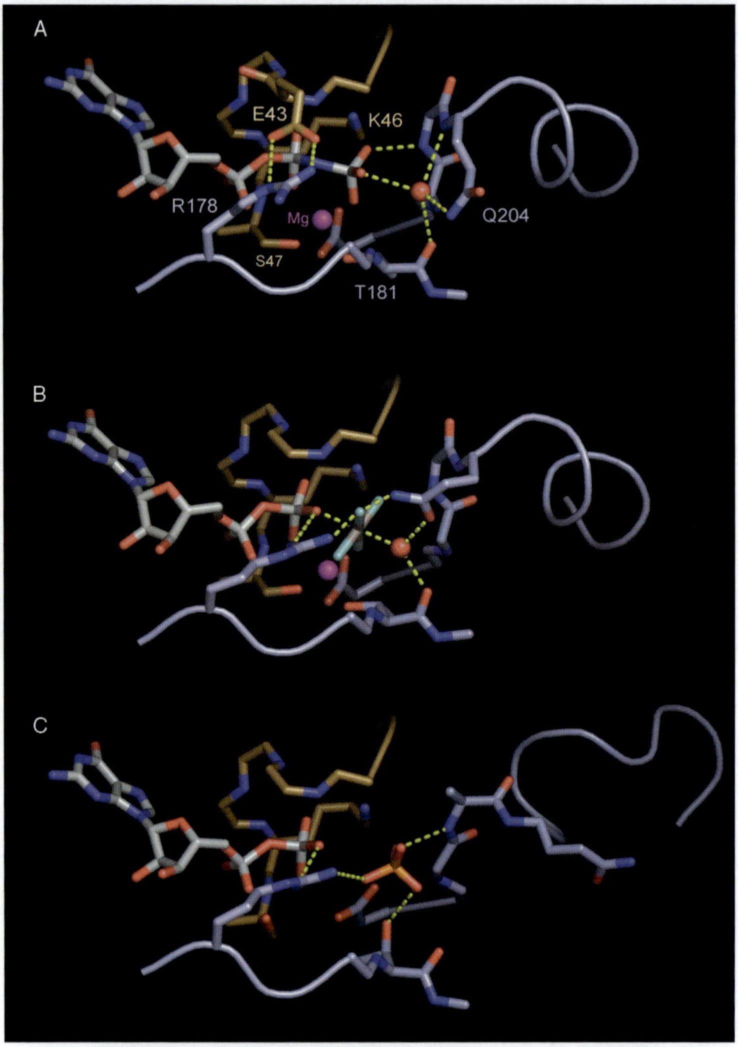


Sprang *et al.*, Chapter 1, Fig. 6. (See Legend in Text.)

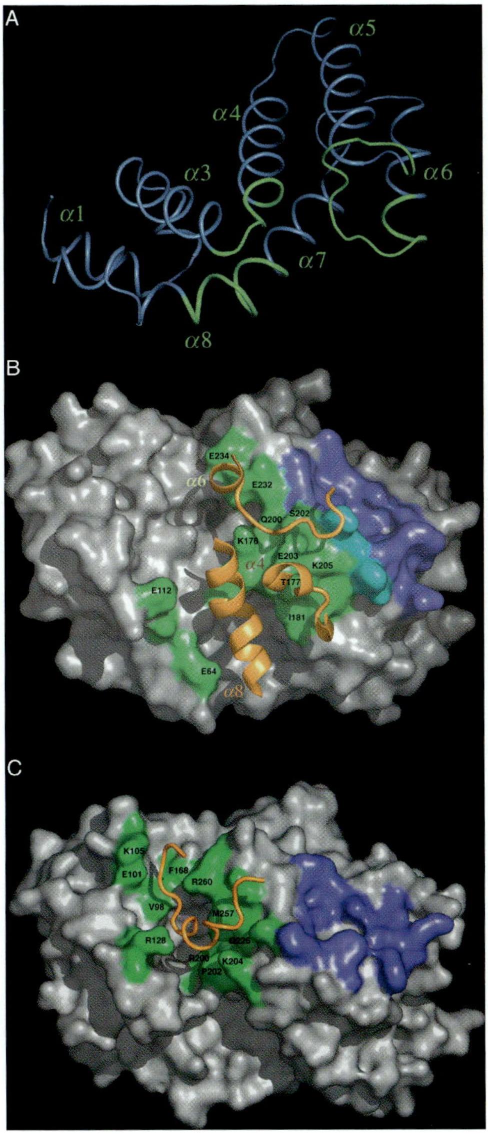


SPRANG *ET AL.*, CHAPTER 1, FIG. 7. (See Legend in Text.)

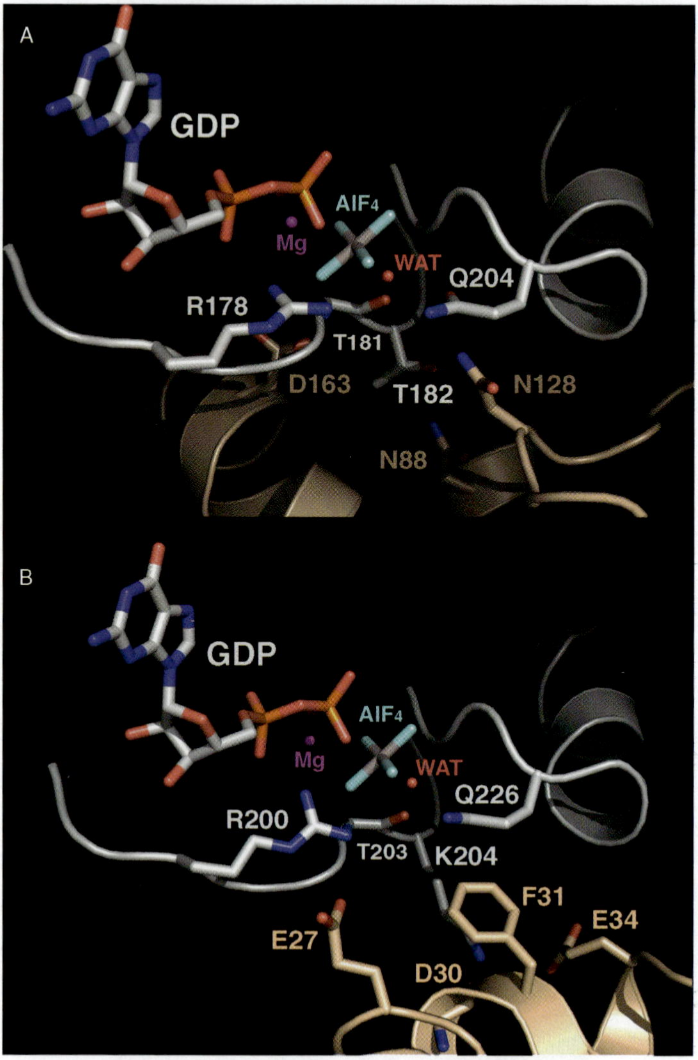


SPRANG *ET AL.*, CHAPTER 1, FIG. 8. (See Legend in Text.)

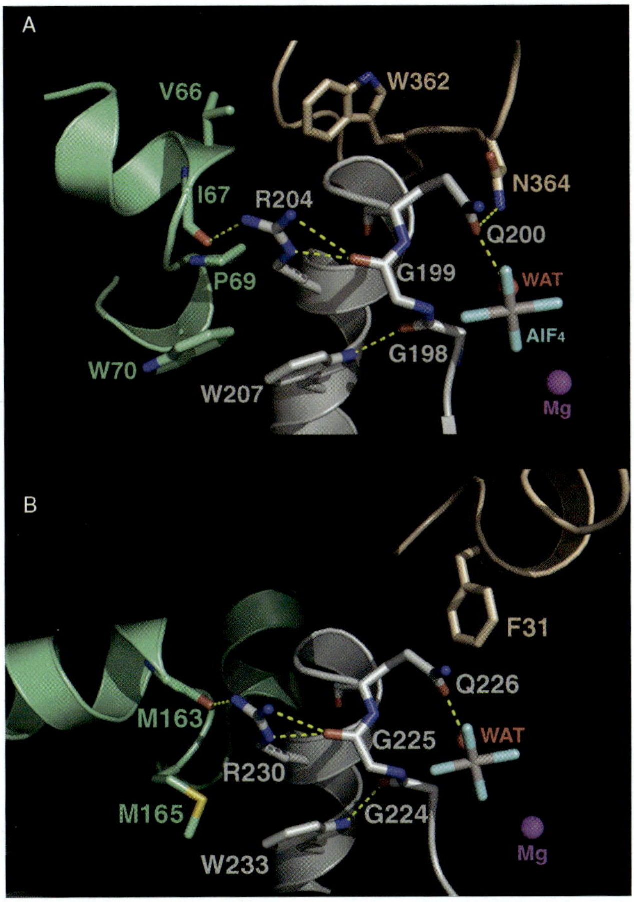


SPRANG *ET AL.*, CHAPTER 1, FIG. 9. (See Legend in Text.)

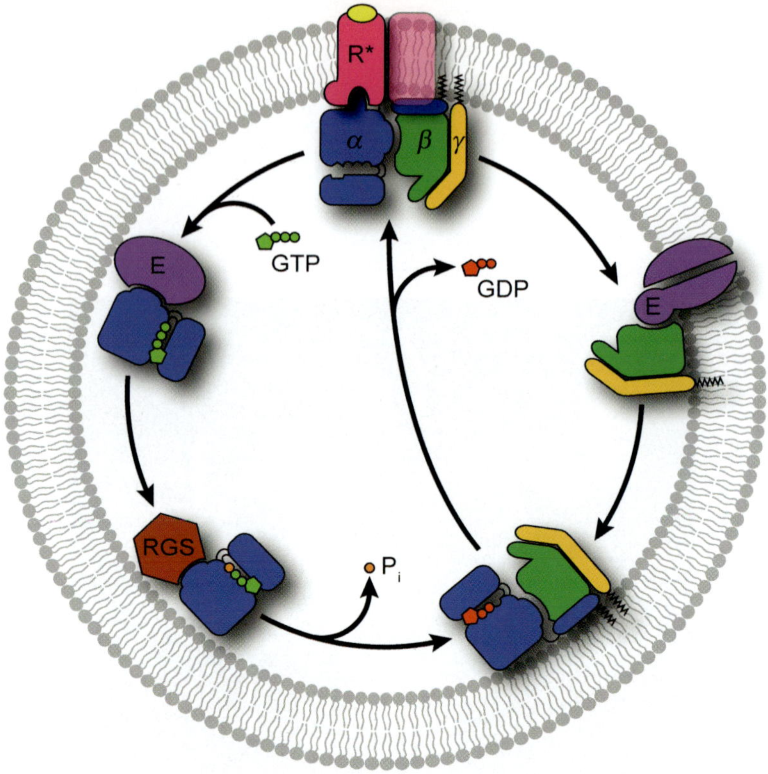


OLDHAM AND HAMM, CHAPTER 2, FIG. 1. (See Legend in Text.)

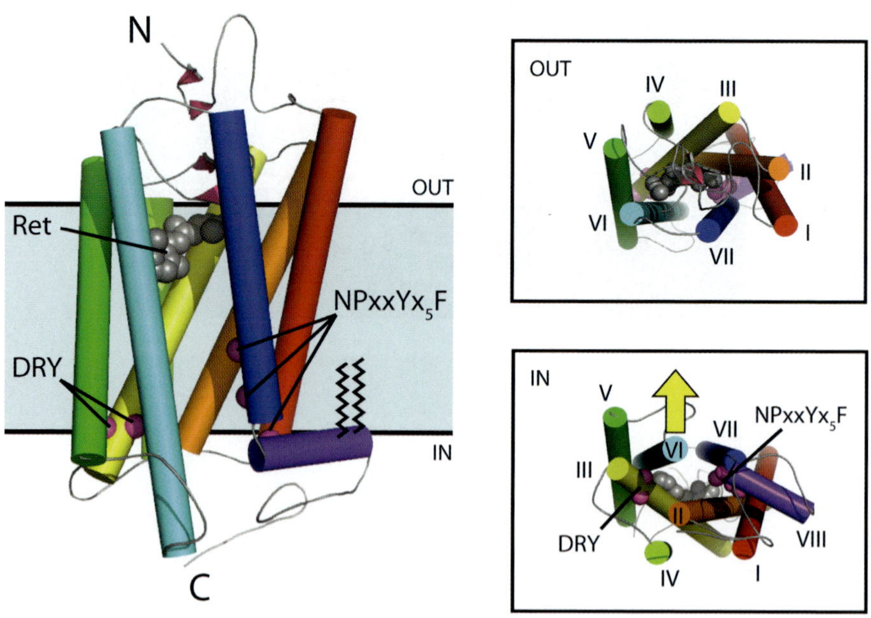


OLDHAM AND HAMM, CHAPTER 2, FIG. 2. (See Legend in Text.)

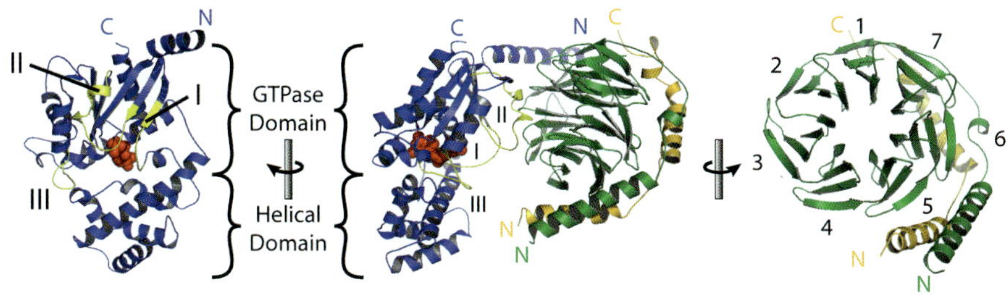


OLDHAM AND HAMM, CHAPTER 2, FIG. 3. (See Legend in Text.)

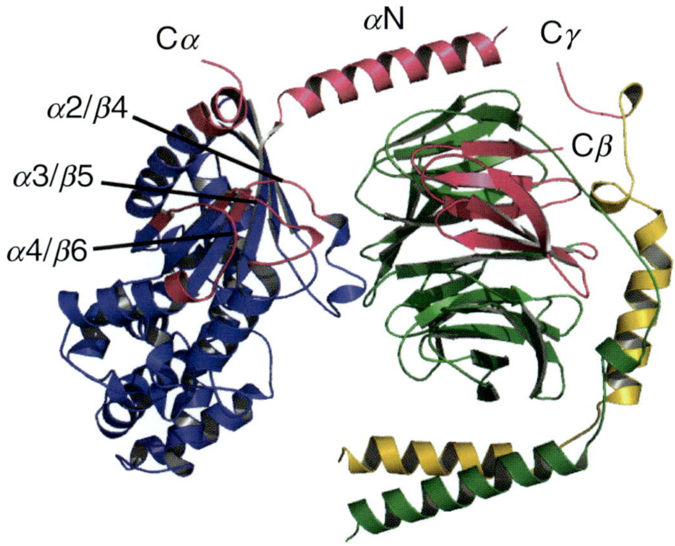


OLDHAM AND HAMM, CHAPTER 2, FIG. 4. (See Legend in Text.)

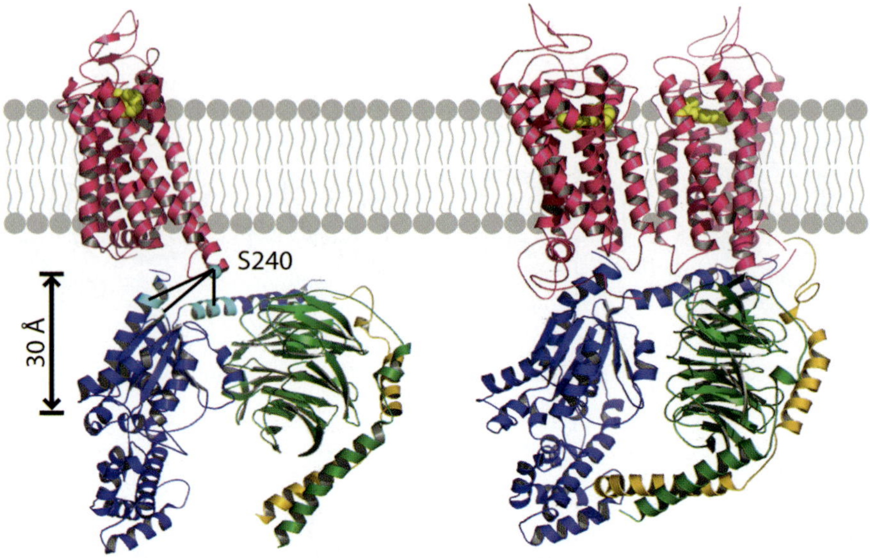


OLDHAM AND HAMM, CHAPTER 2, FIG. 5. (See Legend in Text.)

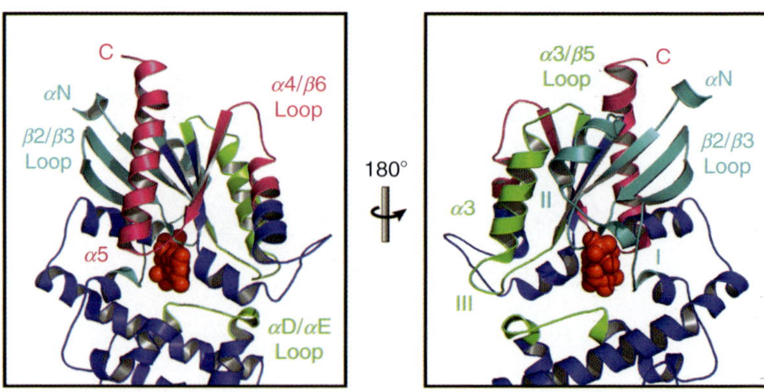


OLDHAM AND HAMM, CHAPTER 2, FIG. 6. (See Legend in Text.)

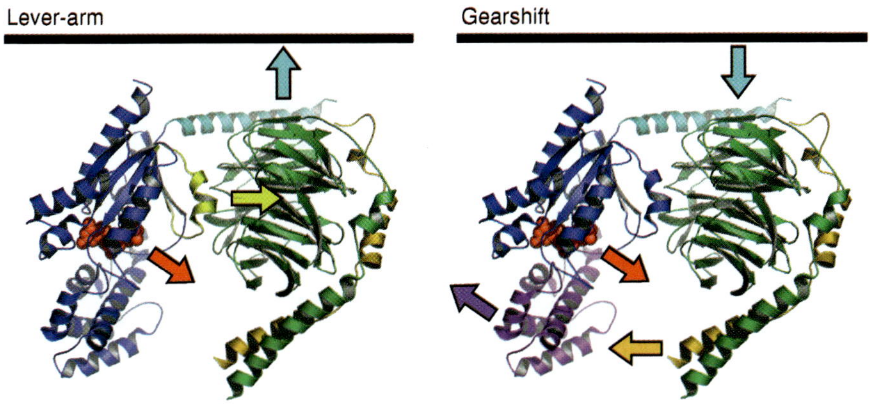


Oldham and Hamm, Chapter 2, Fig. 7. (See Legend in Text.)

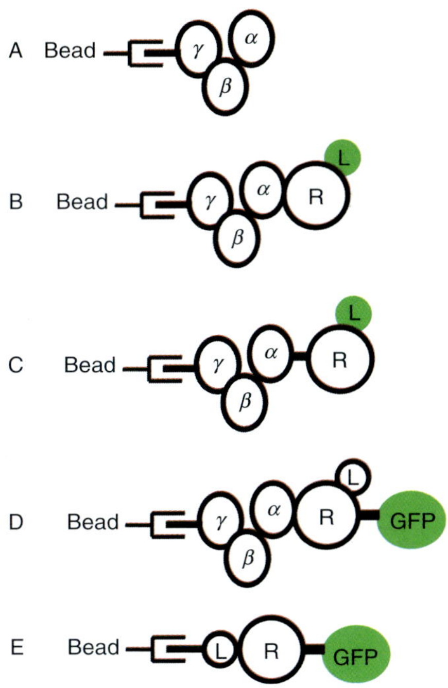


Buranda *et al.*, Chapter 3, Fig. 1. (See Legend in Text.)

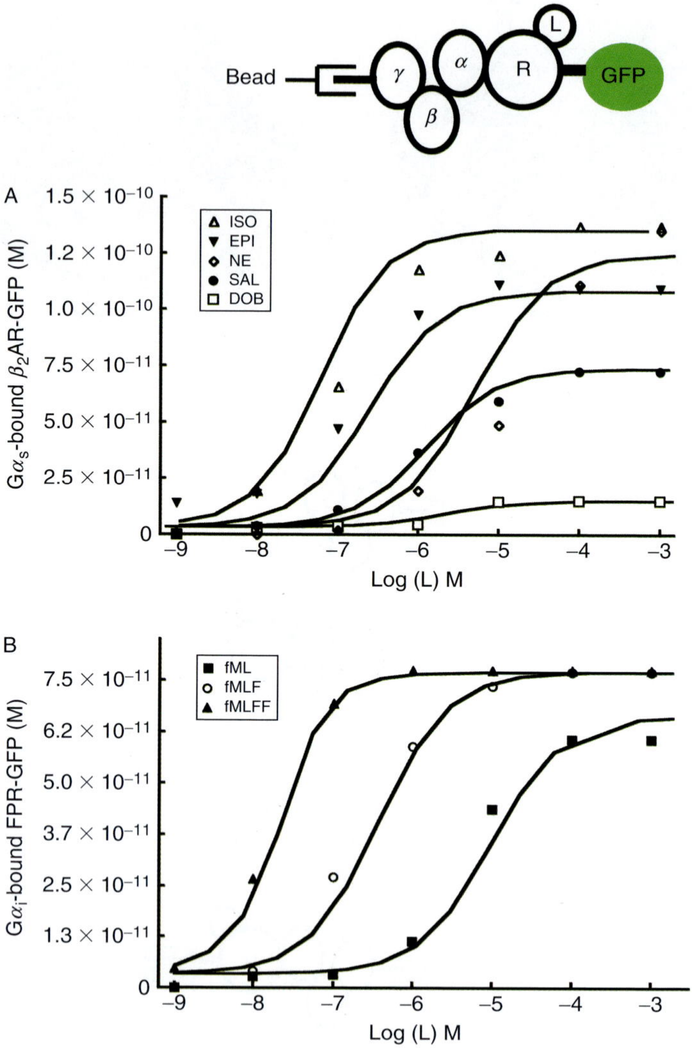


BURANDA *ET AL.*, CHAPTER 3, FIG. 3. (See Legend in Text.)

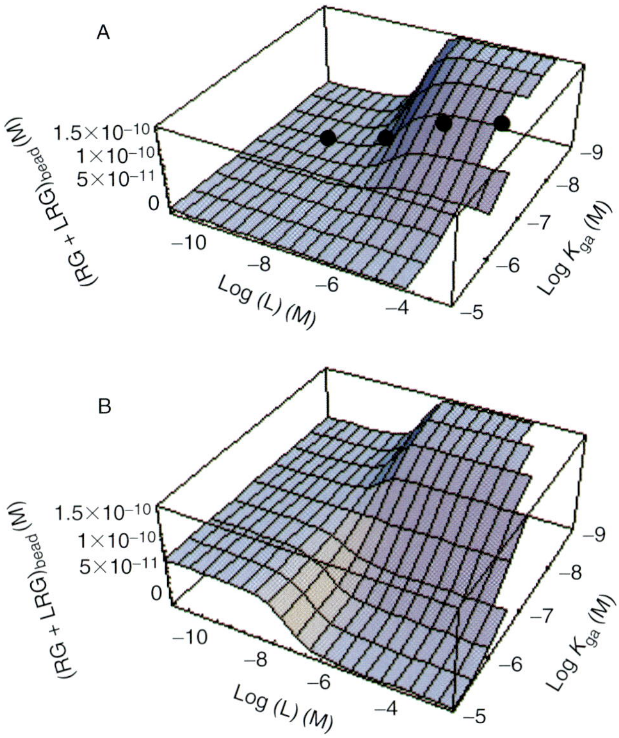


BURANDA *ET AL.*, CHAPTER 3, FIG. 4. (See Legend in Text.)

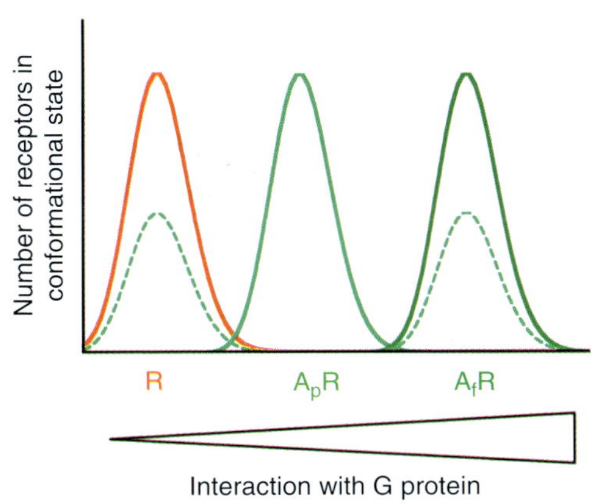


BURANDA *ET AL.*, CHAPTER 3, FIG. 5. (See Legend in Text.)

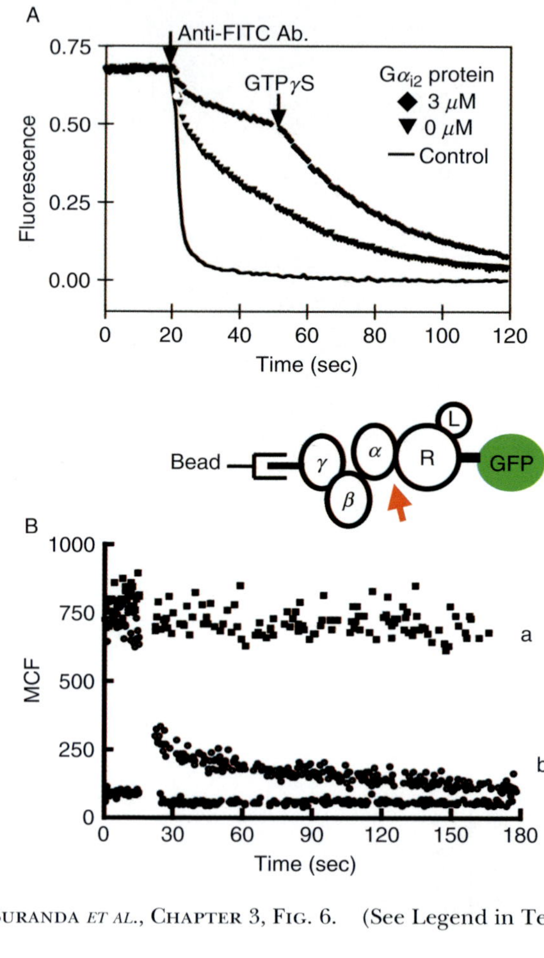


BURANDA *ET AL.*, CHAPTER 3, FIG. 6. (See Legend in Text.)

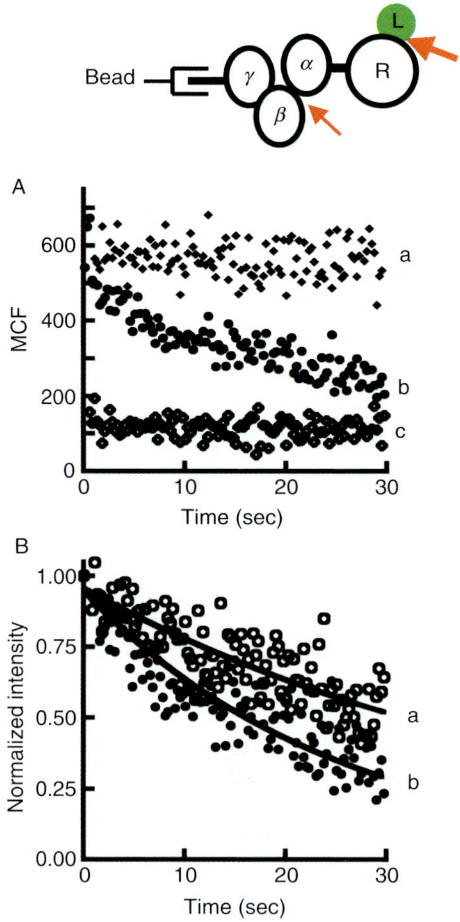


BURANDA *ET AL.*, CHAPTER 3, FIG. 7. (See Legend in Text.)

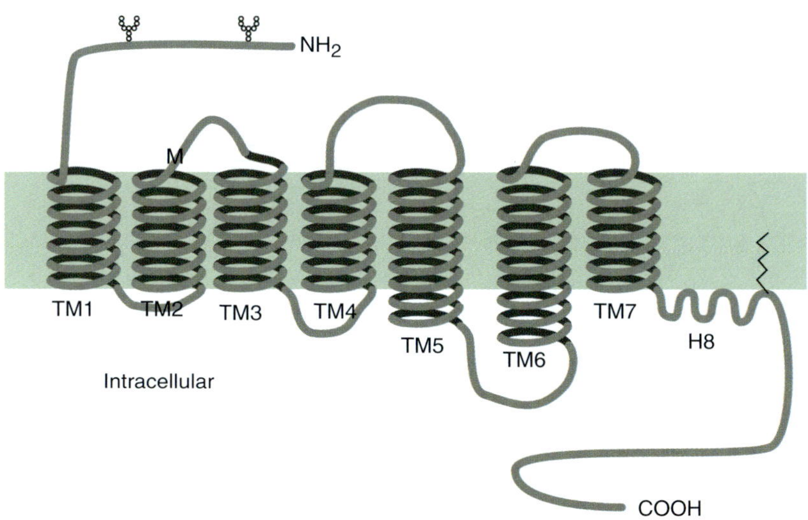


DEUPI AND KOBILKA, CHAPTER 4, FIG. 1. (See Legend in Text.)

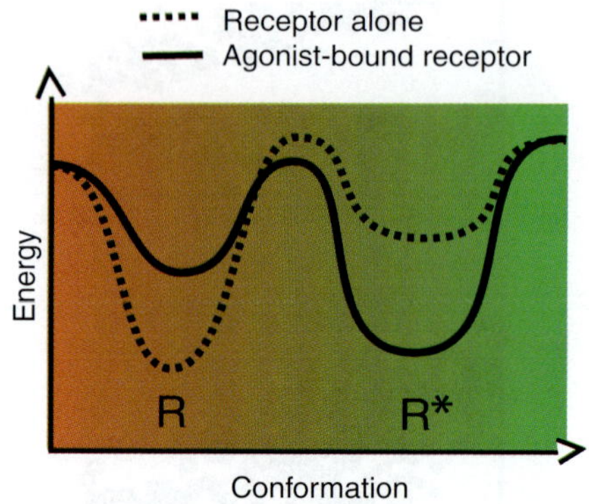


Deupi and Kobilka, Chapter 4, Fig. 2. (See Legend in Text.)

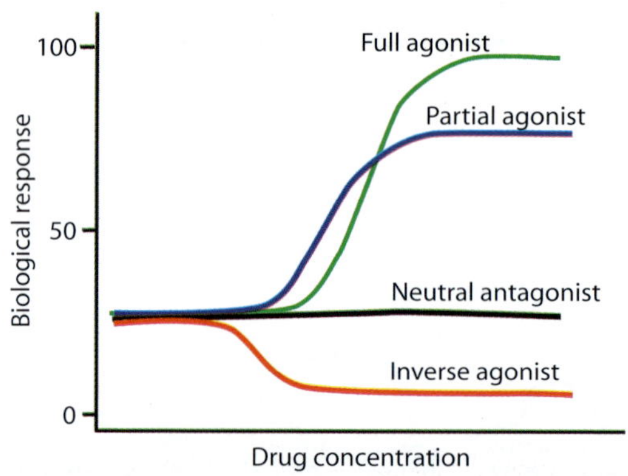


Deupi and Kobilka, Chapter 4, Fig. 3. (See Legend in Text.)

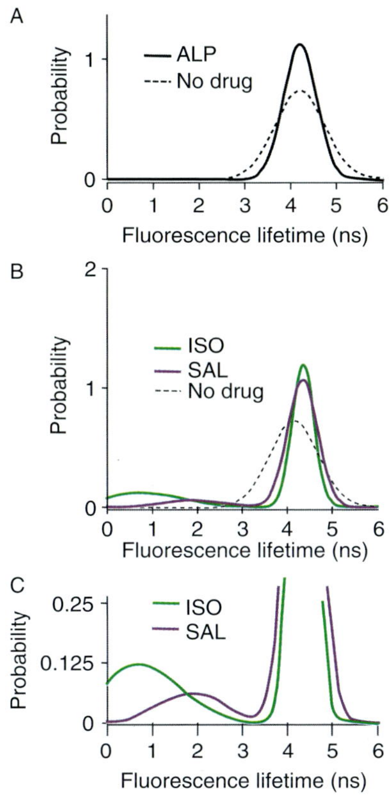


DEUPI AND KOBILKA, CHAPTER 4, FIG. 4. (See Legend in Text.)

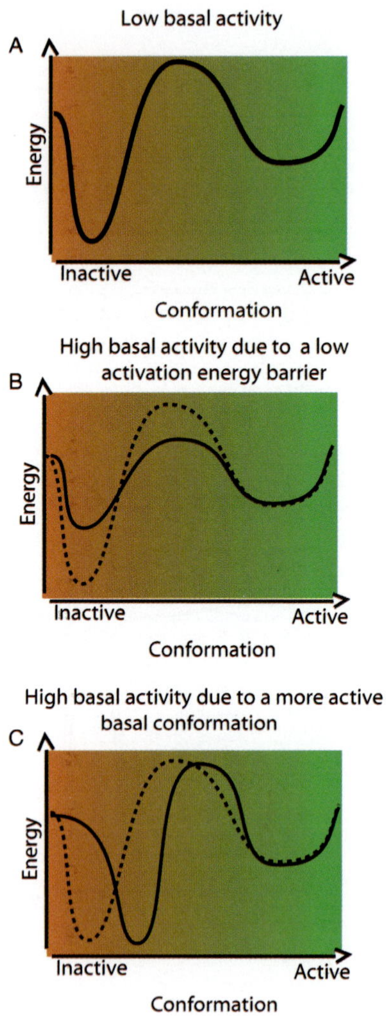


Deupi and Kobilka, Chapter 4, Fig.5. (See Legend in Text.)

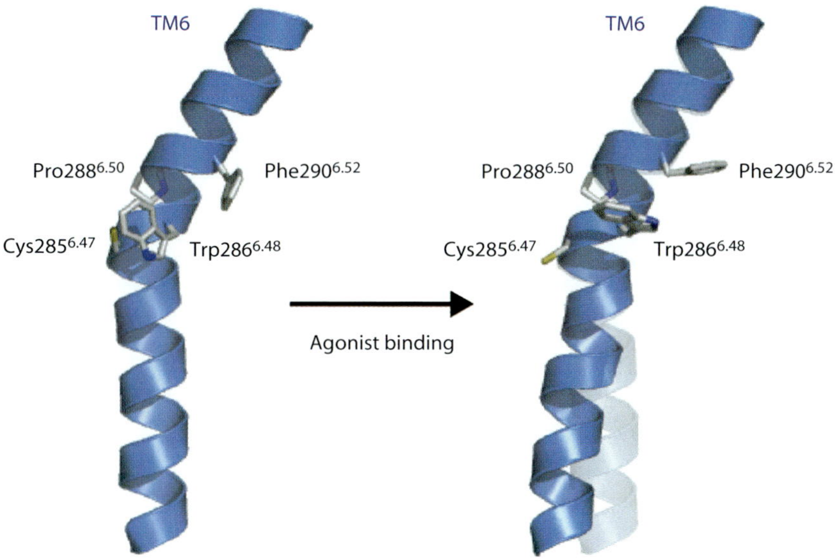


DEUPI AND KOBILKA, CHAPTER 4, FIG. 6. (See Legend in Text.)

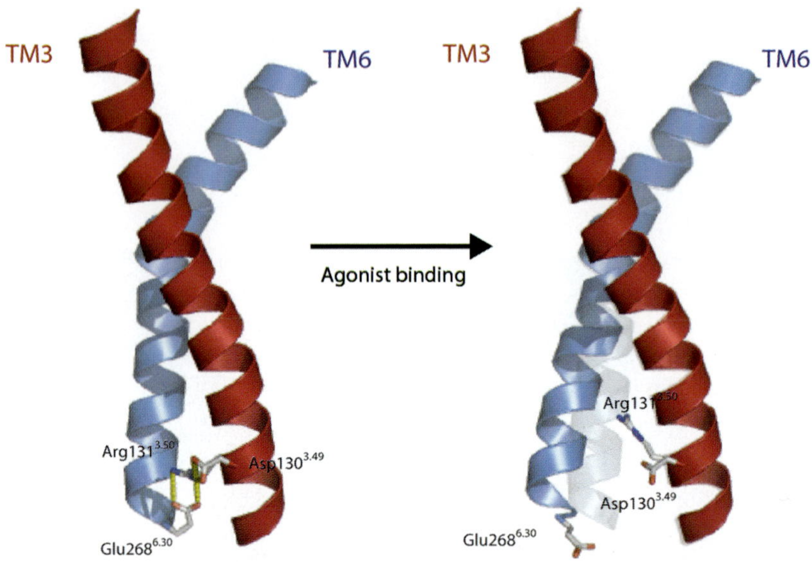


Deupi and Kobilka, Chapter 4, Fig. 7. (See Legend in Text.)

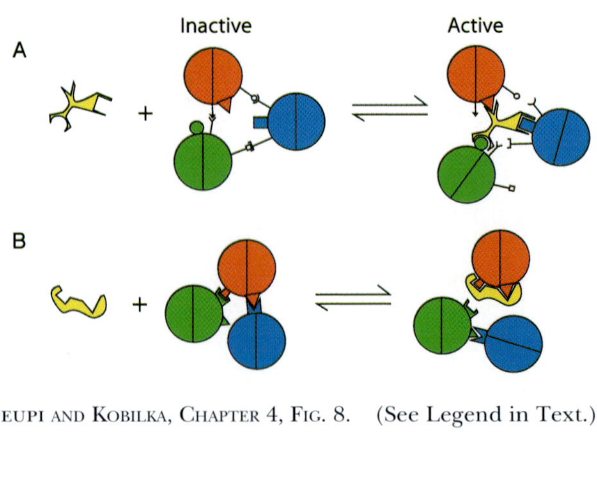


Deupi and Kobilka, Chapter 4, Fig. 8. (See Legend in Text.)

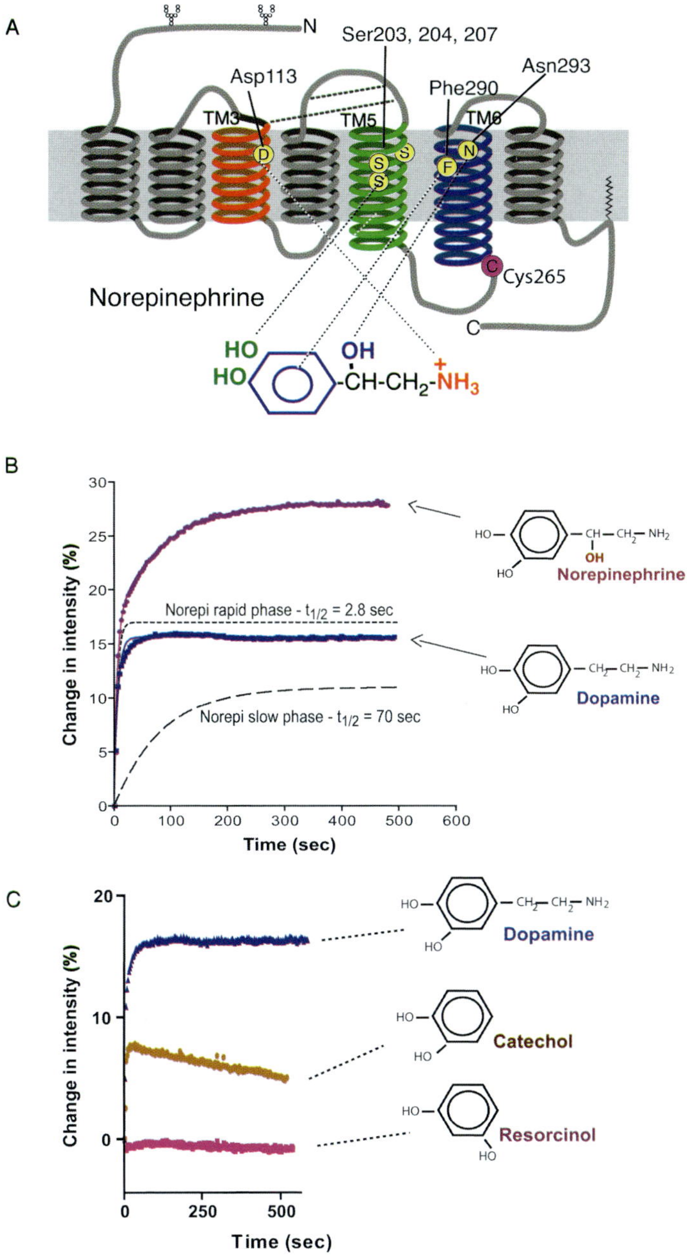


DEUPI AND KOBILKA, CHAPTER 4, FIG. 9. (See Legend in Text.)

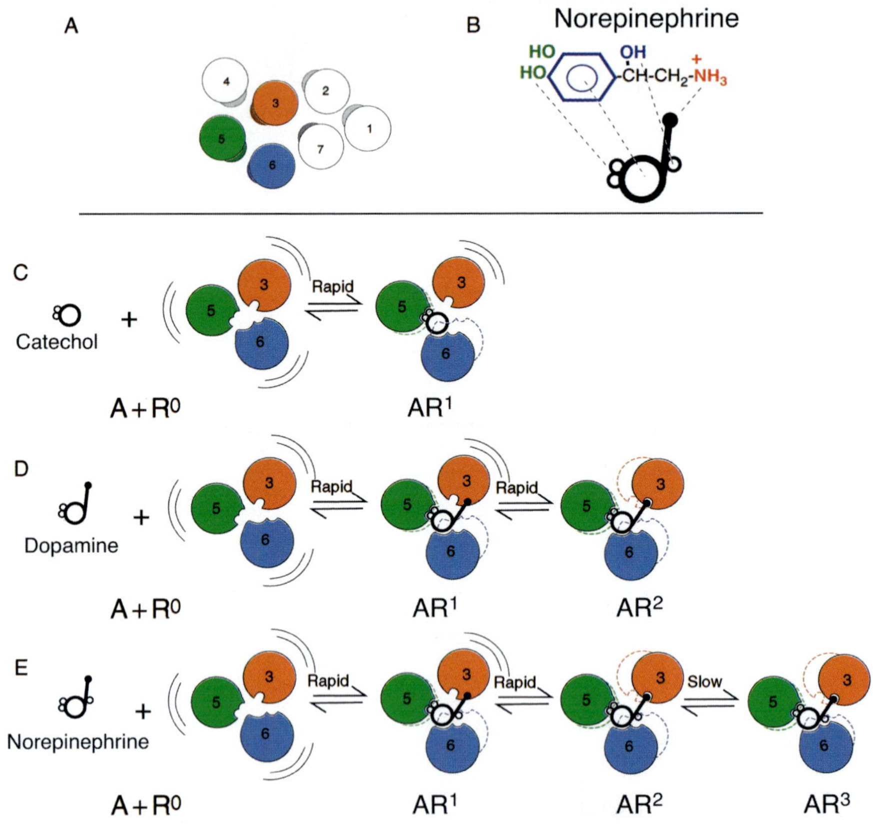


DEUPI AND KOBILKA, CHAPTER 4, FIG. 10. (See Legend in Text.)

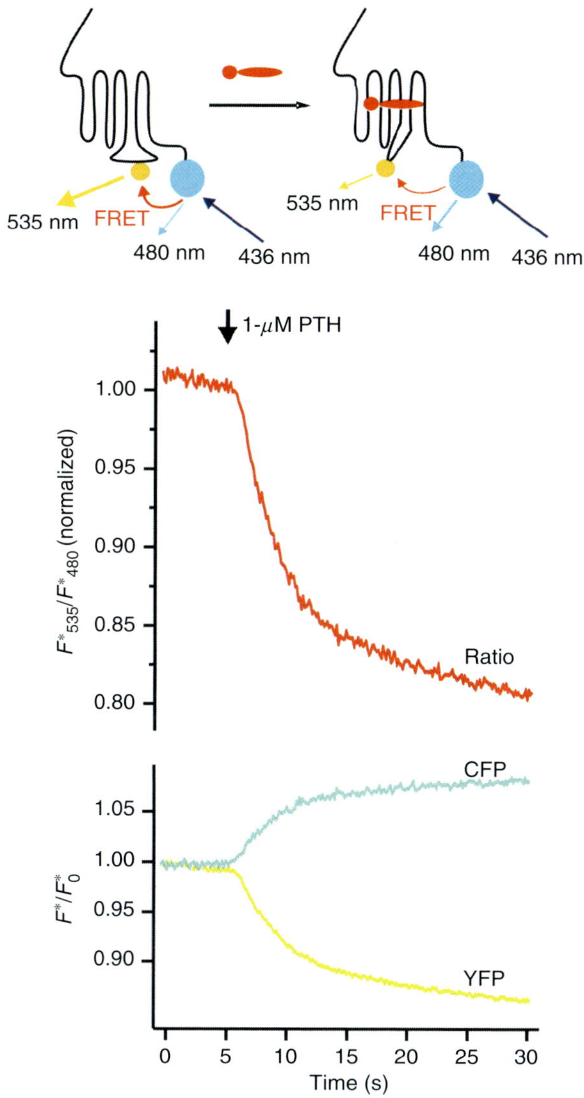


LOHSE *ET AL.*, CHAPTER 5, FIG. 1. (See Legend in Text.)

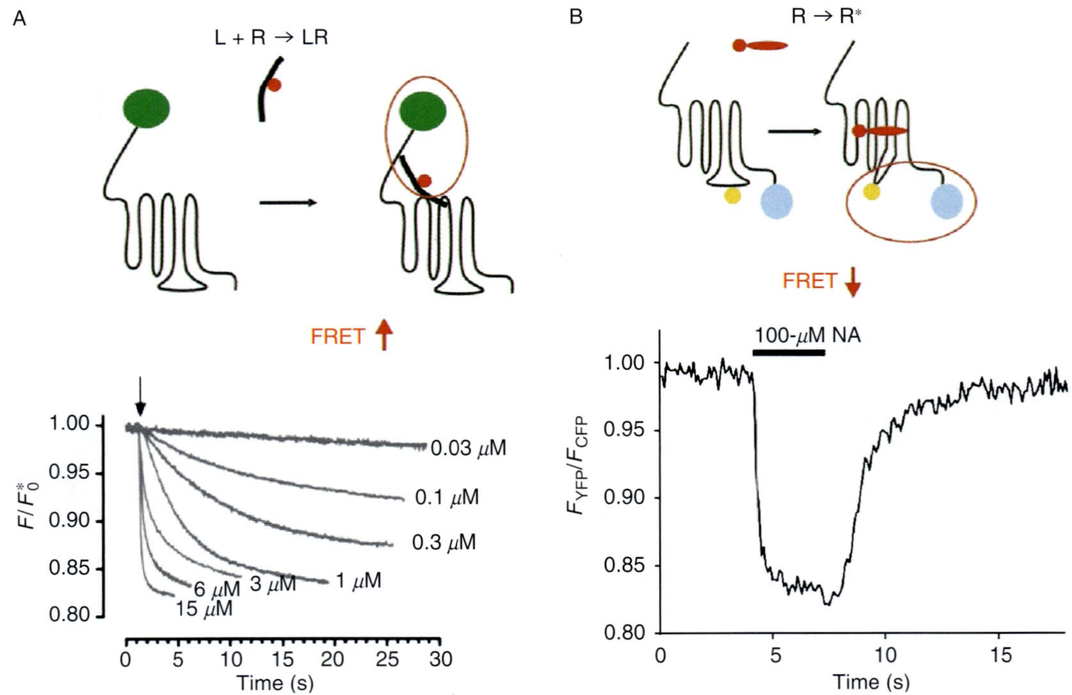
A
L + R → LR
FRET
F/F*0
0.03 μM
0.1 μM
0.3 μM
1 μM
3 μM
6 μM
15 μM
Time (s)
B
R → R*
FRET
100-μM NA
FYFP/FCFP
Time (s)

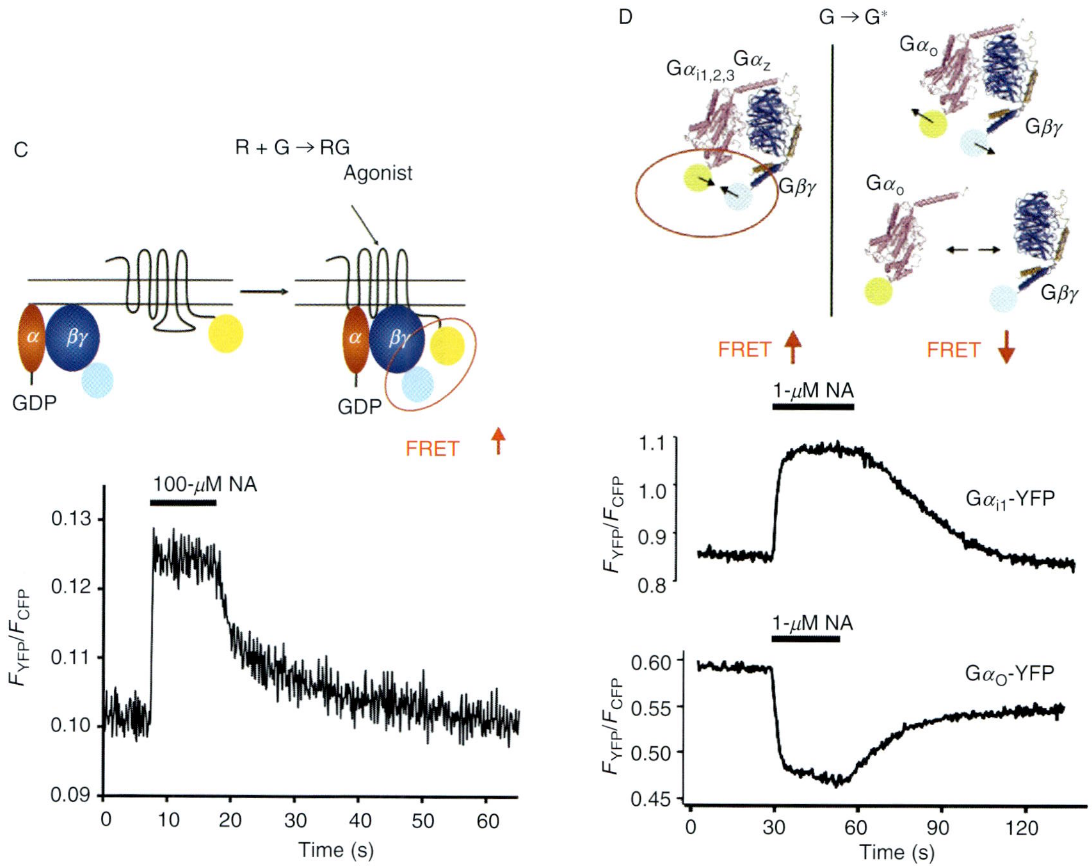


LOHSE *ET AL.*, CHAPTER 5, FIG. 2. (See Legend in Text.)

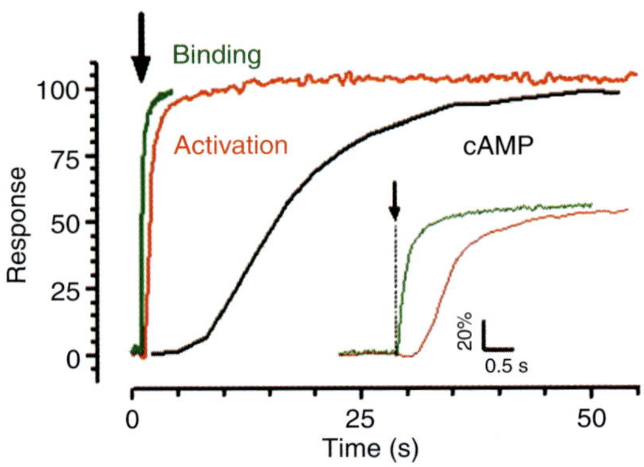


LOHSE *ET AL.*, CHAPTER 5, FIG. 3. (See Legend in Text.)